H. Przuntek and T. Müller (eds.)

Diagnosis and Treatment of Parkinson's Disease – State of the Art

Springer-Verlag Wien GmbH

Prof. Dr. H. Przuntek
Priv.-Doz. Dr. T. Müller

Department of Neurology, St. Josef Hospital

Ruhr University Bochum, Federal Republic of Germany

Originally published by Springer-Verlag Wien New York in 1999
Softcover reprint of the hardcover 1st edition 1999

Typesetting: Best-set Typesetter Ltd., Hong Kong

Graphic design: Ecke Bonk

Printed on acid-free and chlorine-free bleached paper
SPIN: 10715089

With 32 (partly coloured) Figures

CIP-data applied for

ISBN 978-3-211-83275-2 ISBN 978-3-7091-6360-3 (eBook)
DOI 10.1007/978-3-7091-6360-3

Preface

The workshop "State-of-the-Art Diagnosis and Treatment of Parkinson's Disease" was held in November 1997 in Budapest.

A panel of expert clinicians and basic scientists with a special interest in Parkinson's disease discussed and reviewed the current state of science and clinical therapeutics of the disease. This volume represents the papers of this meeting.

Collectively, these articles represent an authorative review of the current state of knowledge regarding preclinical course and symptomatology, subtypes with their impact on the pathology, genetic alterations, novel mechanisms of neuronal cell death, diagnostic tools and old and novel therapeutic approaches with respect to neuroprotection and neuroregeneration in Parkinson's disease. Particular emphasis has been placed on a novel antiparkinsonian drug called budipine with various modes of action also influencing altered non dopaminergic systems in Parkinson's disease.

It became evident, that many questions on the cause, course and treatment of Parkinson's disease are still unanswered and therefore the ideal way to treat a parkinsonian patient remains to be defined.

We wish to acknowledge with thanks the sponsorship and financial support of Promonta Lundbeck in enabling this meeting to take place.

Bochum, February 1999

H. Przuntek
T. Müller

Contents

Post mortem studies in Parkinson's disease — is it possible to detect brain areas for specific symptoms?

K. A. Jellinger

Ludwig Boltzmann Institute of Clinical Neurobiology, Vienna, Austria

Summary. Parkinson's disease (PD) is characterized by progressive neuronal loss associated with Lewy bodies in many subcortical nuclei leading to multiple biochemical and pathophysiological changes of clinical relevance. Loss of nigral neurons causing striatal dopamine deficiency is related to both the duration and clinical stages (severity) of the disease. The clinical subtypes of PD have different morphological lesion patterns: a) The akinetic-rigid type shows more severe cell loss in the ventrolateral part of substantia nigra zona compacta (SNZC) that projects to the dorsal putamen than the medial part projecting to caudate nucleus and anterior putamen, with negative correlation between SNZC cell counts, severity of akinesia-rigidity, and dopamine loss in the posterior putamen. Reduced dopaminergic input causes overactivity of the GABA ergic inhibitory striatal neurons projecting via the "indirect loop" to SN zona reticulata (SNZR) and medial pallidum (GPI) leading to inhibition of the glutamatergic thalamo-cortical motor loop and reduced cortical activation. b) The tremor-dominant type shows more severe neuron loss in medial than in lateral SNZC and damage to the retrorubral field A8 containing only few tyrosine hydroxylase and dopamine transporter immunoreactive (IR) neurons but mainly calretinin-IR cells. A8 that is rather preserved in rigid-akinetic PD (protective role of calcium-binding protein?) projects to the matrix of dorsolateral striatum and ventromedial thalamus. Together with area A10 it influences the strial efflux via SNZR to thalamus and from there to prefrontal cortex. Rest tremor in PD is associated with increased metabolism in the thalamus, subthalamus, pons, and premotor-cortical network suggesting an increased functional activity of thalamo-motor projections. In essential tremor, no significant pathomorphological changes but overactivity of cerebello-thalamic loop have been observed. c) In the akinetic-rigid forms of multisystem atrophy, degeneration is more severe in the lateral SNZC with severe loss of calbindin-IR cells reflecting initial degeneration of the striatal matrix in the caudal putamen with transsynaptic degeneration of striatonigral efferences that remain intact in PD. This fact and loss of striatal D2 receptors — as in advanced stages of PD — are reasons for negative response to L-dopa substitution. These data suggest different pathophysiological mechanisms of the clinical subtypes of PD that have important therapeutic implications. d) Involvement of extranigral structures in PD includes the mesocortical dopaminergic system, the noradrenergic locus coeruleus, dorsal vagal nucleus and medullary nuclei, serotonergic dorsal raphe, nucleus basalis of Meynert and other cholinergic brainstem nuclei, e.g. Westphal-Edinger nucleus (controlling pupillomotor function), posterolateral hypothalamus and the limbic system, e.g. amygdaloid nucleus, part of hippocampal formation, limbic thalamic nuclei with prefrontal projections, etc. Damage to multiple neuronal systems by the progressing degenerative process causing complex biochemical changes may explain the variable clinical picture of PD including vegetative, behavioural and cognitive dysfunctions, depression, pharmacotoxic psychoses, etc. Future comparative clinico-morphological and pathobiochemical studies will further elucidate the pathophysiological basis of specific clinical symptoms of PD and related disorders providing a broader basis for effective treatment strategies.

Parkinson's disease (PD) is characterized by progressive degeneration of the nigrostriatal dopaminergic system and other subcortical neuronal systems leading to striatal dopamine deficiency and other biochemical deficits related to the variable clinical signs and symptoms of the disorder. While the pathogenesis and aetiology of PD are still unknown, the recent elucidation of the morphological and pathophysiological substrates of several clinical dysfunctions in PD has provided a better insight into the course of the disease and important implications for treatment (Jellinger, 1998).

Morphological lesion pattern in PD

Histopathology of PD is featured by the presence of Lewy bodies (LB) in association with neuronal loss in pigmented nuclei of the midbrain and many subcortical nuclei. There is severe depletion of melanized neurons (45–66%) (Hirsch et al., 1988; Pakkenberg et al., 1991; Ma et al., 1996) and tyrosine hydroxylase (TH) immunoreactive (IR) neurons (60–85%, mean 75%) in the A9 group of substantia nigra zona compacta (SNZC) (Halliday et al., 1996), particularly in the ventrolateral tier (area alpha 97%), followed by the medioventral and dorsal area (gamma) (Fearnley and Lees, 1994) (Fig. 1). The most severely involved ventral and caudal parts of SNZC are composed of melanin-rich cells with weak immunoreactivity (IRy) of the calcium binding protein calbindin (CAB), (Hirsch et al., 1992). Loss of dopaminergic neurons shows an inverse pattern to the CAB contents: it is most severe in the caudal and ventrolateral parts (up to 98%) and shows progressive course to medial and rostral parts and in caudorostral direction (Damier et al., 1996). The degeneration of SNZS is closely related to the duration and severity of clinical features i.e. neuronal loss increases with the duration and severity of clinical symptoms (Paulus and Jellinger, 1991; Ma et al., 1997). It has been estimated that clinical symptoms occurring when about 50% of SN neurons are lost, are preceded by a long preclinical phase which correlates with "Incidental Lewy body disease" (Fearnley and Lees, 1994). In PD, the non-dopaminergic, GABAergic neurons of the SN zona reticulata (SNZR) are

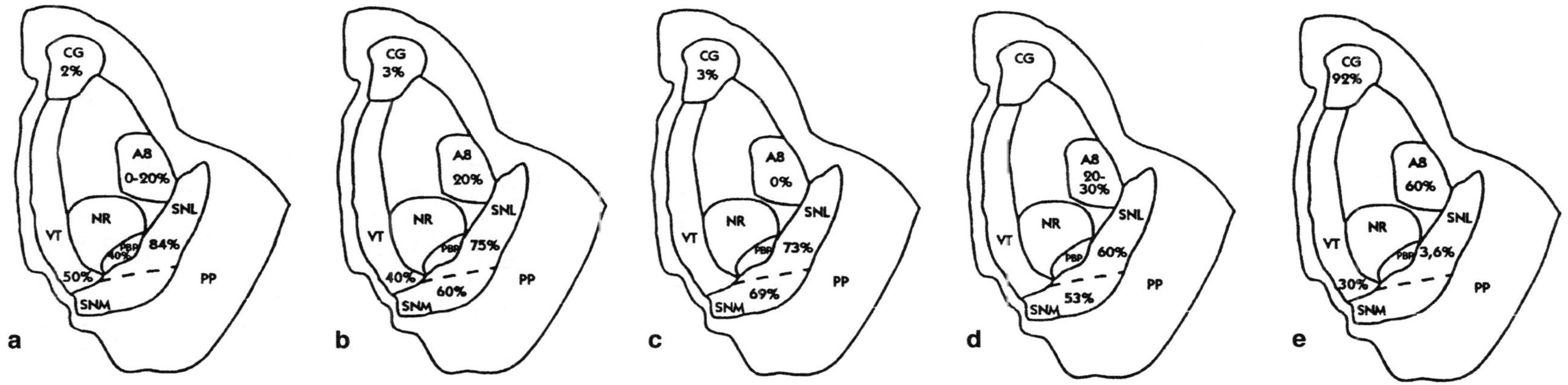

Fig. 1. Neuronal loss in catecholaminergic nuclei of midbrain in Parkinson's disease (modified from Halliday et al., 1996; Hirsch et al., 1992; Jellinger, 1993). **a.** Distribution of melanized, TH-IR neurons in normal controls; **b.** Percent loss of TH-IR neurons in Parkinson's disease (mean values); **c.** % loss of TH-IR neurons in rigid-akinetic IPD; **d.** % loss of TH-IR neurons in tremor-dominant IPD; **e.** % loss of calbindin-IR neurons in IPD. *A-8* perirubral field; *CG* central periaqueductal gray; *SNM*, *SNL* substantia nigra medial, lateral part; *VT* ventral tegmentum; *PP* pes pedunculi; *PBP* parabrachial pigmented nucleus

preserved, but in terminal stages of the disease show loss of paralbumin-IRy (a calcium binding protein) (Halliday et al., 1996). Progressive supranuclear palsy (PSP) shows about 70% nerve cell loss in SNZR (Hardman et al., 1997a). There is much less involvement of the A10 group (ventral tegmentum, nucleus parabrachialis and nucleus parabrachialis pigmentosus) projecting to cortical and limbic areas (mesocortico-limbic dopaminergic system). These nuclei suffer from 40–50% cell loss (McRitchie et al., 1997), while the periretrorubral A8 region containing only few TH-IR but CAB rich neurons, as well as the central periventricular gray either show no definite degeneration (McRitchie et al., 1997) or about 20 to 32% cell loss in A8 (Hirsch et al., 1988; Hirsch et al., 1997) (Fig. 1). This lesion pattern is similar to that produced by the neurotoxic agent MPTP (1-Methyl-4-phenyl-1,2,3,6-tetrahydropyridin) (Varastet et al., 1994) and differs considerably from age-related lesions of the SNZC with cell loss mainly in the dorsal tier that is involved only in late stages of PD (Fearnley and Lees, 1994; Halliday et al., 1996). The particular pattern of lesions in PD and MPTP-induced parkinsonism could be related to a selective vulnerability of neurons with high expression of dopamin transporter (DAT) messenger (m) RNA (Uhl et al., 1994; Joyce et al., 1997; Uhl, 1998), rich in neuromelanin with low content of CAB and other calcium binding proteins, suggesting a protective role of the latter by prohibiting calcium influx into the cells (German et al., 1993; Hirsch et al., 1992) and a weaker neurotrophin support of the more vulnerable lateral SN dopamine neurons (Nishio et al., 1998). Although other data do not support the hypothesis that differential regional DAT gene expression underlies the selective vulnerability of certain SN dopaminergic neurons in PD (Counihan and Penney, 1998), it is supported by the preservation of the subthalamic nucleus (STN) in PD that is rich in calcineurin and parvalbumin-IR neurons (Hardman et al., 1997b), whereas there is selective involvement of the parabrachial nucleus of area A10 containing neurons rich in TH-IRy and GABA but poor in neuromelanin (McRitchie et al., 1997).

Cellular degeneration is preceded by loss of neurofilament (NF) proteins, of neuronal TH-IRy, TH, DAT and NF mRNS and TH- and DAT proteins and cytochrome oxidase C indicating functional neuronal damage (Gai et al., 1994; Hill et al., 1993; Kastner et al., 1993; Joyce et al., 1997; Itoh et al., 1997). It is accompanied by distribution of melanin with uptake into macrophages, astroglial reaction, and proliferation of microglia indicating active neuron degeneration, toxic and immunological reactions (McGeer et al., 1988). Increased nitric oxide (NO) synthase produced by astroglia may be both a cause or sequela of dopaminergic neuronal degeneration (Hunot et al., 1996) that has been related to a cascade of multiple noxious factors including formation of free radicals, lipid peroxidation, oxidative stress, melanin-iron interaction with increased iron content in SN, mitochondrial dysfunction, disorders of calcium homeostasis, excitatory amino acids, and inhibition or loss of neuroprotective mechanisms (Gerlach et al., 1996; Good et al., 1997; Hirsch et al., 1997; Jenner and Olanow, 1996; Kienzl et al., 1995; Schapira, 1995; Linert et al., 1996; Burke and Kolodilov, 1998).

Lewy bodies and neuronal cell death

The diagnostic hallmark of PD are LBs that are seen in two different types; the classical brainstem and cortical type. Classical LBs are intraneuronal cytoplasmic spherical inclusions ranging from 8 to 30μ in diameter with a hyaline eosinophilic core and a narrow pale-stained halo. They are composed of radially arranged 7 to 20 nm intermediate filaments associated with a granular electron-dense coating material and vesicular structures; the core contains densely packed filaments associated with dense granular material. Cortical LBs are eosinophilic, rounded, angular or reniform structures without an obvious halo. Ultrastructurally, they are composed of felt-like arranged intermediate filaments associated with granular material (Lowe et al., 1997; Pollanen et al., 1993). The main components of LBs are phosphorylated NF proteins, ubiquitin and proteins involved in ubiquitin metabolism. NF proteins are phosphorylated prior to incorporation into LBs and subsequently modified by proteolysis in the tail components of heavy and medium components (NFH and NFM) (Pollanen et al., 1993). Phosphorylated NF are present in both core and peripheral regions of LB. The aggregation of intermediate filaments in LBs, may be mediated through αB-crystallin; LBs also contain tubulin, microtubule-associated proteins and β-amyloid precursor protein (Lowe et al., 1997), and the recently detected presynaptic protein alpha-synuclein (Spillantini et al., 1997, 1998; Galvin et al., 1997; Baba et al., 1998; Irrizary et al., 1998; Trojanowski and Lee, 1998), while they do not contain microtubule-associated tau proteins (Trojanowski and Lee, 1994). Immunoreaction for cyclin-dependent kinase 5 suggests the involvement of this enzyme in phosphorylation of NF, a major step in the formation of insoluble fibrils (Sun et al., 1996). Ubiquitin, a heat-shock or cell stress protein, usually targeting proteins for breakdown, is another major component of both classical and cortical LBs, and enzymes associated with ubiquitin mediated proteolysis (ubiquitin carboxyl-terminal hydrolase and multicatalytic protease) are also seen in LBs (Lowe et al., 1997). Immunoreactivity for synaptophysin and chromogranin A suggests that the vesicular structures seen in LBs may represent degenerating nerve endings (Nishimura et al., 1994). It is unknown whether LB are cytotoxic or harmless side products or markers of cell damage, while the involvement of the ubiquitin proteolytic system suggests that they may be the structural manifestation of a cytoprotective response designed to eliminate damaged cellular elements. The selective accumulation of insoluble alpha-synuclein in LBs, degenerating neurities in PD and cytoplasmic inclusions in multiple system atrophy (MSA) (Arima et al., 1998) suggest that a reduction in its solubility may induce it to form filaments that aggregate into cytoplasmic inclusions. The deposits of proteinaceous fibrils may contribute to dysfunction or death of the involved cells (Trojanowski and Lee, 1998). However, the biological significance of these cytoplasmic inclusions, especially the role they might play in neurodegeneration, is still enigmatic, since Tompkins and Hill (1997), in the majority of SN neurons showing DNA fragmentation (positive TUNEL reaction) did not find somal LBs, suggesting that their presence does

Table 1. Distribution of Lewy bodies in Parkinson's disease (modified from Braak et al., 1996; Jellinger, 1989)

Affected Region[a]	Major (putative) neuromediators[b]	Frequent	Rare
Cerebral cortex	Multiple		+
Anterior cingulate gyrus		++	
Temporal, insular		+	
Allocortex			+
Amygdala, central accessory cortex nucleus		++	
Basal nuclei			+
Nucleus basalis of Meynert	Acetylcholine	++	
Thalamus, midline nuclei		++	
Hypothalamus, lateral nucleus		++	
Lateral posterior nucleus	Norepinephrine		+
Paraventricular nucleus	Multiple		+
Tuberomamillar nucleus		++	
Subthalamic nucleus	Dopamine		+
Periaqueductal gray	Multiple		+
Substantia nigra zona compacta	Dopamine	++	
Nucleus parabrachialis pigmentosus	Dopamine	++	
Nucleus paranigralis	Dopamine	++	
Westphal-Edinger nucleus	Acetylcholine	+	
Darkschewitsch nucleus	Acetylcholine		+
Supratrochlear nucleus	Serotonin	+	
Nucleus tegmenti pedunculopontinus	Acetylcholine	+	
Central pontine gray	Multiple		+
Locus coeruleus	Norepinephrine	++	
Nucleus subcoeruleus	Norepinephrine	+	
Nucleus pontis centralis oralis	Serotonin	+	
Central superior nucleus of raphe	Serotonin		+
Processus griseum pontis supralemniscalis	?	+	
Dorsal motor nucleus of vagus	Norepinephrine	++	
Nucleus of Roller	?		+
Nucleus gigantocellularis	Serotonin		+
Nucleus paragigantocelluaris laterialis	?		+
Nucleus medullae oblongatae centralis	Serotonin		+
C1 and C2 groups in medulla oblongata (Gai et al., 1993)	Epinephrine	++	
Spinal cord, intermediolateral column	Multiple		+
Spinal cord, intermediomedial column	Acetylcholine		+
Spinal cord, anterior horn			+
Autonomic (sympathetic) ganglia	Catecholamines	++	
Enteric nerve plexuses	Catecholamines		+
Adrenal medulla	Catecholamines		+

[a] Nomenclature of the brain stem nuclei according to Olszewski and Baxter (1982)
[b] Nieuwenhuys et al. (1988)
+ mild; ++ severe

not predispose a neuron to undergo cell death. On the other hand, LBs being the sequelae of frustranous proteolytic degradation of abnormal cytoskeletal elements may represent — similar to other cellular inclusions in neurodegenerative disorders, e.g. neurofibrillary tangles, Pick bodies, glial cytoplasmic inclusions in MSA, etc, — end products or reactions to hitherto unknown neuronal degeneration processes that are associated with disturbances of axonal protein transport and finally lead to cell death.

LBs are found in surviving neurons in SN, locus ceruleus (LC), dorsal motor vagal nucleus, thalamus, hypothalamus, substantia innominata, in the mesocorticolimbic system, raphe nuclei, pedunculopontine, and Westphal-Edinger nuclei, the intermediolateral columns of the spinal cord, olfactory bulb, in sympathic and parasympathic neurons, enteric nervous plexuses, and in adrenal medulla (Jellinger, 1991; Lowe et al., 1997) (Table 1). Cortical LBs are seen in the neurons of layers V and VI of temporal, insular, and cingulate regions (Braak et al., 1996). Pale bodies, rounded intracytoplasmic areas of granular pale-staining eosinophilic material in neurons, have been purposed as precursors of LBs and show similar immuno histochemical profile (Dale et al., 1992). Both LBs and pale bodies are diagnostic hallmarks for both PD and Lewy body dementia (LBD) but are not specific for these disorders; they have been described in a variety of conditions as a secondary pathology, e.g. multiple system atrophy (MSA), PSP, LB dysphagia, corticobasal degeneration, motor neuron disease, Hallervorden-Spatz disease, ataxia telangiectasia, sporadic and familiar Alzheimer's disease, Down syndrome, Meige syndrome, subacute sclerosing panencephalitis, and normal aging (Lowe et al., 1997). A distinctive neuritic degeneration with abnormal ubiquitin-IR neurites considered as "intraneuritic" LBs has been observed in the CA2–3 region of the hippocampus, in the amygdala and in many brainstem nuclei suggesting a degenerative process involving multiple neuronal systems (Gai et al., 1995). Absence of TH-IR suggests that many of these neuritic processes are not derived from dopaminergic neurons.

While the causes of neuronal death in PD are still unknown, several mechanisms have been discussed, i.e. apoptosis, cell necrosis, and autophagic degeneration (Anglade et al., 1997). Apoptosis of nigral neurons can be experimentally induced by striatal dopaminergic denervation (Jackson-Lewis et al., 1989). However, in nigral neurons of human PD subjects DNA fragmentation indicating cell death is only rarely seen (Mochizuki et al., 1997; Tompkins et al., 1997; Jellinger, 1999), more frequent in glial cells but without relation to microglial activation (Kösel et al., 1997; Banati et al., 1998). Nigral neurons only very rarely show the morphologic signs of apoptosis (less than 1%) (Kösel et al., 1997; Mochizuki et al., 1997; Tompkins et al., 1997), while recent studies showed no significant expression of apoptosis-related or stress proteins, suggesting that mechanisms distinct from classical apoptosis may play a role in the pathogenesis of PD and related neurodegenerative disorders (Jellinger, 1999).

Pathophysiology of basal ganglia in PD

The progressive degeneration of the nigrostriatal system causes dopaminergic denervation of the striatum. Due to the reciprocal innervation pattern of this system SN cell losses correlate with a ventromedial gradient of loss of TH-IR fibers and reduction of dopamine (DA), its enzymes and metabolites: more severe losses are seen in putamen than in caudate nucleus (Bernheimer et al., 1973) and in the caudal than rostral putamen with most severe reduction in rostral parts of caudate nucleus (Kish et al., 1988). The sensorimotor striatum, i.e. the dorsolateral putamen, severely involved in PD, is innervated by the ventral tier of SNZC (Fig. 2), while the limbic or ventral striatum is supplied by the less affected dorsal tier of SNZC and area 10 (ventral tegmentum) (Gerfen, 1992; Lynd-Balta and Haber, 1994). Even in early stages of PD, severe loss of dopa uptake and of the dopamine transporter (DAT) bound to presynaptic endings involves the posterior putamen more than the caudate (Booij et al., 1997; Rinne et al., 1997; Rinne et al., 1995b; Tissingh et al., 1997; Guttman et al., 1997). In hemiparkinsonism there is bilateral reduction with more severe loss contralateral to clinical symptoms (Marek et al., 1996; Rinne et al., 1997; Rinne et al., 1995b; Tissingh et al., 1998). Immunohistochemical data show strong reduction of TH- and DAT-IR fibers and endings in putamen, with relative preservation of its matrix and very little involvement of periventricular parts of caudate and nucleus accumbens (Miller et al., 1997). Decrease in dopa uptake in striatum correlates with the density of SNZC neurons, the loss of which is also related to both duration of illness and the clinical staging of PD (Ma et al., 1997). In early stages of illness, DA uptake in caudal and rostral putamen are reduced by about 50% with little involvement of ventral putamen. With progression of PD, DA loss progresses to the ventrorostral putamen which, however, even in late stages of illness shows only about 30% reduction of DA uptake. This suggests a focal progression of the degeneration with early involvement of the dorsal putamen in preclinical stages and later of the ventral putamen (Morrish et al., 1996). Onset of clinical PD symptoms occurs when striatal dopa uptake is reduced by 57 to 80% (Bernheimer et al., 1973; Kish et al., 1988; Morrish et al., 1996), while the threshold level for DAT loss is 56%, indicating that about 50% of dopaminergic striatal innervation are sufficient for normal function (Guttman et al., 1997).

TH and DAT are frequently co-localized in synaptic vesicles and plasma membranes in striatum. This suggests that transporters facilitate the uptake of DA and neurotoxines into distal dendritic and axonal processes of nigrostriatal neurons (Nirenberg et al., 1996). The predominant loss of DAT protein IRy in the metenkephalinergic striosomes of putamen with preservation of its matrix and of periventricular caudate islands also suggests that the DAT-richest endings are most sensitive towards degenerative changes in PD and MPTP parkinsonism (Miller et al., 1997). The presynaptic D2 autoreceptors regulating DA release and synthesis in midbrain and putamen are much more sensitive than postsynaptic DA receptors (Elsworth and Roth, 1996). Striatal D1 receptors are predominantly

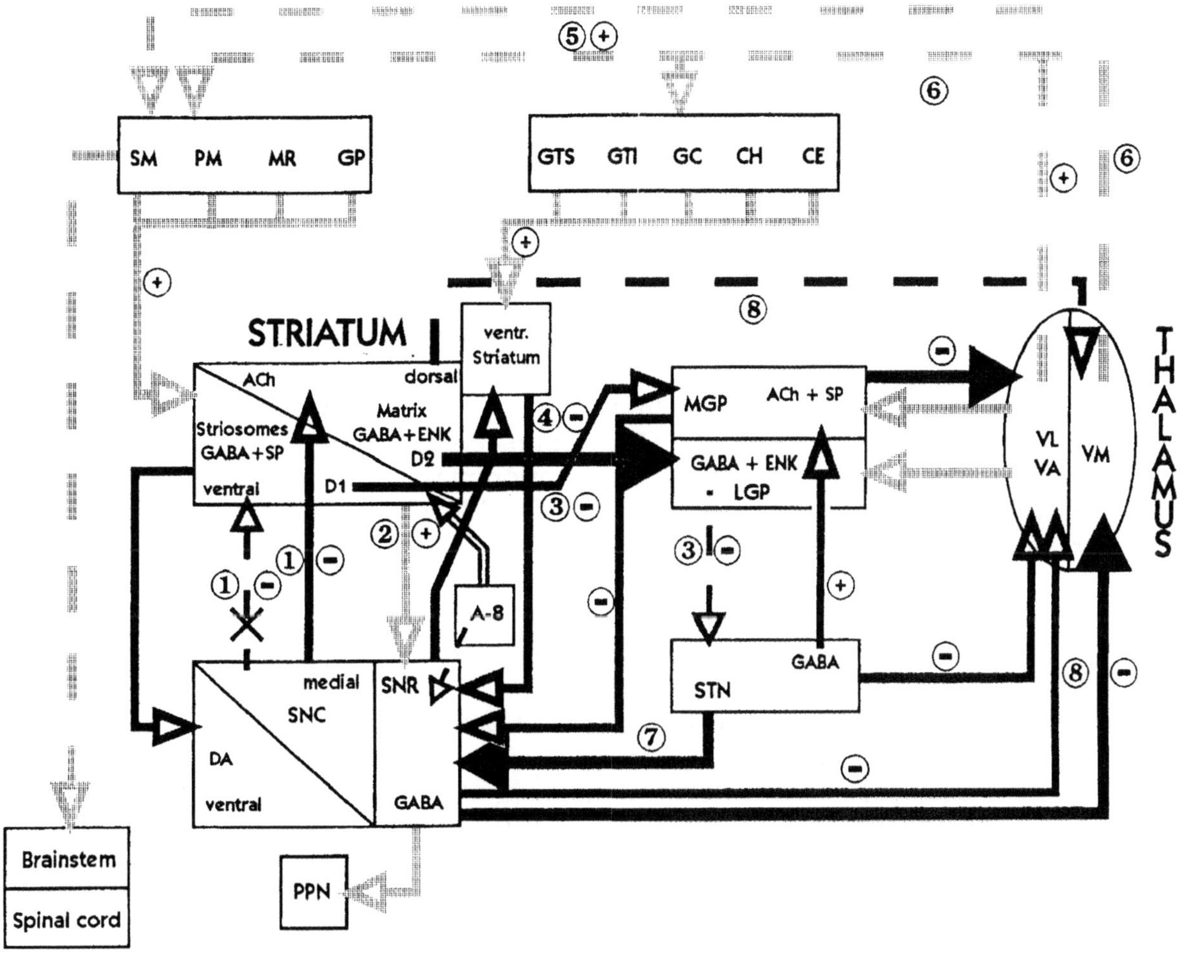

Fig. 2. Schematic diagram of basal ganglia-thalamocortical circuitry under normal conditions and in Parkinson's disease (modified after Albin, 1995 and Jellinger, 1996). Cortex: *SM* supplementary motor field, *PM* premotor field, *MR* motor cortex, *GP* postcentral gyrus, *GC* gyrus cinguli, *CH* hippocampus, *CE* entorhinal cortex. Basal ganglia: *SNC* substantia nigra zona compacta, *SNR* substantia nigra reticulata, *STN* subthalamic nucleus, *A-8* retrorubral field, *GPM* medial globus pallidus, *GPL* lateral globus pallidus, *PAN* peduriculopontine nucleus, *VL*, *VLM* ventro-lateral/medial thalamus, *VM* medioventral thalamus, *CS* superior colliculus, *MPT* mesopontine tegmentum, *ACh* acetyl choline, *SP* substance P, *ENK* enkephalin, *DA* dopamine. *1* nigrostriatal dopaminergic pathway *2* striato-nigral pathway *3* "indirect" loop *4* "direct" loop *5* motor or complex loop *6* thalamocortical pathway *7* pallido-subthalamic pathway; + excitatory, − inhibitory; Glutamat ⟶ GABA ⟶ dopamine ⟶ unchanged ---→ reduced ⟶ increased

located at striatonigral projection neurons ("direct pathway"), while D1 receptors are also present at endings of GABA neurons of SNZR that regulate DA release via dendrites of nigrostriatal neurons (Robertson, 1992).

Dopaminergic denervation of the striatum causes a severe loss of dendrites on type I medium spiny neurons, the principal goal of dopaminergic afferences form SN (Neill et al., 1988) which suggests transsynaptic degeneration. Ultrastructural analyses of caudate biopsies in patients with progressed PD showed dystrophic axons and neurites indicating transsynaptic degeneration of postsynaptic neurites due to dying back disturbances of nigrostriatal axon transport (Lach et al., 1992). These data and progressive loss of TH- and DAT-IR nigrostriatal fibers in striatum indicate transsynaptic degeneration as a possible morphological substrate for the severity of the disease. Progressive loss of initially preserved islands of dopaminergic innervation with little loss of DAT regulating the storage and release of L-Dopa, may contribute to decreased efficacy of L-dopa substitution in late stages of PD (Miller et al., 1997). Putamen D1 binding shows a negative correlation with duration of disease and L-dopa treatment (Turjanski et al., 1997; Stoessl and Ruth, 1998), while progressive loss of striatal D2 receptors preserved for a long time even with high degrees of presynaptic denervation may also contribute to dopa-resistance in late stages of PD as well as in MSA and PSP (Brooks, 1993; Gilman et al., 1996; Hierholzer et al., 1998).

Symptoms — specific lesion patterns in PD?

Progressive neuronal degeneration in PD causes functional disturbances in the complex regulatory loops of the basal ganglia as the basis of clinical signs and symptoms. Damage to the nigrostriatal system caused by SN cell loss produces dopaminergic denervation of the striatum whose efflux systems via pallidum, SNZR and thalamus to the cortex and the parallels striato-nigro-pallido-cortical loop to premotor cortex (Fig. 2) remain intact in early stages of PD. In addition to correlations between SN cell loss and both duration of PD and severity of symptoms (Ma et al., 1997), the major clinical subtypes of PD show specific morphological lesion patterns of pathophysiological relevance (Paulus and Jellinger, 1991; Jellinger, 1993, 1996):

a) In the *rigid-akinetic* type the ventrolateral part of SNZC which projects to the dorsal putamen degenerates more severely than the medial part projecting to the caudate nucleus and anterior putamen (Fig. 3a,b). There is a negative correlation between SNZC neuron counts, severity of rigidity-akinesia, and dopamin loss in striatum (Bernheimer et al., 1973; Paulus and Jellinger, 1991; Ma et al., 1997). Projections to the severely involved ventrolateral SNZC come from met-enkephalin-substance P rich, acetylcholinesterase poor striosomes of the putamen, the organization of which remains intact despite loss of dopaminergic influx, as are the CAB-positive, somatostatin rich matrix neurons that project to GABA-neurons of SNZR and further to motor thalamus (Gerlach et al., 1996). This is supported by selective increase of somatostatin mRNA expression in the basal ganglia in PD (Eve et al., 1998). The substance P-IR endings are in contact with D2 receptor IR dendrites and enkephalin-IR endings with D1 receptor IR dendrites by which the axon collaterals forming the "direct loop" are in synaptic contact with neurons of the "indirect loop" and vice versa. Thus, both information loops of the basal

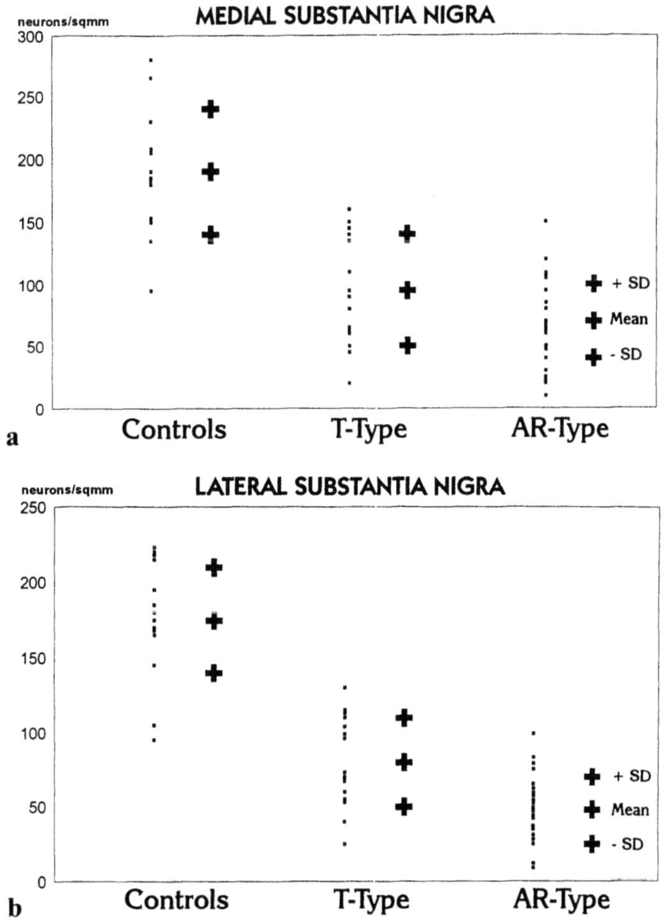

Fig. 3. Neuronal counts per mm² in medial (**a**) and lateral (**b**) substantia nigra zona compacta comparing controls (mean age 76.4 years), tremordominant (*T*) and akinetic-rigid (*AR*) cases of Parkinson's disease (mean age 77.7 vs. 77.2 years; duration 8.9 vs. 8.7 years). Symbols represent individual cases. Differences between the groups are significant ($p > 0.02$). Modified from (Paulus and Jellinger, 1991)

ganglia in the neostriatum are in mutual synaptic contact (Yung et al., 1996). Despite progressive dopaminergic denervation, in PD and Guam Parkinson dementia complex, the CAB-positive neurons of the striatal matrix and, thus, the striatal efferences remain intact (Ito et al., 1992). By contrast, in MSA and PSP, loss of calcineurin IR neurons in caudal and lateral putamen and of methionin-, substance P and CAB-IR in dorsolateral putamen, ventrolateral pallidum and lateral SNZC cause dysfunction of striatal efferences to internal pallidum and ventrolateral SNZR due to subregional deafferentation of striatal efflux nuclei (Goto et al., 1996; Ito et al., 1996). These disorders show lesion of both striatal afferences and efferent projection systems ("motor loop") (Fig. 4b,c). In the rigid-akinetic forms of MSA, nigral atrophy is more severe in the lateral SNZC with severe loss of calbindin-IR cells reflecting initial degeneration of the striatal matrix in the caudal putamen with

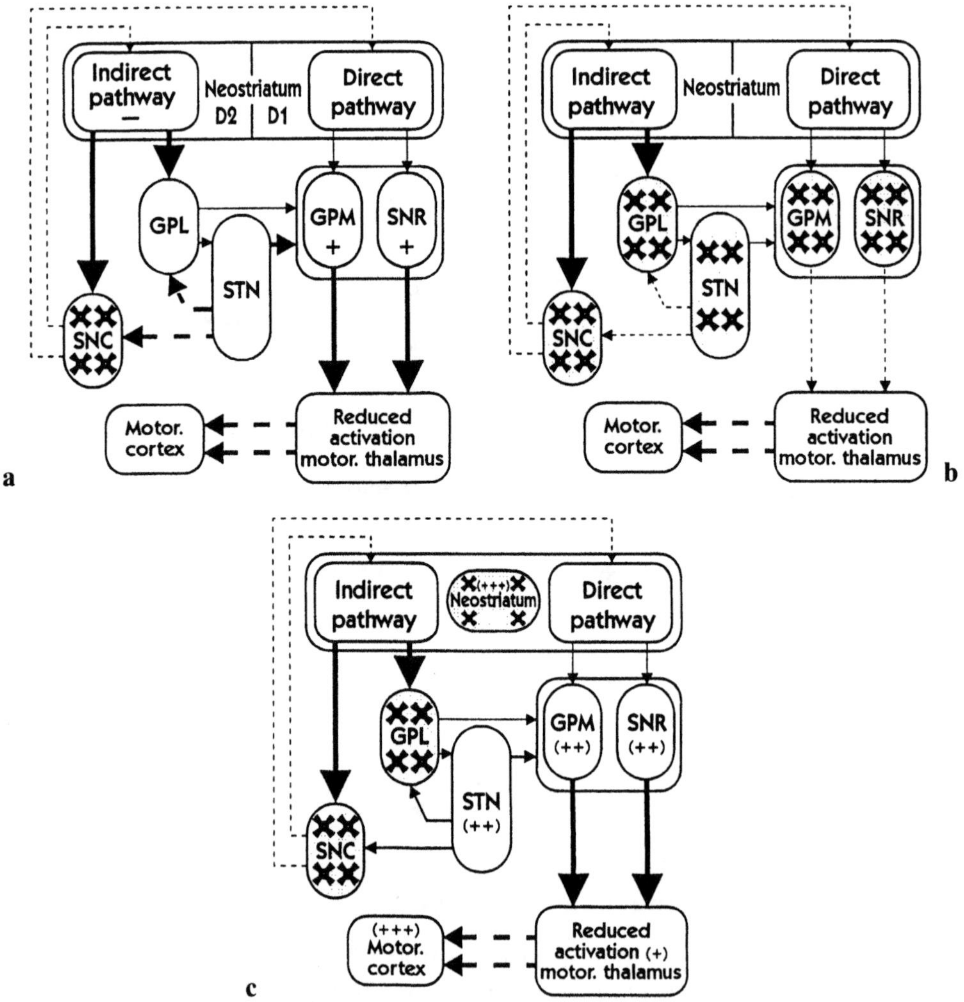

Fig. 4. Schematic diagram of basal ganglia-thalamocortical circuitry in Parkinson's disease (**a**), Progressive supranuclear palsy (**b**), and multiple system atrophy (**c**); xx degenerated; a and b partly modified from (Halliday et al, 1996). ---▸ reduced, ➞ increased, C(+), (++), (+++) frequency of inclusion bodies in MSA

transsynaptic degeneration of striatonigral efferences that remain intact in PD (Goto et al., 1996; Kume et al., 1996). Recent studies on the functional effect of striatal dysfunction in PD provide evidence for the hypothesis that, on the level of the striatum, motor impairment — rigidity and akinesia — results from altered DA neuronal integrity in the putamen but not in the caudate causing increased GABA activity of the "indirect" striatal efferent loop via SNZR and GPI leading to increased GABA output to the ventrolateral thalamus (Holthoffdetto et al., 1997). The marked inhibition of thalamic nuclei projecting to the cortex (thalamocortical motor loop) leads to reduced cortical activation by glutamate and thus to an akinetic-rigid syndrome that can be explained by overactivity of the STN and GPI (Fig. 4a). Since pallidotomy has been shown to diminish cortical inhibition in PD (Strafella et al., 1997), ablation of the sensorimotor portion of GPI is a highly affective treatment for advanced PD with pronounced effects on both tremor and dyskinesia (Fernandez and Dujovny, 1997; Gross et al., 1997; Johansson et al., 1997; Kraus et al., 1997; Taha et al., 1997).

Nigrostriatal denervation in PD patients and MPTP monkeys further causes loss of glutamate decarboxylase mRNS activity in SNZR and superior colliculus that receives inhibitory influx from GABA neurons in SNZR and is involved in eye motor control (Vila et al., 1996). Overactivity of inhibitory nigrotectal pathways may be responsible for oculomotor and cognitive disturbances in PD that are hardly influenced by L-dopa substitution. An imbalance between DA and glutamatergic activities with upregulation of the inhibitory striopallidal indirect loop causes desinhibition of the STN that, via glutamatergic fibers, activates the GPI and SNZR causing glutamatergic overactivity of cortical afferences (Fig. 2). Under such circumstances, the subunit composition and/or phosphorylation state of glutamatergic receptors of the N-methyl-D-aspartate (NMDA) subtype expressed on the dendritic spines of medium-sized striatal neurons changes in ways that compromise motor performance. Although levodopa acts, after conversion to DA, to reverse these changes by restoring striatal dopaminergic transmission, significant differences exist between the normally functioning DA system and the restoration of function provided by standard levodopa therapy. The nonphysiologic stimulation of DA receptors on striatal spiny neurons associated with current levodopa regimens now appears to contribute to the motor response complications that ultimately affect most parkinsonian patients. Current evidence suggests that alterations in signaling systems linking dopaminergic and glutamatergic receptors within these GABAergic efferent neurons induce NMDA receptor modification. Functionally, the resultant long-term change in glutamatergic synaptic efficacy leads to alterations in spiny neuron output, favoring the appearance of motor complications (Chase et al., 1998). Due to relative preservation of the strio-cortical efferences in PD, restitution of dopaminergic transmission by L-dopa substitution maintains the function of the "motor loop", while progressive degeneration with transgression to non-dopaminergic neuronal systems in later stages of PD cause loss and reduced stimulation of postsynaptic D2 and muscarinic cholinergic receptors in the striatum by dopamin with unchanged antagonist binding as a major cause of both drug resistance of rigidity and akinesia and adverse effects of L-dopa (fluctuation, dyskinesia, etc). Similar loss of striatal D2 receptors is seen in MSA and PSP explaining their frequent L-dopa resistance (Brooks, 1993; Gilman et al., 1996). In PSP, there is additional severe damage to pallidum, SNZR and STN causing dysfunction of these inhibitory nuclei projecting to motor thalamus (Hardman et al., 1997a; Hardman et al., 1997b; Litvan et al., 1996) (Fig. 4b).

b) The *tremor dominant type* of PD shows less severe neuronal loss in the total SNZC than the akinetic-rigid form (mean 52.8 vs 69%) and also less severe depletion in the lateral than in the medial SNZC (Paulus and Jellinger, 1991) (Fig. 3a,b). This possibly correlates with an inverse relation of dopa-uptake between caudate and putamen to the severity of tremor, while such an index does not exist in relation to the degree of bradykinesia and rigidity

(Otsuka et al., 1996). While progressive loss of striatal dopa-uptake causes increasing rigidity and akinesia, there is no relationship to the degree of rest tremor in PD. Hirsch et al. (1992) also observed less severe SNZC cell loss in tremor-dominant PD (60 vs 68%), but considerable cell losses in the periretrorubal field A8 in the lateral reticular formation (Fig. 2) which is usually preserved in akinetic-rigid PD (McRitchie et al., 1997). The area A8 that, in contrast to A9 and A10, contains only few TH-IR and DAT-IR but mainly calretinin-IR neurons, is widely independent of striatal influences (Gerfen, 1992; Lynd-Balta and Haber, 1994), but projects to the matrix of the dorsolateral striatum and ventromedial thalamus (Deutch et al., 1988). Both the A8 and A10 areas directly influence the striatal efflux via SNZR to thalamus and from there to the prefrontal cortex (Gerfen, 1992; Percheron et al., 1994). Tremor synchronous electric activity has been reported in STN, internal pallidum and ventral intermedial thalamus (VIM) (Taha et al., 1997; Obeso et al., 1997) and recent PET studies revealed increased expression of a metabolic network comprising the thalamus, pons, and premotor-cortical regions in PD patients with tremor indicating an increased functional activity of ventral thalamic projections to cortical motor regions (Antonini et al., 1998). In essential tremor, a hyperactivity in thalamus and contralateral cerebellum, pallidum and nucleus ruber has been observed by functional MRI (Bucher et al., 1997), while in both essential and primary orthostatic tremor, bilateral overactivity of cerebellar connections has been reported (Wills et al., 1996). In autopsy cases of essential tremor, no significant pathological changes have been observed. PET findings showing unchanged F-dopa uptake in striatum argue against relations to PD (Brooks, 1993), whereas in isolated resting tremor, reduced F-dopa uptake in striatum may indicate some subtype of PD (Chang et al., 1995). While data in experimental parkinsonism in primates suggest severe DA loss in putamen and/or pallidum to be an important factor of tremor in early PD (Bagmen et al., 1994), neurophysiological data suggesting neurons with 4–6 Hertz tremor activity in internal pallidum point to its role in the genesis of rest tremor in PD (Hutchison et al., 1997). These data, differences in biopterin content of ventricular CSF between akinetic-rigid and tremor dominant forms of PD (Furukawa et al., 1991), and in the relation of dopa uptake between caudate and putamen (Otsuka et al., 1996) suggest different pathophysiological mechanisms of these two major clinical subtypes of PD that have important therapeutic consequences for both drug and surgical treatment — i.e. stereotactic thalamotomy, with nucleus stimulation, pollidotomy and subthalamotomy — (Benabid et al., 1996; Boecker et al., 1997; Fernandez and Dujovny, 1997; Johansson et al., 1997; Koller et al., 1997; Kraus et al., 1997; Taha et al., 1997; Tasker et al., 1997; Dormont et al., 1998; Limousin et al., 1998; Kupsch and Earl, 1999; Ondo et al., 1998).

Involvement of extranigral systems

PD appears to be a multisystem disorder involving a variety of subcortical neuronal systems and causing multiple neuromediator dysfunctions that result

Table 2. Subcortical ascending systems in Parkinson's disease (modified after Jellinger, 1991)

Neuronal system	Reduction vs. controls %
1. Mesocortical dopaminergic systems	
VTA: loss of melanin+. TH-IR neurons	40–86
Ventral mesencephalon: dopamine/TH-IR loss	75
Limbic areas and neocortex: dopamine loss	40–60
2. Noradrenergic system	
Locus coeruleus: pigmented cell loss	40–50
Motor vagal nucleus: neuron loss (TH-IR, SP-IR)	5–77
Supraoptic, paraventricular nucleus: cell loss	0
Neocortex, limbic areas: norepinephrine loss	40–75
3. Serotonergic system	
Dorsal raphe nucleus: neuron loss	20–40
Striatum, neocortex: serotonin loss	20–60
Striatum, neocortex: 5-HT S-1, S-2 binding sites	reduced
4. Cholinergic system	
Nucleus basalis of Meynert: neuron loss	32–93
Neocortex, hippocampus: ChAT, AChE loss	50–60
Neocortex, hippocampus: nicotonic receptors	30–55
Nucleus tegmenti pedunculopontinus: neuron loss	36–57
Westphal-Edinger nucleus: neuronal loss	~50
5. Peptidergic systems	
Cholecystokinin-IRy s. nigra	30–40
Met-enkephalin-nigra, putamen, pallidum	50
Substance P-nigra, globus pallidus	30–40
IRy-nigra, pallidum	0
brainstem, nuclei, cell loss	57–85
Somatostatin: cortex, hippocampus (PD + AD)	30–60
Neuropeptide Y-medulla; neuronal loss	70
IR: cortex, hippocampus	10–30

in complex patterns of functional and clinical deficits (Braak et al., 1996; Jellinger, 1988, 1989, 1991, 1993). Most of the lesions are not random, but region-specific, affecting not all neurons containing a specific transmitter or harboring LBs (Tables 1, 2).

1. Mesocortical dopaminergic system

In PD, there is considerable damage to the mesocorticolimbic DA system which originates in the ventral tegmental area (VTA) and medial SNZC, both projecting to limbic and prefrontal areas and to the upper brainstem (Nieuwenhuys et al., 1988; Parent and Hazrati, 1995). In demented PD subjects, the medial portion of SNZC shows greater neuronal loss than the lateral regions (Paulus and Jellinger, 1991; Rinne et al., 1989; Zweig et al., 1993), whereas Duyckaerts et al. (1993) found no differences in SNZC neuronal density between subjects with and without dementia. Dementia in PD patients without concurrent AD has been associated with greater cell depletion in VTA (Zweig et al., 1993), 40–60% loss of DA in limbic and prefrontal cortex (Agid et al., 1990), diffuse reduction of TH-IRy in the prefrontal cortex (Gaspar et al., 1991), and neocortical monoamine terminal loss (Marie et al., 1995). Experimental evidence confirms that damage to DA neurons in VTA and medial SNZC and their prefrontal terminals resulting in a mesocorticolimbic DA deficiency is related to cognitive and behavioural dysfunctions in PD (Pillon et al., 1994; Sawaguchi and Goldman-Rakic, 1991; Jellinger, 1977; Pahwa et al., 1998; Monza et al., 1998).

2. Noradrenergic system

The locus coeruleus (LC, area A-6), the main source of noradrenergic innervation, with the rostral mesencephalon, hypothalamus, motor vagal nucleus, hippocampus, and neocortex as major projection fields (Nieuwenhuys et al., 1988), has an internal regional topographic organization (Loughlin et al., 1986). In PD, LC cell loss averages 40 to 50% with predominant involvement of the more caudal, compact parts projecting to cerebellum and spinal cord. LC cell loss is less severe in non-demented PD subjects than in those with depression and/or dementia (Gaspar and Gray, 1984) in whom it approaches the values seen in Alzheimer's disease (AD) (Chan-Palay, 1993; Zweig et al., 1993). However, other workers describe overlap between controls and PD subjects with and without dementia (Duyckaerts et al., 1993). Recent image morphometric studies of the LC displayed a 39% decrease of large pigmented neurons with a 44% (though not significant) increase of small unpigmented cells in PD patients without depression or dementia, but an 82% loss of large pigmented neurons and 39% decrease of small unpigmented cells in AD (Hoogendijk et al., 1995). Here, neuron depletion in the rostral parts of LC projecting to temporal cortex and hippocampus significantly correlates with the density of neuritic AD lesions indicating a retrograde degeneration of subcortical projection nuclei (German et al., 1992) with both cell death and shrinkage due to loss of synapses in their cortical target areas which is an early morphological feature in AD (Heinonen et al., 1995; Masliah, 1995). In contrast, LC in PD brains shows not only cell loss, but also cell shrinkage and a loss of phenotype without relationship to cortical pathology, suggesting primary degeneration of this nucleus (Jellinger, 1991; Hoogendijk et al., 1995). LC damage is most significant in severely demented PD-patients without concurrent AD and involves all anatomical levels (Chan-Palay, 1993; Zubenko et al., 1990). It results in severe deprivation of noradrenergic innervation in the forebrain and neocortex (Gaspar et al., 1991; Javoy-Agid et al., 1989) the effect of which has been related to dementia and depression (Zubenko et al., 1992) and to autonomic dysfunction (Gerlach et al., 1994).

In PD the melanin-pigmented neurons of the arcuate and periventricular hypothalamic nuclei are preserved (Matzuk and Saper, 1985), but a 25–35% decrease in nucleolar volume and cytoplasmic RNA and a loss of TH-IR cells in the related hypothalamic paraventricular and supraoptic nuclei (Mann et al., 1983) indicate reduction in norepiphrenin (NE) hypothalamic input. The decline in functional capacity of hypothalamic neurons as a result of degeneration of the ascending NE pathways based on the LC is more severe in demented PD cases (Mann et al., 1983). Central NE deficiency caused by degeneration of the ascending LC system has been related to mental changes and depression in PD (Agid et al., 1987; Zubenko et al., 1990) although recent studies showed no differences in the decrease of cerebral NE and the activation of LC neurons between depressed and nondepressed AD patients suggesting that the remaining LC neurons are hyperactive to compensate for decreased cerebral NE levels (Hoogendijk et al., 1998).

The noradrenergic dorsal vagal nucleus has been reported to show no or only very little changes in aged controls, PD, and PSP (Mann et al., 1983), with 5–17% cell loss, but frequent occurrence of LBs. Halliday et al. (1990a) reported a 77% loss in substance P (SP) immunoreactive preganglionic cells of the dorsal motor vagal nucleus, whereas the noradrenergic TH-IR neurons, although often harboring LBs, are not severely affected (<5% reduction). These data confirm earlier findings of degeneration of the motor preganglionic neurons with relative sparing of the pigmented ones (Jellinger, 1991), although the presence of many LBs in noncatecholaminergic neurons may indicate that the degeneration of the motor vagal nucleus is caused by a primary process and is not secondary to damage to pathways that synapse with these neurons (Halliday et al., 1990a). Loss of SP-IR vagal neurons, involvement by LBs of peripheral autonomic neurons in enteric plexuses (Wakabayashi et al., 1988), and reduced TH-IRy in adrenal medulla (Riederer et al., 1978), with significant decrease of DA (Carmichael et al., 1988), indicate damage to the sympathetic noradrenergic system that may contribute to vegetative symptoms in PD (Agid et al., 1990; Gerlach et al., 1994; Halliday et al., 1990b). Morphometric studies of the adrenergic nuclei A-1 and A 2 in the medulla oblongata containing poorly pigmented neurons revealed no cell loss in PD or PSP, but considerable degeneration of these nuclei in MSA (Malessa et al., 1990), while a loss of NE synthesizing cells was observed on the C1 area of the medulla (Gai et al., 1993).

3. Serotonergic system

The dorsal raphe nucleus (DRN) or nucleus supratrochlearis (Olszewski and Baxter, 1982) and the central superior (raphe) nucleus, corresponding to cell groups B6–8, the large neurons of which are serotonin-immunoreactive (Zweig et al., 1989b), give rise to the ascending serotonergic pathways and are bidirectionally connected with many central nervous system (CNS) centers (Nieuwenhuys et al., 1988). In PD, the DRN suffers a severe decrease of TH-IR neurons, whereas the phenylalanine hydroxylase (PH-8) immunostained serotoninergic neurons are unaffected. By contrast, there is about 60% reduction of PH-8^+ serotonin-synthesizing neurons in caudal midbrain and pons (Halliday et al., 1990b). Many of the remaining neurons contain LBs or Lewy neurites (Gai et al., 1995). Central serotonergic deficiency that is reflected by reduction of serotonin, its metabolises and receptors in the striatum and medial frontal cortex (Ruberg and Agid, 1988; Agid et al., 1990) has been related to cognitive disorders and depression in patients with PD.

4. Cholinergic system

Degeneration of the cholinergic system is characterised by cell loss and shrinkage in the magnocellular posterior parts of the Ch_4 region of the nucleus basalis of Meynert (NBM) projecting to the neocortex (Nieuwenhuys et al., 1988; Steriade and Biesold, 1990), in the pedunculopontine tegmental nucleus (Zweig et al., 1989a), the Westphal-Edinger nucleus, and other cholinergic brainstem nuclei (Jellinger, 1991). In PD, NBM cell depletion averages 30 to 40% without correlation to age or duration of illness, and is much higher in demented PD patients (where it approaches the values seen in AD (50 to 70%) than in those without dementia (0–40%) who show neuronal losses only slightly higher than normal aged controls (Table 2) In other studies, however, considerable NBM cell loss (about 50%) was also seen in non-demented PD subjects (Gaspar and Gray, 1984), with no relationship between cell counts and mental status (Tagliavini et al., 1984; Paulus and Jellinger, 1991). LBs in NBM are seen in 100% of cases and NFT in 30 to 65% of all PD brains, whereas neuritic plaques are rare in non-demented cases (Jellinger, 1991). Severe depletion of NBM with 75 to 80% loss of large cholinergic neurons is also observed in LBD (Jellinger and Bancher, 1996). This is accompanied by a decrease of cholinergic innervation of the cortex and hippocampus that may or may not correlate with the severity of NBM cell loss and with mental status (Gaspar and Gray, 1984; Duyckaerts et al., 1993), although both deficits are usually higher in demented PD (Ruberg and Agid, 1988; Perry et al., 1993). Comparative studies in 50 autopsy cases of PD revealed that in non-demented patients NBM cell loss ranging from 15 to 62% was associated with no or little cortical Alzheimer pathology, while in demented subjects, NBM cell depletion ranged from 64 to 90% and was often, but inconsistently, accompanied by severe cortical AD lesions (Jellinger, 1993; Jenner and Olanow, 1996). These findings and the demonstration of both frontal cholinergic deficiency and NBM cell loss in PD patients without cognitive deficits (Gaspar and Gray, 1984) suggest that degeneration of the ascending cholinergic system may precede the onset of mental changes and that there is a critical threshold of 75 to 80% neuronal loss and/or shrinkage within the NBM with equivalent cortical cholinergic denervation before dementia becomes apparent (Jellinger, 1989). The loss of cholinergic neurons within the substantia innominata and in the dorso-lateral tegmental nucleus of the brainstem may affect hippocampal and prefrontal structures, either directly or indirectly via the dorsomedial thalamic nucleus (Groenewegen et al., 1993) and may thus contribute to cognitive impairment and subcortical behavioural changes in PD (Dubois et al., 1994; Reid et al., 1992). Although the variable correlations between neuropsychological, morphological and biochemical changes in PD suggest that degeneration of the cholinergic forebrain system often may not be sufficient to induce severe dementia (Duyckaerts et al., 1993), some severely demented PD cases without concurrent AD show considerable degeneration of the NBM with neuronal loss and LBs (Jellinger, 1997). The heterogeneity of degeneration of cholinergic neurons in basal forebrain (Lehericy et al., 1993), and the variability in NBM cell depletion and loss of cholinergic markers in neocortex and hippocampus, irrespective of cortical pathology, suggest a primary degenerative process of the cholinergic forebrain system in PD, while secondary retrograde degeneration is proposed for AD (Saper et al., 1985) and has been confirmed by defective retrograde transport of nerve growth factor to NBM in AD (Mufson et al., 1995).

The nucleus tegmenti pedunculopontinus, pars compacta (PPNc) is a cholinergic nucleus in the dorsolateral part of the caudal mesencephalic tegmentum, the cells of which react with AChE, ChAT, SP, and NADPH diaphorase, and probably belong to the cholinergic cell group Ch5 (Halliday et al., 1990b). It is recognized as an important loop nucleus, receiving fibers and providing major projections to the thalamus, SNZC, STN, striopallidum, pontine tegmentum, basal forebrain, and minor projections to widespread cortical areas (Hallanger et al., 1987; Scarnati et al., 1987). The PPNc is considered an extrapyramidal center for affecting the balance between cholinergic and dopaminergic functions of the basal ganglia (Graybiel et al., 1990; Steriade and Biesold, 1990). Severe cell loss in the PPNc (75–80%) has been observed in PSP (Hirsch et al., 1987; Zweig et al., 1989a) and in patients with PD, in whom it ranges from 36 to 57%, with loss of 57% of the SP-IR cells (Jellinger, 1988), and strong correlation to neuron loss in the SNZC, but negative correlations to the patients' age, duration of illness, and LB counts. These data suggest a severe involvement of the putative cholinergic and SP^+ PPNc neurons in PSP and PD. Overactivity of the NPPc in experimental MPTP parkinsonism (Mitchell et al., 1990) could indicate dysfunctions in the tegmento-nigro-subthalamo-cortical pathways, although the clinical significance of the nucleus in PD remains unclear. Recent findings that the parameters of cholinergic transmission in thalamus and STN remain unaltered in PD and AD suggest that the involvement of the PPNc is a secondary retrograde phenomenon rather than part of a systemic cholinergic fiber degeneration (Xuereb et al., 1990). Damage to the PPNc may contribute to disorders of locomotor activities, abnormalities of gait and posture (Coles et al., 1989), coordination of the sleeping-waking cycle, or cognitive disturbances in PD and related disorders (Zweig et al., 1989a).

The Westphal-Edinger nucleus, a visceral subdivision of the oculomotor complex, giving rise to cholinergic fibers to the ciliary ganglion regulating pupilloconstriction (Bender, 1980), suffers a 54% neuronal loss in PD, and 2–3% of the cells are affected by LBs or NFTs (Hunter, 1985). Among 50 PD brains, LBs and NFTs were observed in 94 and 29% respectively (Jellinger, 1991). In PSP, these and other cholinergic mesencephalic nuclei suffer a significant decrease in the number of neurons with ChAT-IRy (69 to 93%), indicating a regionally selective destruction of cholinergic neurons (Juncos et al., 1991). Damage to this and other brainstem nuclei including periaqueductal gray, nucleus pontis oralis, and nucleus interstitialis of Cajal may explain neuroophthalmic dysfunctions in PD, and/or PSP (Guiloff et al., 1980; Rascol et al., 1989; Juncos et al., 1991; Leigh and Zee, 1991).

5. Pathological lesions in the amygdala

This nuclear complex, a primary limbic structure, is anatomically interconnected with the prefrontal cortex, hippocampus, basal forebrain, brainstem, and other areas regulating behavioural, mnestic, endocrine and autonomic functions (Sims and Williams, 1990; Vereecken et al., 1994). The amygdala is affected differently in PD and AD. In PD, LBs and ubiquitinated Lewy neurites mainly involve the accessory cortical and lateral nuclei, with less severe lesions in the basal and lateral nuclei (Braak et al., 1996). The central accessory nucleus integrates information from hypothalamus and brainstem to the basal forebrain and centers controlling endocrine or autonomic functions. Since it is involved in "alerting" behaviour and cognition, this damage may contribute to relevant clinical changes in PD. However, since PD-associated amygdala damage has been found to be unrelated to mental impairment, it appears not to be a major cause of dementia.

6. Peptidergic systems

Neuropeptides are thought to modulate the excitability of dopaminergic neurons in the extrapyramidal system (Graybiel et al., 1990; McGeer and McGeer, 1989). Whereas some neuropeptides are severely decreased, others are unchanged (Jellinger, 1991) (Table 2).

There is considerable reduction of Met- and Leu-enkephalin and of somatostatin in striatum and SN in PD and, in lesser degree, in "incidental Lewy body disease" (presymptomatic PD), while SP is only diminished in putamen. This reduction of peptides in the basal ganglia is considered an intragral part of PD and not secondary to either loss of dopaminergic neurons or prolonged dopa therapy (Fernandez et al., 1996). While some experimental models of PD show increased glutamatergic activity in striatum (Blandini et al., 1996), in human PD brain glutamate and aspartate are unchanged, GABA is mildly reduced in medial thalamus, and the NMDA receptors are decreased in caudate which may lead to new considerations of the current models of excitatory and inhibitory transmitters in the thalamocortical loop in PD (Gerlach et al., 1996).

In conclusion, damage to multipe neuronal systems causing complex biochemical changes and pathophysiological disturbances may represent the basis for the variable clinical picture of PD including motor, vegetative, behavioural, and cognitive dysfunctions, depression, pharmacotoxic psychoses, and other symptoms that usually increase with progressive stages of the disease. The pathogenesis of both the striatonigral and extranigral lesions responsible for the aggravation and complications of the disease and their mutual interrelations are still poorly understood. Further clarification of the morphological, pathophysiological and molecular substrates of clinical signs and symptoms of PD and related disorders is a major challenge for modern neurosciences in order to provide a broader basis for effective treatment strategies.

References

Agid Y, Javoy-Agid F, Ruberg M (1987) Biochemistry of neurotransmitters in Parkinson's disease. In: Marsden CD, Fahn ST (eds) Movement disorders 2. Butterworth, London, pp 166–230

Agid Y, Graybiel AM, Ruberg M et al (1990) The efficacy of levodopa treatment declines in the course of Parkinson's disease. Do non-dopaminergic lesions play a role? Adv Neurol 53: 83–100

Albin RL (1995) The pathophysiology of chorea, ballism and parkinsonism. Parkinsonism Rel Disord 1: 2–133

Anglade P, Vyas S, Javoy-Agid F et al (1997) Apoptosis and autophagy in nigral neurons of patients with Parkinson's disease. Histol Histopathol 12: 25–31

Antonini A, Moeller JR, Nakamura T, Spetsieris P, Dhawan V, Eidelberg D (1998) The metabolic anatomy of tremor in Parkinson's disease. Neurology 51: 803–810

Arima K, Ueda K, Sunohara N, Arakawa K, Hirai S, Nakamura M, Tonozuka-Ueahra H, Kawai M (1998) NACP/α-synuclein immunoreactivity in fibrillary components of neuronal and oligodendroglial cytoplasmic inclusions in the pontine nuclei in multiple system atrophy. Acta Neuropathol 96: 439–444

Baba M, Nakajo S, Tu PH, et al. (1998) Aggregation of α-synuclein in Lewy bodies of sporadic Parkinson's disease and dementia with Lewy bodies. Am J Pathol 152: 879–884

Bagmen T, Carmine B, De-Long MR (1994) Parkinsonian tremor is associated with low frequency oscillations in selective loops of the basal ganglia. Adv Behav Biol 41: 317–325

Banati RB, Daniel SE, Path MRC, Blunt SB (1998) Glial pathology but absence of apoptotic nigral neurons in long-standing Parkinson's disease. Mov Disord 13: 221–227

Benabid AL, Pollak P, Gao DM et al (1996) Chronic electrical stimulation of the ventralis intermedius nucleus of the thalamus as a treatment of movement disorders. J Neurosurg 84: 203–214

Bender MB (1980) Brain control of conjugate horizontal and vertical eye movements. Brain 103: 25–69

Bernheimer H, Birkmayer W, Hornykiewicz O et al (1973) Brain dopamine and the syndromes of Parkinson and Huntington: Clinical, morphological and neurochemical correlations. J Neurol Sci 20: 415–455

Blandini F, Porter RHP, Greenamyre JT (1996) Glutamate and Parkinson's disease. Mol Neurobiol 12: 73–94

Boecker H, Wills AJ, Ceballos-Baumann A et al (1997) Stereotactic thalamotomy in tremor-dominant Parkinson's disease — An (H2O)-O-15 PET motor activation study. Ann Neurol 41: 108–111

Booij J, Tissingh G, Boer GJ et al (1997) [123I]FP-CIT SPECT shows a pronounced decline of striatal dopamine transporter labelling in early and advanced Parkinson's disease. J Neurol Neurosurg Psychiatry 62: 133–140

Braak H, Braak E, Yilmazer D et al (1996) Pattern of brain destruction in Parkinson's and Alzheimer 's diseases. J Neural Transm 103: 455–490

Brooks DJ (1993) Functional imaging in relation to parkinsonian syndromes. J Neurol Sci 115: 1–17

Bucher SF, Seelos KC, Dodel RC et al (1997) Activation mapping in essential tremor with functional magnetic resonance imaging. Ann Neurol 41: 32–40
Burke RE, Antonelli M, Sulzer D (1998) Glial cell line-derived neurotrophic growth factor inhibits apoptotic death of postnatal substantia nigra dopamine neurons in primary culture. J Neurochem 71: 517–525
Carmichael SW, Wilson RJ, Brimijoin WS et al (1988) Decreased catecholamines in the adrenal medulla of patients with parkinsonism. N Engl Med J 319: 254
Chan-Palay V (1993) Depression and dementia in Parkinson's disease: Catecholaminergic changes in the locus ceruleus. A basis for therapy. Adv Neurol 60: 438–446
Chang MH, Chang TW, Lai PH, Sy CG (1995) Resting tremor only — a variant of Parkinson's disease or of essential tremor. J Neurol Sci 130: 215–219
Chase TN, Oh JD, Blanchet PJ (1998) Neostriatal mechanisms in Parkinson's disease. Neurology 51(2) [Suppl 2]: 30–35
Coles SK, Iles JF, Nicolopoulos-Stournaras S (1989) The mesencephalic centre controlling locomotion in the rat. Neuroscience 28: 149–157
Counihan TJ, Penney JB, Jr (1998) Regional dopamine transporter gene expression in the substantia nigra from control and Parkinson's diseased brains. J Neurol Neurosurg Psychiatry 65: 164–169
Dale GE, Probst A, Luthert P, et al (1992) Relationship between Lewy bodies and pale bodies in Parkinson's disease. Acta Neuropathol 83: 525–529
Damier P, Hirsch EC, Agid Y (1996) Patterns of cell loss in the substantia nigra in Parkinson's disease. Neurology 46: A442
Deutch AY, Goldstein M, Baldino FJ, Roth RH (1988) Telencephalic projections of the A8 dopaminergic cell group. Ann NY Acad Sci 537: 27–50
Dormont D, Cornu P, Pidoux B et al (1997) Chronic thalamic stimulation with three-dimensional MR stereotactic guidance. Am J Neuroradiol 18: 1093–1107
Dubois B, Malapani C, Verin M et al (1994) Cognitive functions and the basal ganglia. The model of Parkinson's disease. Rev Neurol 150: 763–770
Duyckaerts C, Gaspar P, Costa C et al (1993) Dementia in Parkinson's disease. Morphometric data. Adv Neurol 60: 447–455
Eadie MJ (1963) The pathology of certain medullary nuclei in parkinsonism. Brain 86: 781–795
Elsworth J, Roth R (1996) Dopamine autoreceptor pharmacology and function: Recent insights. In: Neve K, Neve R (eds) The dopamine receptors. Humana Press, Totowa, NJ, pp 223–265
Eve DJ, Nisbet AP, Kingsburg AE et al (1997) Selective increase in somatostatin mRNA expression in human basal ganglia in Parkinson's disease. Mol Brain Res 50: 59–70
Fearnley JM, Lees AJ (1994) Pathology of Parkinson's disease. In: Calne DB (ed) Neurodegenerative diseases. Saunders, Philadelphia, pp 545–554
Fernandez A, Deceballos ML, Rose S et al (1996) Alterations in peptide levels in Parkinson's disease and incidental Lewy body disease. Brain 119: 823–830
Fernandez PM, Dujovny M (1997) Pallidotomy — Editorial review. Neurol Res 19: 25–34
Furukawa Y, Kondo T, Nishi K et al (1991) Total biopterin levels in the ventricular CSF of patients with Parkinson's disease: A comparison between akineto-rigid and tremor types. J Neurol Sci 103: 232–237
Gai WP, Geffen LB, Dehoroy L, et al (1993) Loss of C1 and C3 epinephrine-synthesizing neurons in the medulla oblongata in Parkinson's disease. Ann Neurol 33: 357–367
Gai WP, Vickers JC, Blumbergs PC, Blessing WW (1994) Loss of nonphosphorylated neurofilament immunoreactivity, with preservation of tyrosine hydroxylase, in surviving substantia nigra neurons in Parkinson's disease. J Neurol Neurosurg Psychiatry 57: 1039–1046
Gai WP, Blessing WW, Blumbergs PC (1995) Ubiquitin-positive degenerating neurites in the brainstem in Parkinson's disease. Brain 118: 1447–1459

Galvin JE, Lee AMY, Baba M, Mann DMA, Dickson DW, Yamaguchi H, Schmidt ML, Iwatsubo T, Trojanowski JQ (1997) Monoclonal antibodies to purified cortical Lewy bodies recognize the mid-size neurofilament subunit. Ann Neurol 42: 595–603

Gaspar P, Gray F (1984) Dementia in idiopathic Parkinson's disease. A neuropathological study of 32 cases. Acta Neuropathol 64: 43–52

Gaspar P, Duyckaerts C, Alvarez C et al (1991) Alterations of dopaminergic and noradrenergic innervations in motor cortex in Parkinson's disease. Ann Neurol 30: 365–374

Gerfen C (1992) The neostriatal mosaic: Multiple levels of compartmental organization. Trends Neurosci 15: 133–139

Gerlach M, Jellinger K, Riederer P (1994) The possible role of noradrenergic deficits in selected signs of Parkinson's disease. In: Briley M, Marien M (eds) Noradrenergic mechanisms in Parkinson's disease. CRC Press, Boca Raton, pp 59–71

Gerlach M, Riederer P, Youdim MBH (1996) Molecular mechanisms of neurodegeneration. Synergism between reactive oxygen species, calcium, and excitotoxic amino acids. Adv Neurol 69: 177–194

German DC, Manaye KF, White CL (1992) Disease specific patterns of locus ceruleus cell loss. Ann Neurol 32: 667–676

German DC, Manaye KE, Sonsalia PK, Brooks BA (1993) Midbrain dopaminergic neurons (nuclei A8, A9, and A10): Three-dimensional reconstruction in the rat. J Comp Neurol 331: 297–309

Gilman S, Frey KA, Koeppe RA et al (1996) Decreased striatal monoaminergic terminals in olivopontocerebellar atrophy and multiple system atrophy demonstrated with positron emission tomography. Ann Neurol 40: 885–892

Good PF, Olanow CW, Perl DP (1997) LAMMA studies of iron, oxidative stress, and neuroprotective strategies in Parkinson's disease. In: Yasui M, Strong MJ, Ota K, Veritiy MA (eds) Mineral and metal neurotoxicology. CRC Press, Boca Raton, pp 379–390

Goto S, Matsumoto S, Ushio Y, Hirano A (1996) Subregional loss of putaminal efferents to the basal ganglia output nuclei may cause parkinsonism in striatonigral degeneration. Neurology 47: 1032–1036

Graybiel AM, Hirsch EC, Agid Y (1990) The nigrostriatal system in Parkinson's disease. Adv Neurol 53: 17–29

Groenewegen JH, Roeling T, Voorn P, Berendse H (1993) The parallel arrangement of basal ganglia-thalamocortical circuits: a neuronal substrate for the role of dopamine in motor and cognitive functions? In: Wolters EC, Scheltens P (eds) Mental dysfunction in Parkinson's disease. Vrije Universiteit Amsterdam 193: 3–18

Gross C, Rougier A, Guehl D et al (1997) High-frequency stimulation of the globus pallidus internalis in Parkinson's disease — a study of seven cases. J Neurosurg 87: 491–498

Guiloff RJ, George RJ, Marsden DC (1980) Reversible supranuclear ophthalmoplegia associated with parkinsonism. J Neurol Neurosurg Psychiatry 43: 352–354

Guttman M, Burkholder J, Kish SJ et al (1997) [C-11]RTI-32 PET studies of the dopamine transporter in early dopa-naive Parkinson's disease — implications for the symptomatic threshold. Neurology 48: 1578–1583

Hallanger AE, Levey AL, Lee HJ et al (1987) The origins of cholinergic and other subcortical afferents to the thalamus in the rat. J Comp Neurol 262: 105–124

Halliday CM, Blumbergs PC, Cotton RCH et al (1990a) Loss of brainstem serotonin- and substance P-containing neurons in Parkinson's disease. Brain Res 510: 104–107

Halliday CM, Blumbergs PC, Cotton RGH et al (1990b) Neuropathology of immunohistochemically identified brainstem neurons in Parkinson's disease. Ann Neurol 27: 373–385

Halliday GM, McRitchie DA, Cartwright HR et al (1996) Midbrain neuropathology in idiopathic Parkinson's disease and diffuse Lewy body disease. J Clin Neurosci 3: 52–60

Hardman CD, McRitchie DA, Halliday GM et al (1996) The substantia nigra pars reticulata in Parkinson's disease. Neurodegeneration 5: 49–55

Hardman CD, Halliday GM, McRitchie DA et al (1997a) Progressive supranuclear palsy effects both the substantia nigra pars compacta and reticulata. Exp Neurol 144: 183–192

Hardman CD, Halliday GM, McRitchie DA, Morris JGL (1997b) The subthalamic nucleus in Parkinson's disease and progressive supranuclear palsy. J Neuropathol Exp Neurol 56: 132–142

Heinonen O, Soininen H, Sorvari H et al (1995) Loss of synaptophysin-like immunoreactivity in the hippocampal formation as an early phenomenon in Alzheimer's disease. Neuroscience 64: 375–384

Hierholzer J, Cordes M, Venz S et al (1998) Loss of dopamine-D2 receptor binding sites in parkinsonian plus syndromes. J Nucl Med 39: 954–960

Hill WD, Arai M, Cohen JA, et al (1993) Neurofilament mRNA is reduced in Parkinson's disease substantia nigra pars compacta neurons. J Comp Neurol 328–336

Hirsch EC, Graybiel AM, Duyckaerts C, Javoy-Agid F (1987) Neuronal loss in the pedunculopontine tegmental nucleus in Parkinson's disease and progressive supranuelear palsy. Proc Natl Acad Sci USA 84: 5976–5980

Hirsch E, Graybiel AM, Agid YA (1988) Melanized dopaminergic neurons are differentially susceptible to degeneration in Parkinson's disease. Nature 334: 345–348

Hirsch EC, Mouatt A, Thomasset M et al (1992) Expression of calbindin D (28K) — like immunoreactivity in catecholaminergic cell groups of the human midbrain; normal distribution and distribution in Parkinson's disease. Neurodegeneration 1: 83–93

Hirsch EC, Faucheux B, Damier P et al (1997) Neuronal vulnerability in Parkinson's disease. J Neural Transm [Suppl] 50: 79–88

Holthoffdetto VA, Kessler J, Herholz K et al (1997) Functional effects of striatal dysfunction in Parkinson's disease. Arch Neurol 54: 145–150

Hoogendijk WJG, Pall CW, Troost D et al (1995) Image analysis-assisted morphometry of the locus ceruleus in Alzheimer's disease, Parkinson's disease, and amyotrophic lateral sclerosis. Brain 118: 131–143

Hunot S, Boissiere F, Faucheux B et al (1996) Nitric oxide synthase and neuronal vulnerability in Parkinson's disease. Neuroscience 72: 355–363

Hunter S (1985) The rostral mesencephalon in Parkinson's and Alzheimer's disease. Acta Neuropathol 68: 326–334

Hutchison WD, Lozano AM, Tasker RR et al (1997) Identification and characterization of neurons with tremor-frequency activity in human globus pallidus. Exp Brain Res 113: 557–563

Irrizary MC, Growdon W, Gomez-Isla T et al (1998) Nigral and cortical Lewy bodies and dystrophic nigral neurites in Parkinson's disease and cortical Lewy body disease contain α-synuclein immunoreactivity. J Neuropathol Exp Neurol 57: 334–337

Ito H, Goto S, Sakamoto S, Hirano A (1992) Calbindin-D28K in the basal ganglia of patients with Parkinsonism. Ann Neurol 32: 543–550

Ito H, Kosaka H, Matsumoto S, Imai T (1996) Striatal efferent involvement and its correlation to levodopa efficiacy in patients with multiple system atrophy. Neurology 47: 1291–1299

Itoh K, Weis S, Mehraein P, Müller-Höcker J (1997) Defects of cytochrome c oxidase in the substantia nigra of Parkinson's disease: An immunohictochemical and morphometric study. Mov Disord 12: 9–16

Jackson-Lewis V, Donaldson D, Przedborski S (1997) Apoptosis and Parkinson's disease (PD). Neurology 48: A323

Javoy-Agid F, Scatton B, Ruberg M et al (1989) Distribution of monoaminergic, cholinergic and gabaergic markers in the human cerebral cortex. Neuroscience 29: 251–269

Jellinger (1988) The pedunculopontine nucleus in Parkison's disease, supranuclear palsy and Alzheimer's disease. J Neurol Neurosurg Psychiatry 51: 540–544

Jellinger K (1989) Pathology of Parkinson's syndrome. In: Calne DB (ed) Drugs for the treatment of Parkinson's disease. Springer, Berlin Heidelberg New York Tokyo, pp 47–112
Jellinger KA (1991) Pathology of Parkinson's disease. Changes other than the nigrostriatal pathway. Mol Chem Neuropathol 14: 153–197
Jellinger K (1993) Pathogenese und Pathophysiologie der Parkinson-Krankheit. Neuropsychiatrie 7: 29–37
Jellinger KA (1996) Die Bewegungsstörungen im höheren Lebensalter. In: Zapotoczky HG, Fischhoff PK (Hrsg) Handbuch der Gerontopsychiatrie. Springer, Wien New York, pp 202–290
Jellinger KA (1997) Morphological substrates of dementia in parkinsonism. A critical update. J Neural Transm [Suppl] 51: 123–147
Jellinger KA (1998) Neuropathology of movement disorders. Neurosurg Clin North Am 9: 237–262
Jellinger KA (1999) Cell death mechanisms in Parkinson's disease. J Neural Transm (in press)
Jellinger KA, Bancher C (1996) Dementia with Lewy bodies. Relationship to Parkinson's and Alzheimer's disease. In: McKeith LG, Perry E et al (eds) Dementia with Lewy bodies. Cambridge University Press, New York, pp 268–286
Jenner P, Olanow CW (1996) Oxidative stress and the pathogenesis of Parkinson' disease. Neurology 56 [Suppl 3] pp 161–170
Johansson F, Malm J, Nordh E, Hariz M (1997) Usefulness of pallidotomy in advanced Parkinson's disease. J Neurol Neurosurg Psychiatry 62: 125–132
Joyce JN, Smutzer G, Whitty CJ, Myers A, Bannon MJ (1997) Differential modification of dopamin transporter and tyrosine hydroxylase mRNAs in midbrain of subjects with Parkinson's, Alzheimer's with parkinsonism, and Alzheimer's disease. Mov Disord 12: 885–897
Juncos JL, Hirsch EC, Malessa S et al (1991) Mesencephalic cholinergic nuclei in progressive supranuclear palsy. Neurology 41: 25–30
Kastner A, Hirsch EC, Agid Y, Javoy-Agid F (1993) Tyrosine hydroxylase protein and messenger RNA in the dopaminergic neurons of patients with Parkinson's disease. Brain Res 606: 341–345
Kienzl E, Puchinger L, Jellinger K et al (1995) The role of transition metals in the pathogenesis of Parkinson's disease. J Neurol Sci 134 [Suppl]: 69–75
Kish SJ, Shannak K, Hornykiewicz O (1988) Uneven patterns of dopamine loss in the striatum of patients with Parkinson's disease. N Engl J Med 318: 876–880
Koller W, Pahwa R, Busenbark K et al (1997) High-frequency unilateral thalamic stimulation in the treatment of essential and parkinsonian tremor. Ann Neurol 42: 292–299
Kösel S, Egensperger R, von Eitzen U et al (1997) On the question of apoptosis in the substantia nigra in Parkinson's disease. Acta Neuropathol 93: 105–109
Kraus JK, Jankovic J, Lai EC et al (1997) Posteroventral medial pallidotomy in Levodopa-unresponsive parkinsonism. Arch Neurol 54: 1026–1029
Kume A, Takahashi A, Hashizume Y (1993) Neuronal cell loss of the striatonigral system in multiple system atrophy. J Neurol Sci 117: 33–40
Kupsch A, Earl C (1998) Neurosurgical interventions in the treatment of idiopathic Parkinson disease: neurostimulation and neural implantation. J Molec Med 77: 178–184
Lach H, Grimes D, Benoit B, Minkiewicz-Janda A (1992) Caudate nucleus pathology in Parkinson's disease. Ultrastructural and biochemical findings in biopsy material. Acta Neuropathol 83: 352–360
Lehericy S, Hirsch EC, Pervera-Plerot P et al (1993) Heterogeneity of the degeneration of choliergic neurons in basal forebrain in patients with Alzheimer's disease. J Comp Neurol 330: 15–31
Leigh JR, Zee DS (1991) The neurology of eye movement, 2nd edn. FA Davis Philadelphia

Limousin P, Krack P, Pollak P et al (1998) Electrical stimulation of the subthalamic nucleus in advanced Parkinson's disease. New Engl J Med 339: 1105–1111
Linert W, Herlinger E, Jameson RF et al (1996) Dopamine, 6-hydroxydopamine, iron, and dioxygen — their mutual interactions and possible implication in the development of Parkinson's disease. Biochem Biophys Acta 1316: 160–168
Litvan I, Hauw JJ, Bartko JJ et al (1996) Validity and reliability of the preliminary NINDS neuropathologic criteria for progressive supranuclear palsy and related disorders. J Neuropathol Exp Neurol 55: 97–105
Loughlin SE, Foote SL, Bloom FE (1986) Efferent projections of nucleus locus coeruleus: Topographic organization of cells of origin demonstrated by three-dimensional reconstruction. Neuroscience 18: 291–306
Lowe J, Lennox G, Leigh PN (1997) Disorders of movement and system degenerations. In: Graham DI, Lantos PL (eds) Greenfield's neuropathology, 6th edn. E Arnold, London, pp 280–366
Lynd-Balta E, Haber SN (1994) Primate striatonigral projections: A comparison of the sensorimotor-related striatum and the ventral striatum. J Comp Neurol 345: 562–578
Ma SY, Rinne JO, Collan Y et al (1996) A quantitative morphometrical study of the neuron degeneration in the substantia nigra in patients with Parkinson's disease. J Neurol Sci 140: 40–45
Ma SY, Röyttä M, Rinne JO et al (1997) Correlation between neuromorphometry in the substantia nigra and clinical features in Parkinson's disease using disector counts. J Neurol Sci 151: 83–87
Malessa S, Hirsch EC, Cerver P et al (1990) Catecholaminergic systems in the medulla obiongata in Parkinsonian syndromes. A quantitative immunohistochemical study in Parkinson's disease, progressive supranuclear palsy, and striatonigral degeneration. Neurology 40: 1739–1742
Mann DMA, Yates PO, Hawkes J (1983) The pathology of the human locus coeruleus. Clin Neuropathol 2: 1–7
Marek KL, Seibyl JP, Zoghbi SS et al (1996) (123I)β-CIT/SPECT imaging demonstrates bilateral loss of dopamine transporters in hemi-Parkinson's disease. Neurology 46: 231–237
Marie RM, Barre L, Rioux P et al (1995) PET imaging of neocortical monoaminergic terminals in Parkinson's disease. J Neural Transm (PD-Dem Sect) 9: 55–71
Masliah E (1995) Mechanisms of synaptic dysfunction in Alzheimer's disease. Histol Histopathol 10: 505–519
Matzuk MM, Saper CB (1985) Preservation of hypothalamic dopaminergic neurons in Parkinson's disease. Ann Neurol 18: 552–555
McGeer EG, McGeer PL (1989) Biochemical neuroanatomy of the basal ganglia. In: Calne DB (ed) Drugs for the treatment of Parkinson's disease. Handbook of experimental pharmacology, vol 88. Springer, Wien-New York, pp 112–148
McGeer PL, Itagaki S, Akiyama H, McGeer EG (1988) Rate of cell death in parkinsonisrn indicates active neuropathological process. Ann Neurol 24: 574–576
McRitchie DA, Cartwright HR, Halliday GM (1997) Specific A10 dopaminergic nuclei in the midbrain degenerate in Parkinson's disease. Exp Neurol 144: 202–213
Miller GW, Staley JK, Heilman CJ et al (1997) Immunochemical analysis of dopamine transporter protein in Parkinson's disease. Ann Neurol 41: 530–539
Mitchell IJ, Aambrook MA, Jackson A et al (1990) Basal ganglia function in experimental movement disorders. In: Franks AJ (ed) Function and dysfunction in the basal ganglia. Manchester University Press, Manchester, NY, pp 94–109
Mochizuki H, Mori H, Mizuno Y (1997) Apoptosis in neurodegenerative disorders. J Neural Transm [Suppl] 50: 125–140
Monza D, Soliveri P, Radice V et al (1998) Cognitive dysfunction and impaired organization of complex motility in degenerative Parkinsonian syndromes. Arch Neurol 55: 372–378

Morrish PK, Sawle GV, Brooks DJ (1996) Regional changes in [18F]dopa metabolism in the striatum in Parkinson's disease. Brain 119: 2097–2103
Mufson EJ, Conner JM, Kordower JH (1995) Nerve growth factor in Alzheimer's disease. Defective retrograde transport to nucleus basalis. Neuroreport 6: 1063–1066
Neill TH, Brown SA, Rafols JA, Shoulson L (1988) Atrophy of medium spiny I striatal dendrites in advanced Parkinson's disease. Brain Res 455: 148–152
Nieuwenhuys R, Voogel J, Van Huizen C (1988) The human central nervous system. A synopsis and atlas, 3rd edn. Springer, Berlin Heidelberg New York Tokyo
Nirenberg MJ, Vaughan RA, Uhl GR et al (1996) The dopamin transporter is localized to dendritic and axonal plasma membranes of nigrostriatal dopaminergic neurons. J Neurosci 16: 436–447
Nishimura M, Tomimoto H, Suenaga T, et al (1994) Synaptophysin and chromogranin A immunoreactivities of Lewy bodies in Parkinson's disease brains. Brain Res 634: 339–344
Nishio T, Furukawa S, Akiguchi I, Sunohara N (1998) Medial nigral dopamine neurons have rich neurotrophin support. NeuroReport 9: 2847–2851
Obeso JA, Guridi J, DeLong M (1997) Surgery for Parkinson's disease. J Neurol Neurosurg Psychiatry 62: 2–8
Olanow CW, Tatton N, Redman R, Perl D, Walker R, Tatton WG (1998) Apoptosis and mitochondrial potential in Parkinson's disease (abstract). Ann Neurol 44: 452
Olszewski J, Baxter D (1982) Cytoarchitecture of the human brain stem 2nd edn. Karger, Basel
Ondo W, Jankovic J, Schwartz K, Almaguer M, Simpson RK (1998) Unilateral thalamic deep brain stimulation for refractory essential tremor and Parkinson's disease tremor. Neurology 51: 1063–1069
Otsuka M, Ichiya Y, Kuwabara Y et al (1996) Differences in the reduced 18F-Dopa uptakes of the caudate and the putamen in Parkinson's disease: correlation with the three main symptoms. J Neurol Sci 136: 169–173
Pahwa R, Paolo A, Tröster A, Koller W (1998) Cognitive impairment in Parkinson's disease. Eur J Neurol 5: 431–441
Pakkenberg B, Møller A, Gundersen HJG et al (1991) The absolute number of nerve cells in substantia nigra in normal subjects and in patients with Parkinson's disease estimated with an unbiased stereological method. J Neurol Neurosurg Psychiatry 54: 30–33
Parent A, Hazrati L-N (1995) Functional anatomy of the basal ganglia. I. The cortico-basal-ganglia-thalamo-corticol loop. Brain Res Rev 20: 91–127
Paulus W, Jellinger K (1991) The neuropathologic basis of different clinical subtypes of Parkinson's disease. J Neuropathol Exp Neurol 50: 143–155
Percheron G, Francois C, Yelnik J et al (1994) The basal ganglia related system of primates: Definition, description and informational analysis. In: Percheron G, McKensie JS, Féger J (eds) The basal ganglia, Vol IV, New ideas and data on structure and function. Plenum Press, New York, pp 3–20
Perry EK, Irving D, Kerwin JM et al (1993) Cholinergic transmitter and neurotrophic activities in Lewy body dementia: similarity to Parkinson's and distinction from Alzheimer disease. Alzheimer Dis Assoc Disord 7: 69–79
Pillon B, Deweer B, Malapani C et al (1994) Explicit memory disorders of demented parkinsonian patients and underlying neuronal basis. In: Korczyn AD (ed) Dementia in Parkinson's disease. Monduzzi, Bologna, pp 265–271
Pollanen MS, Dickson DW, Bergeron C (1993) Pathology and biology of the Lewy body. J Neuropathol Exp Neurol 52: 183–191
Rascol O, Clanet M, Motastruc JL et al (1989) Abnormal ocular movements in Parkinson's disease. Brain 112: 1193–1214
Reid WJG, Broe, GA, Morris JGL (1992) The roile of cholinergic deficiency in neuropsychological deficits in idiopathic Parkinson's disease. Dementia 3: 114–120

Riederer P, Rausch WD, Birkmayer W et al (1978) CNS modulation of adrenal tyrosine hydroxylase in Parkinson's disease and metabolic encephalopathies. J Neural Transm [Suppl] 14: 121–133
Rinne JO, Rummukainen J, Paljärvi J, Rinne UK (1989) Dementia in Parkinson's disease is related to neuronal loss in the medial substantia nigra. Ann Neurol 26: 47–50
Rinne JO, Burn DJ, Mathias CJ et al (1995a) Positron emission tomography studies on the dopaminergic system and striatal opioid binding in the olivopontocerebellar atrophy variant of multiple system atrophy. Ann Neurol 37: 568–573
Rinne JO, Leihinnen A, Ruottinen H et al (1995b) Increased densitiy of dopamine D2 receptors in the putamen, but not in the caudate nucleus in early Parkinson's disease — A PET study with [C-11]raclopride. J Neurol Sci 132: 156–161
Rinne JO, Kuikka JT, Berström MA et al (1997) Striatal dopamine transporter in Parkinson's disease; a study with a new radioligand, (123I)B-CIT-FP. Parkinsonism Rel Disord 3: 77–81
Robertson H (1992) Dopamine receptor interactions: Some implications for the treatment of Parkinson's disease. Trends Neurosci 15: 201–206
Ruberg M, Agid Y (1988) Dementia in Parkinson's disease. In: Iversen LL, Iversen SD, Snyder SH (eds) Handbook of psychopharmacology, vol 20, Psychopharmacology of aging nervous system. Plenum Press, New York, pp 157–206
Saper CD, German DC, White CL (1985) Neuronal pathology in the nucleus basalis of Meynert and associated cell groups in senile dementia of the Alzheimer's type. Possible role of cell loss. Neurology 35: 1089–1095
Sawaguchi T, Goldman-Rakic PS (1991) D1 dopamine receptors in prefrontal cortex: involvement in working memory. Science 251: 247–250
Scarnati E, Casbarri A, Campana E, Pacitti C (1987) The organization ot the nucleus tegmenti pedunculopontine neurons projecting to basal ganglia and thalamus. Neurosci Lett 79: 11–16
Schapira AHV (1995) Oxidative stress in Parkinson's disease. Neuropathol Appl Neurobiol 21: 3–9
Sims KS, Williams RS (1990) The human amygdaloid complex. Neuroscience 36: 449–472
Spillantini MG, Schmidt ML, Lee VMY, Trojanowski JQ, Jakes R, Goedert M (1997) α-synuclein in Lewy bodies. Nature 388: 839–840
Spillantini MG, Crowther RA, Jakes R et al (1998) α-synuclein in filamentous inclusions of Lewy bodies from Parkinson's disease and dementia with Lewy bodies. Proc Natl Acad Sci USA 95: 6469–6473
Steriade M, Biesold D (1990) Brain cholinergic systems. Oxford University Press, Oxford
Stoessl AJ, Ruth TJ (1998) Neuroreceptor imaging: new developments om PET and SPECT imaging of neuroreceptor binding (including dopamine transporters, vesicle transporters and post synaptic receptor sites). Curr Opin Neurol 11: 327–333
Strafella A, Ashby P, Munz M et al (1997) Inhibition of voluntary activity by thalamic stimulation in humans — relevance for the control of tremor. Mov Disord 12: 727–737
Sun D, Leung CL, Liem RKH (1996) Phosphorylation of the high molecular weight neurofilament protein (NF-H) by cdk5 and p53. J Biol Chem 271: 14245–14251
Tagliavini F, Pilleri G, Bouras C, Constantinidis J (1984) The basal nucleus of Meynert in idiopathic parkinson's disease. Acta Neurol Scand 69: 20–28
Taha JM, Favre J, Baumann TK, Burchiel KJ (1997) Tremor control after pallidotomy in patients with Parkinson's disease — correlation with microrecording findings. J Neurosurg 86: 642–647
Tasker RR, Lang AE, Lozano AM (1997) Pallidal and thalamic surgery for Parkinson's disease. Exp Neurol 144: 35–40
Tissingh G, Booij J, Winogrodzka A et al (1997) IBZM- and CIT-SPECT of the dopaminergic system in parkinsonism. J Neurol Transm [Suppl] 50: 31–37

Tissingh G, Bergmans P, Booij J et al (1998) Drug-naive patients with Parkinson's disease in Hoehn and Yahr stages I and II show a bilateral decrease in striatal dopamine transporters as revealed by [1-123]β-cit spect. J Neurol 245: 14–20

Tompkins MM, Hill WD (1997) Contribution of somal Lewy bodies to neuronal death. Brain Res 775: 24–29

Tompkins MM, Basgall EJ, Zamrini E, Hill WD (1997) Apoptotic-like changes in Lewy body-associated disorders and normal aging in substantia nigral neurons. Am J Pathol 150: 119–131

Trojanowski JQ, Lee VM-Y (1994) Phosphorylation of neuronal cytoskeletal proteins in Alzheimer's disease and Lewy body dementia. Ann NY Acad Sci 747: 92–109

Trojanowski JQ, Lee VMY (1998) Aggregation of neurofilament and α-synuclein proteins in Lewy bodies — Implications for the pathogenesis of Parkinson's disease and Lewy body dementia. Arch Neurol 55: 151–152

Turjanski N, Lees AJ, Brooks DJ (1997) In vivo studies on striatal dopamine D_1 and D_2 site binding in l-dopa treated Parkinson's disease patients with and without dyskinesias. Neurology 49: 717–723

Uhl GR (1998) Hypothesis: the role of dopaminergic transporters in selective vulnerability of cells in Parkinson's disease. Ann Neurol 43: 555–560

Uhl GR, Walther D, Mash D et al (1994) Dopamine transporter messenger RNA in Parkinson's disease and control substantia nigra neurons. Ann Neurol 35: 494–498

Varastet M, Riche D, Maziere M, Hantraye P (1994) Chronic MPTP treatment reproduces in baboons the differential vulnerability of mesencephalic dopaminergic neurons in Parkinson's disease. Neuroscience 63: 47–56

Vereecken ThHLG, Vogels OJM, Nieuwenhuys R (1994) Neuron loss and shrinkage in the amygdala in Alzheimer's disease. Neurobiol Aging 15: 45–54

Vila M, Herrero MT, Levy R et al (1996) Consequences of nigrostriatal denervation on the γ-aminobutyric acidic neurons of substantia nigra pars reticulata and superior colliculus in parkinsonian syndromes. Neurology 46: 502–509

Wakabayashi K, Takahashi H, Takeda S et al (1988) Parkinson's disease: The presence of Lewy bodies in Auerbach's and Meissner's plexuses. Acta Neuropathol 76: 217–221

Wills AJ, Thompson PD, Findley LJ, Brooks DJ (1996) A positron emission tomography study of primary orthostatic tremor. Neurology 46: 747–752

Xuereb JH, Perry EK, Candy JM et al (1990) Parameters of cholinergic neurotransmission in the thalamus in Parkinson's disease and Alzheimer's disease. J Neurol Sci 99: 185–197

Yung KKL, Smith AD, Levey AL, Bolam JP (1996) Synaptic connections between spiny neurons of the direct and indirect pathways in the neostriatum of the rat — Evidence from dopamine receptor and neuropeptide immunostaining. Eur J Neurosci 8: 861–869

Zubenko GS (1992) Biological correlates of clinical heterogeneity in primary dementia. Neuropsychopharmacology 6: 72–93

Zubenko GS, Moossy J, Kopp U (1990) Neurochemical correlates of major depression in primary dementia. Arch Neurol 47: 209–214

Zweig RM, Jankel WR, Hedreen JC et al (1989a) The pedunculopontine nucleus in Parkinson's disease. Ann Neurol 26: 41–46

Zweig RM, Ross CA, Hedreen JC et al (1989b) Neuropathology of aminergic nuclei in Alzheimer's disease. In: Iqbal K, Wisniewski HM (eds) Alzheimer's disease and related disorders. Liss, New York, pp 353–365

Zweig RM, Cardilio JE, Cohen M, Giere S, Hedreen JC (1993) The locus ceruleus and dementia in Parkinson's disease. Neurology 43: 986–991

Author's address: Prof. Dr. Kurt Jellinger, Ludwig Boltzmann Institute of Clinical Neurobiology, Psychiatric Hospital, Baumgartner Höhe 1, A-1140 Vienna, Austria

Pathogenesis and preclinical course of Parkinson's disease

P. Foley and **P. Riederer**

Clinical Neurochemistry, Department of Psychiatry, University of Würzburg, Federal Republic of Germany

Summary. Idiopathic parkinsonism (IP) is defined by its classic symptomology, its responsiveness to therapies which elevate dopamine levels, and by the failure to identify a specific etiological factor. The progressive and irreversible degeneration of dopaminergic neurons projecting from the substantia nigra pars compacta (SNc) to the striatum and the presence of SNc Lewy bodies are regarded as the essential pathological bases of IP, but neither the initiator(s) nor the nature of the degeneration have been determined, nor its relationship with degenerative changes in other parts of the IP brain. This paper discusses the various hypotheses that have been proposed to explain these phenomena, arguing that IP be regarded as a multisystem disorder, both at the level of individual neurons and at the whole brain level. It is probable that IP is the result of a multifactorial process, and that a cascade of interacting and overlapping biochemical mechanisms determine the course of the disease.

Introduction

Idiopathic parkinsonism (IP) is defined by its classic symptomology (tremor, bradykinesia and rigidity), its responsiveness to a therapy which elevates brain dopamine (DA) levels, and, significantly, by the failure to identify a specific etiological factor, thus distinguishing it from a range of parkinsonian syndromes, such as manganese toxicity. The progressive and irreversible degeneration of the pigmented dopaminergic neurons which project from the substantia nigra pars compacta (SNc) to the dorsal striatum, accompanied by the presentation of Lewy bodies (LB) in the SNc, is generally regarded as the essential pathological basis of IP, but neither the initiator(s) nor the nature of this degeneration have been determined, nor has its relationship with degenerative changes seen in other parts of the IP brain been clarified (Braak et al., 1995; Jellinger, 1995). These uncertainties persist despite the ever increasing sophistication of the approaches applied to the investigation of the pathophysiological basis of IP.

Current therapeutic strategies ameliorate the symptoms of the disease, at least temporarily, by compensating for the declining striatal DA levels associated with the loss of the nigral input, but are not capable of arresting the further progression of the disorder. In contrast, a "neuroprotective" strategy would aim to protect vulnerable, surviving neurons from whatever neurotoxic processes are responsible for the observed neuropathological changes, thereby preserving their function. The development of such a neuroprotective strategy, however, requires an understanding both of the processes involved and of the preclinical course of IP.

A number of pathogenetic processes have been proposed as underlying the degeneration of the SNc in IP. The current paper will discuss first the phenomena which an "explanation" of IP must take into account, and will then evaluate the various explanations which have been proposed. The implications of this discussion for the development of a theory of the etiology of IP will then be considered, as will the consequences for the development of novel therapeutic approaches.

The pathophysiology and early course of IP

The course of IP can be divided into two phases: a presymptomatic stage and a clinical stage. This is, of course, an arbitrary division; retrospective studies of IP patients have often revealed the presence of subtle disturbances of motor function years before the "symptomatic" phase (Koller and Montgomery, 1997). In fact, little is securinly known about the preclinical course of IP, and it is this phase which is most critical in determining the origin of the disease. It is recognized that several forms of parkinsonism have long "latent" phases (Koller and Montgomery, 1997), and the length of the preclinical stage for IP has been variously estimated as lying between five (based on a comparison of normal and PD brains; Fearnley and Lees, 1991) and 40 years (assuming a constant rate of nigral neuronal loss throughout the course of IP; Sherman et al., 1989).

But most IP patients also exhibit symptoms not related to motor performance. Olfactory dysfunction is an early, non-progressive feature of most IP case histories, often apparent years prior to the onset of motor symptoms (Doty et al., 1988). It has also been suggested that there is a premorbid "parkinsonian personality", including depression, which is manifest before the appearance of motor symptoms (Poewe et al., 1983), as are a variety of autonomic nervous system dysfunctions (Birkmayer and Riederer, 1985). Taken together, these phenomena suggest the presence of brain changes prior to the commencement of the clinical phase, changes almost certainly not restricted to the SNc. Any of these symptoms alone would be too unspecific to allow a decisive diagnosis of IP, but the development of a "PD Battery" (incorporating assessments of motor function, olfaction and mood) as a tool for the early detection of IP is currently being examined (Koller and Montgomery, 1997).

The brain which the neuropathologist examines generally represents the endstage of the disease and can therefore only provide hints with respect to what biological events are responsible for the neurological changes presented. The most striking alterations observed in the brain of the IP patient at autopsy — a brain which on the whole, before closer examination, appears normal — are:

- The selective degeneration of the pigmented dopaminergic cells of the brainstem. The loss is greatest in the SNc (76%), significant in the SN pars lateralis (34%), ventral tegmental area (55%) and the catecholaminergic cell group A (31%), and minimal in the central gray substance (7%; Hirsch et al., 1988). It is thought that at least 60% of the SNc neurons must be lost before clinical manifestation of the disorder (Riederer and Wuketich, 1976); but the loss of neurons in the SNc itself is also highly selective, with greater losses in the ventral (97%) than the dorsal regions (57%), and an apparent preference for melanized and calbindin-poor cells (Hirsch et al., 1997; see, however Gibb, 1992). The various clinical subtypes of IP differ in the specific morphological patterns presented (see reviews by Jellinger, 1991; Braak, 1995).
- The presence of Lewy bodies (LBs) or Lewy neurites in specific brain regions, both in the SNc and in extranigral nuclei, especially the amygdala and its projection regions, but also in the cortex, autonomic ganglia and the olfactory bulbs (Gibb, 1989; Braak et al., 1995).

Neurodegeneration. It is important to recognize that the well-known nigrostriatal neurodegeneration does not represent the total picture of IP neuropathophysiology; Braak et al. (1994, 1995, 1997) have emphasized that in IP all major structures of the limbic system, including the deep layers of the entorhinal cortex, the CA2 area of the hippocampus and that portion of the cingulate gyrus projecting to the amygdala, are characteristically afflicted by a thick network of Lewy neurites. All regions of the amygdala itself exhibit severe, IP-specific damage (including the presence of LBs), especially the central core. The multifarious projections of the amygdala mean that this is of consequence for both the motor and nonmotor symptoms of IP; this region exercises considerable influence on the frontal association cortex, important for decision-making in humans, as well as on the vegetative and neurosecretory nuclei of the hypothalamus. Braak et al. (1997) has indeed suggested that the vital steering function of the amygdala may be significantly impaired early in the course of IP.

And not all losses are of catecholaminergic neurons: the loss of serotonin- and substance P-containing neurons in the brainstem is documented (Halliday et al., 1990), and a massive loss of cholinergic cells in the nucleus basalis Meynert has been reported in IP associated with dementia (Jellinger, 1991). It is not clear whether the changes in the SNc precede those in other brain regions, none of which are directly connected with the SNc or the dorsal striatum, or whether the brainstem changes are in themselves sufficient to explain the major clinical features of the disorder. It is, however, assumed that the loss of striatal DA consequent to nigrostriatal neuronal loss leads to a dominance of the inhibitory output from the basal ganglia to the cortex, provoking the characteristic motor symptoms of IP (Gerlach and Riederer, 1993).

Neuroimaging has also been applied to the early detection of nigrostriatal changes in "prospective" IP patients. The most commonly reported feature linked to IP has been a decreased uptake of [^{18}F]-L-DOPA on one side of the brain in relatives of IP sufferers or subjects who on closer examination exhibited subtle motor abnormalities (Brooks, 1991; Sawle et al., 1992; Snow, 1996). Similar PET studies have followed the progression of the disease during its clinical phase; the authors interpreted their findings as indicating that the preclinical phase lasted 3.1 years (Morrish et al., 1996); Lee et al. (1994) concluded that nigral cell loss was greatest at the onset of clinical symptoms, subsequently slowing to the rate seen in normal ageing. This dynamic is consistent with the findings of a number of other workers (Scherman et al., 1989; Fearnley and Lees, 1991).

The Lewy body. Braak and his coworkers (1995) have argued that IP is essentially a cytoskeletal disorder affecting specific cell groups, and see the LB as the critical histopathological entity in the disease. The LB, a cytoplasmic inclusion first identified in the brainstem in 1912 (Lewy, 1912), is generally regarded to be the result of altered neurofilament metabolism and/or transport (Gibb et al., 1991; Hansen and Galasko, 1992); whether it is the result or cause of neuronal damage and degeneration, however, is not known. The brainstem LB, or "classical" LB, consists ultrastructurally of a dense core of circular structures with a rim of radiating filaments, the so-called "halo" (Duffy and Tennyson, 1965; Hansen and Galasko, 1992; Forno, 1995); LBs in the cortex appear similar, but are less tightly arranged, and are not as closely associated with neurodegeneration (Kosaka, 1978). Immunohistochemical studies have demonstrated the presence in brainstem LBs of phosphorylated forms of all species of neurofilament protein, with the interesting exception of the tau protein, centrally involved in Alzheimer's disease (AD), and of ubiquitin; the latter is also associated with neurodegeneration in neurofibrillary tangles and Pick bodies, and is a highly conserved protein which is often found conjugated with proteins destined for proteolysis (Goldman et al., 1983; Bancher et al., 1989; Jellinger, 1990; Hill et al., 1991; Forno, 1995). A structure similar to the LB, the "pale body" (PB), lacks a dense core and halo, but is thought to represent an early, immature form of the LB (Leigh et al., 1989). Both LBs and PBs are found only in neural cells; that is, they do not occur in glial cells (Jellinger, 1990), and are more common in recently diagnosed IP cases than in longer term or slowly progressing patients (Forno, 1995).

LBs have not been definitely identified in species other than man. Similar structures have been observed in older squirrel monkeys treated with MPTP (see below), but these more closely resemble the PB described above; they may, in these cases, indicate cellular stress, and are reversible (Forno et al., 1995, 1996). Interestingly, however, they occur most commonly in the periamygdaloid cortex, a finding which stimulated the reexamination of brains from IP patients; this led to the recognition of the involvement of extranigral structures in IP pathology.

Interestingly, it is believed that about 5 to 10% of asymptomatic individuals over 60 years of age possess LBs ("incidental Lewy body disease"; ILBD), mainly in the SN (Hansen and Galasko, 1992; Forno and Langston, 1993), a fact which has been interpreted as the presence of preclinical IP in these subjects (Gibb and Lees, 1989; Smith and Prayson, 1996). While LBs are not seen in many other forms of parkinsonism (Taussig and Plante-Bordeneuve, 1997), they occur in other neurodegenerative disorders, including AD, and can thus not be regarded as specific to IP (Gibb and Lees, 1989; Perry et al., 1990). Of particular interest in the last few years has been the recognition of "diffuse (or cortical) Lewy body disease" (DBLD), a disorder which more closely resembles AD than IP. The age of disease onset is lower than for IP, and, although parkinsonian symptoms usually develop during the course of the disease, an AD-style dementia dominates the clinical picture (Gibb et al., 1985). The role of the LB in IP is thus by no means clear; nevertheless, the LB is such a characteristic feature for IP, that it seems a logical object for investigation of the biochemical nature of this disorder.

The investigator is therefore confronted with two basic questions when addressing the etiology of IP: what role do the observed changes play in the manifestation of the disease, and why/how do they arise? In this framework, the following clinical features must also be taken into account:

— The only well-established risk factor for sporadic IP is old age: presentation of the disease before the fifth decade is rare, and the risk rises to a peak of 0.6–0.8% in the eighth decade. Gender does not appear to be significant, and, although a number of other risk factors have been suggested, no coherent link amongst them has led to the identification of a common factor (Martilla and Rinne, 1981; Román et al., 1995).
— The disease is characterized by its progressive nature — that is, the symptoms are not simply the acute response to an immediate physiological challenge, but rather a process which unfolds inexorably over a period of years.
— Genetic factors probably play a role, but the disease generally manifests itself sporadically, although family clusters are also known, accounting for about 10% of all IP cases. Ethnicity now seems also to be uninvolved (Román et al., 1995).

Current explanations for IP

The MPTP model of IP. A major problem for the investigation of IP remains the lack of a totally satisfactory animal or human model of the disease. A number of models for individual aspects of the disease are available, but few reproduce the essentially progressive and irreversible nature of the disease (see review: Gerlach and Riederer, 1996). The most fruitful contribution in this regard was the recognition in the early 1980s of the toxic effects of 1-methyl-4-phenyl-1,2,3,6-tetrahydropyridine (MPTP) and its MAO-B metabolite 1-methyl-4-phenylpyridine (MPP^+), which provided the impulse for the development of an etiological hypothesis based on toxin-induced oxidative stress (Langston et al., 1983; Gerlach et al., 1991; Tipton and Singer, 1993; Gerlach and Riederer, 1996; Langston, 1996). In humans, MPTP provokes a progressive parkinsonism which includes all important clinical features of IP. MPTP toxicity has also been examined in other primates and in rodents, but its effects appear to be highly species-specific; despite eliciting parkinsonian symptoms in most test animals, only in higher primates (such as the baboon and rhesus monkey) is the syndrome of a permanent nature.

The precise mechanism of the toxic action of MPP^+ is still not entirely clear, but it was recognized early to inhibit the mitochondrial complex I (NADH dehydrogenase), causing a decline in ATP production and an elevated generation of free radicals, both of which ultimately lead to oxidative damage to the cell membrane and other cell structures (Singer et al., 1987; Adams and Odunze, 1991). Evidence for oxidative stress in the SNc in IP has been identified, as was a deficit in mitochondrial function, so that MPTP/MPP^+ toxicity has become the most studied human and animal model of the disease. It must, however, be borne in mind that the relevance of the MPTP model is still disputed, as is the essential basis of its toxicity. Despite sharing a number of significant features with IP, important differences also exist: the lack of LBs and the lack of involvement of extranigral regions in the histopathology have lead to it being described as a model of "pure nigral parkinsonism" (Davis et al., 1979). Nevertheless, it provided the spur which led to the recognition that a single event could lead to an immensely complex chain of events and to the neurodegeneration of a specific brain region, and thus afforded valuable insights into the pathophysiology of IP.

Oxidative stress

Free radical production. Despite criticism, the hypothesis that oxidative stress (OS) is responsible for the degeneration of nigral neurons remains one of the most investigated hypotheses for the etiology of IP (reviewed in Gerlach et al., 1996; Jenner and Olanow, 1996a; Olanow, 1997). It proposes that free radicals, such as the superoxide radical ($^{\bullet}O_2^-$), nitric oxide ($NO^{\bullet}$) and, of particular significance, the hydroxyl radical ($^{\bullet}OH$), damage essential components of the dopaminergic neuron, including DNA, mitochondria, protein structures and the cell membrane, resulting in the disruption of their function and ultimately in cell death (Halliwell, 1992). Such radicals, especially $^{\bullet}OH$, are highly reactive molecules which are produced during normal cellular metabolism, but specific cellular detoxification systems normally minimize their damaging effects. Oxygen radicals are produced during the reduction of molecular oxygen to water in the oxidative phosphorylation process:

$$O_2 \xrightarrow{+e^-} {}^{\bullet}O_2^- \xrightarrow[+2H^+]{+e^-} H_2O_2 \xrightarrow{+e^-} {}^{\bullet}OH + OH^- \xrightarrow[+2H^+]{+e^-} H_2O_2$$

Most oxidative damage in vivo is attributed to $^{\bullet}OH$, as $^{\bullet}O_2^-$ is mostly reductive in nature, its significance lying in its being the precursor of H_2O_2, although its reaction with $NO^{\bullet}$ may also be important (see below). H_2O_2 is an oxidant, but its clearance by cellular dismutases renders it comparatively harmless; in the presence of transition metals, however, its conversion to $^{\bullet}OH$ can become significant.

It is not possible to directly measure free radicals in a living organism, but the concentration of their reaction products and their damage which they cause are used as indirect indices of their levels, and there is evidence that these are elevated in the SN in IP, either as the result of increased production or reduced clearance by the radical-metabolizing enzymes. For example, Jenner's group found evidence of increased membrane peroxidation in the SNc of IP patients and widespread oxygen species-induced protein modification, with protein carbonyl levels elevated in all regions of the IP brain (Dexter et al., 1989, 1994; see also Yoritaki et al., 1996). Further, elevated levels of thiobarbituric acid-reactive substances (a measure of the secondary products of lipid peroxidation) with decreased levels of polyunsaturated fatty acids (the substrate for peroxidation) have been reported (Dexter et al., 1989), as have increased nigral levels of 8-hydroxy-2′deoxyguanosine, a measure of free radical-mediated DNA damage; such damage has also been detected in other areas of the IP brain (Sanchez-Ramos et al., 1994; Alam et al., 1997).

DA and oxidative stress. L-DOPA itself, the most commonly administered agent for the treatment of IP, induces oxidative stress and degeneration of dopaminergic neurons in vitro (Spencer et al., 1994; Spencer-Smith et al., 1994; Walkinshaw and Waters, 1995). Although this effect is not observed in vivo (in rodents — Hefti et al., 1981; Perry et al., 1984 — although it may elicit such effects in lesioned animals — Blunt et al., 1993), the possibility of locally elevated oxygen species concentrations due to treatment rather than disease must be borne in mind. It is unclear whether the generalized DNA damage in the IP brain, for instance, should be attributed to the disease process itself or to the effects of L-DOPA (Alam et al., 1997). On the other hand, Olanow (1997) has argued that "a little" L-DOPA-induced OS may be neuroprotective, and it is only when this level is exceeded that damage outweighs benefit.

DA metabolism can also lead to oxidative stress (reviewed in Halliwell, 1992; Götz et al., 1994). Autooxidation of DA leads to the production of toxic semiquinone species ($SQ^{\bullet}$), which via a metabolic cascade may be polymerized to form neuromelanin:

$$DA + O_2 \longrightarrow SQ^{\bullet} + {}^{\bullet}O_2^- + H^+$$
$$DA + {}^{\bullet}O_2^- + 2H^+ \longrightarrow SQ^{\bullet} + H_2O_2$$

There is, however, no evidence for an increased activity in this pathway. The enzymatic metabolism of DA generates hydrogen peroxide (H_2O_2), an in itself fairly harmless compound which is inactivated by glutathione peroxidase and its cofactor glutathione (GSH; see below):

$$DA + O_2 + H_2O \xrightarrow{MAO} DOPAC + NH_3 + H_2O_2$$

It has, however, been suggested that DA turnover increases during the preclinical phase of IP, perhaps in an effort to compensate for the death of nigrostriatal neurons (Scherman et al., 1989), leading to excessive production of H_2O_2 and its conversion to $^{\bullet}OH$ via the iron-mediated Fenton reaction; this would be particularly problematic if the GSH system were itself already impaired or deficient (evidence for which is discussed below). Evidence of increased oxidative stress during elevated DA turnover in rats has been provided by the measurement of elevated levels in vivo of oxidized glutathione (GSSG; Spina and Cohen, 1989); such an increase is not found, however in IP.

A word of caution is, however, necessary with respect to overestimating the significance of DA oxidation in IP: the pattern of neurodegeneration follows more closely the distribution of LBs than of DA. While DA may play a local role in the problems of the nigrostriatal system, it cannot be held responsible for all the damage which occurs in the IP brain; this is confirmed by the fact that we have many animal models in which dopaminergic function is altered, but none which reproduces the pattern of damage seen in PD.

Iron in IP. The normal basal ganglia are rich in iron, with highest levels in the SN, pallidum and putamen (Dexter et al., 1987). Absolute iron levels are increased by about 35% in the IP SNc, but not in other parts of the basal ganglia, accompanied by a shift in the iron(II)/iron(III) ratio from almost 2:1 to 1:2 (Sofic et al., 1988; Riederer et al., 1989; Good et al., 1992; Gerlach et al., 1994); the latter is consistent with an increased iron(II)-catalyzed conversion of H_2O_2 to $^{\bullet}OH$, and favors a greater turnover in the Haber-Weiss cycle. As noted earlier, the superoxide radical is not as great an oxidative threat to the cell as the hydroxyl radical; but it can react with iron(III) in the Haber-Weiss reaction to produce the hydroxyl radical:

$$
\begin{array}{c}
^{\bullet}O_2^{-} + Fe^{3+} \longrightarrow O_2 + Fe^{2+} \\
\underline{H_2O_2 + Fe^{2+} \longrightarrow {}^{\bullet}OH + OH^{-} + Fe^{3+}} \\
^{\bullet}O_2^{-} + H_2O_2 \longrightarrow O_2 + {}^{\bullet}OH
\end{array}
$$

The second reaction in this cycle is the Fenton reaction, whereby the hydroxyl radical can be directly produced by the interaction of a reduced metal ion (in the brain, usually iron(II)) with hydrogen peroxide. The availability of soluble iron complexes in the neuron for participation in these reactions is probably a major determinant of the toxicity of oxygen radicals.

Similar increases in iron concentrations are also seen in other neurodegenerative disorders (Riederer and Youdim, 1993). Jellinger et al. (1990) demonstrated that the increased levels are localized to the SNc and not the SN pars reticulata, where iron levels are known to increase with normal ageing. Further, using energy dispersion X-ray analysis, they localized the ferric iron to NM, not the LB (Jellinger et al., 1992; although some iron was detected in the LB halo in another study: Jellinger et al., 1990); other workers (Hirsch, et al., 1991; Mann et al., 1994), however, found it in both NM and LB deposits.

However, free iron levels in the SNc are tightly regulated, in that ferric iron is normally bound either by ferritin (about 90%) or by NM. Iron participates in free radical-generating reactions only when in its free, ferrous form; it is unclear as to whether the increased iron burden in the nigra is free or bound, which would reduce its availability for catalysis of radical production. Work in our laboratory, however, has shown that a variety of molecules with an o-dihydroxyphenyl structure (including 1,2,4-trihydrobenzene, 6-OHDA and DA), but not the two dopaminergic neuron toxins MPP^+ or 1-trichloromethyl-1,2,3,4-tetrahydro-β-carboline (TaClo), effectively release ferritin-bound iron in vitro and thereby stimulate lipid peroxidation (Double et al., 1997). The iron-releasing properties of 6-OHDA (Jellinger et al., 1995; Double et al., 1997) would also explain another phenomenon: the iron chelator desferrioxamine protects rats against 6-OHDA-induced loss of striatal DA and DA-related spontaneous motor activity; 6-OHDA toxicity is attributed to iron-catalyzed generation of free radical species from H_2O_2, with increased striatal iron levels (Ben-Shachar et al., 1991). Infusion of iron(III) into the SN leads to a dose-dependent and progressive reduction of nigral cell numbers, mostly in the SNc, and of striatal DA concentration (Ben-Shachar and Youdim, 1991; Sengstock et al., 1992, 1994; Wesemann et al., 1994); conversely, nigral iron levels can be elevated by remote lesioning of nigral tracts (Oestreicher et al., 1994; Temlett et al., 1994). The nigral damage following iron(III) infusion is particularly interesting, in that it increased over a period of six weeks; this was the first demonstration of a toxin-induced chronic and progressive neurodegeneration in an animal model (Wesemann et al., 1994).

The source of the increased iron levels in IP is not known; either reduced ferritin levels or increased cellular uptake of iron would be plausible explanations, but the reports here are contradictory. Ferritin levels have been reported as being increased in IP (Riederer et al., 1989), and Jellinger et al. (1990) found a dramatic increase in ferritin-positive microglia; but dramatically reduced levels in the entire parkinsonian brain (Dexter et al., 1990, 1991) and unchanged levels (Mann et al., 1994; Connor et al., 1995) have also been described. The discrepancies might be attributable to the specificities of the different antibodies employed. With respect to uptake mechanisms, there is a similar low concentration of [^{125}I]-transferrin binding sites in both the normal and IP human SN (Faucheaux et al., 1993); increased levels of the ferritin receptor in the striatum have, however, been reported (although not specifically for nigrostriatal neurons; Faucheaux et al., 1995), so that increased nigral iron levels may be the result of increased retrograde transport from this region. Lactoferrin concentrations are elevated in the IP SN (Leveugle et al., 1996), and lactoferrin receptor immunoreactivity is specifically increased on microvessels and neurons of the SN in IP, providing yet another avenue for entry; this change was greatest in the most severely affected patients (Faucheaux et al., 1995).

However, there is evidence that the increase in iron concentration is a later or secondary change in IP; for instance, increased iron levels are not reported for incidental Lewy body disease (ILBD), regarded as a preclinical form of IP (Dexter et al., 1994), and our laboratory has detected elevated levels only in the most severe cases of IP (Riederer et al., 1989). It thus appears likely that iron is not involved early in the pathogenesis of IP, but its potent ability to generate free radicals means that it may nonetheless play a significant role in the progression of the disease. This is especially true, as it is accompanied by a dramatic rise in aluminium levels (Hirsch et al., 1991; Good et al., 1992), which would exacerbate the oxidative damage by iron (Gutteridge et al., 1985).

Nitric oxide as an inducer of OS. It has also been suggested that cytotoxic OS might result from glial, activated microglial or macrophage activity; in particular, nitric oxide (NO) release by these cells is seen as possibly being involved in such activity (see reviews: McCall and Vallance, 1991; Dawson et al., 1992). NO can act as an effective free radical scavenger (Rubanyi et al., 1991), but is also a highly reactive free radical in its own right; it can also react with superoxide radicals to form the peroxynitrite anion ($^{\bullet}ONOO^-$), a potently oxidative radical, which in turn can decompose to $^{\bullet}OH$ and NO_2, potent initiators of lipid peroxidation (Beckman et al., 1990; Radi et al., 1991):

$$O_2^- + NO^{\bullet} \longrightarrow ONOO^-$$
$$ONOO^- + H^+ \longrightarrow ONOOH \longrightarrow {}^{\bullet}OH + NO_2^{\bullet}$$

NO itself may also act directly at several sites of the electron transport chain, including complex I, whose activity is specifically reduced in the IP SNc (see below; NO, however, has a preference for complex IV: Cleeter et al., 1994), and has also been shown to liberate ferritin-bound iron, thereby promoting lipid peroxidation (Reif and Simmons, 1990; Radi et al., 1991). There is evidence of increased expression of inducible NO synthase (NOS) activity in the SNc in IP, possibly secondary to reactive gliosis (Hunot et al., 1996); interestingly, Mufson and Brandabur (1994) reported that NOS-containing cells are spared in IP, perhaps because of the concurrently elevated Mn-SOD levels. Jenner and Olanow (1996b) reported evidence of increased NO-induced damage in IP, as indicated by 3-nitrotyrosine (3-NT) levels in the core (but not the halo) of the LB. Finally, it has been demonstrated that NO increases the toxicity of MPTP, at least in part by rendering irreversible the inhibition of the mitochondrial complex I by MPP^+ (Cleeter et al., 1992).

Free radical defences in IP. In summary, it appears that the IP brain exhibits both specific and nonspecific signs of oxidative damage. Such injuries could arise either through increased free radical production or as a consequence of reduced clearance. Even if exposure to free radicals is increased in IP, it would be expected that the detoxification systems would be adequate to cope with the problem. The major enzymes which protect the cells from free radicals are as follows:

Superoxide dismutase (SOD):

$$^{\bullet}O_2 + {}^{\bullet}O_2 + 2H^+ \longrightarrow H_2O_2 + O_2$$

Glutathione peroxidase (GPO):

$$2\ GSH + H_2O_2 \longrightarrow GSSG + 2H_2O$$

However, the activity of such systems declines even in normal ageing. The activities of catalase (which also converts H_2O_2 to water) and glutathione peroxidase appear to be normal with respect to the age of the patient in the IP SNc (Martilla et al., 1988; Sian et al., 1994). But Damia et al. (1993) noted that the vulnerable cell groups in the SNc have lower levels of glutathione peroxidase than their neighbours; in IP, similar (Martilla, et al., 1988; Sian et al., 1994) or slightly reduced (Ambani et al., 1975; Kish et al., 1985) levels are measured, and the levels of the free radical scavengers ascorbate and α-tocopherol are also preserved (Riederer et al., 1989; Dexter et al., 1992). But two elements of the radical detoxification system do appear to be significantly compromised.

The activity of superoxide dismutase (SOD) is increased in discrete brain regions in various neurodegenerative disorders (for example, see Marklund et al., 1985); in IP, it is specifically and uniquely increased in the SNc (Martilla et al., 1988; Saggu et al., 1989). SOD occurs in two forms, a mitochondrial Mn-SOD and a cytoplasmic Cu/Zn-SOD; it is the Mn-SOD form which is increased in IP (Yoshida et al., 1994), suggesting increased OS at the level of the mitochondrion (see below), although elevated levels of the Cu/Zn SOD mRNA have also been reported (Martilla et al., 1988), indicating cytosolic OS. This increase might be interpreted as a neuroprotective response to increased local oxidative stress, as SOD converts free radicals to H_2O_2; SOD is reported, for instance, to protect rat nigral cells against MPTP toxicity (Przedborski et al., 1992). On the other hand, the increased H_2O_2 flux might be deleterious, as there is no indication of increased catalase or glutathione peroxidase activity to process it further.

Nigral glutathione levels are low compared to those in other brain regions, and both total and reduced glutathione are further reduced in the IP SNc, without an increase in the oxidized form (GSSG) (Perry et al., 1982; Riederer et al., 1989; Sofic et al., 1992). The decline is specific both to this region and to the disorder, with clear consequences for H_2O_2 clearance. Jenner and Olanow (1996a) note that, although the GSH loss occurs in all nigral cell types (Pearce et al., 1997), the magnitude of the decline indicates that the greater part must be attributed to glial cells. Interestingly, the decline in GSH levels is as great in ILBD patients as in IP (Dexter et al., 1994), although in the former patients altered iron levels and mitochondrial function are not evident, so that the decline in GSH levels may be an early indicator of the disorder. Other components of the GSH system are unaltered, with the exception of γ-glutamyl transpeptidase (GGT), associated with GSH translocation and degradation, whose levels are doubled (Sian et al., 1994) — once more specific to IP and the SNc. The reason for this increase has not been investigated, and could be either a reaction to or the reason for the reduced GSH levels. It has been demonstrated in rats that an artificially induced 40–60% loss of GSH, similar to that seen in IP, does not alone lead to major cell loss, nor to changes in mitochondrial SOD activity (see also Seaton et al., 1996), but the sensitivity to the toxicity of both 6-OHDA and MPP^+ is enhanced (Pileblad et al., 1996; Wüllner et al., 1996; Toffa et al., 1997). Jenner and Olanow (1996a) interpreted this as indicating that GSH depletion represents the "first blow" which renders the cell vulnerable to a toxin-induced degeneration at a later stage, noting that the incidence of ILBD exceeds that of IP by a factor of ten; the reduced GSH levels may by themselves be innocuous, but render the subject sensitive to an environmental toxin at a later stage.

Persistent OS might be invoked to explain this decline in GSH levels (analogous to the response to chronic OS in hepatic tissue: DeLeve and Kaplowitz, 1991), or impaired mitochondrial function. Whatever the reason for its decline, this phenomenon may be important in the pathogenesis of IP, either by directly inducing OS, or by rendering the SNc susceptible to attack by another toxin.

Neuromelanin. Would an OS model of IP explain the apparently selective vulnerability of neuromelanin (NM)-containing neurons in the SNc? NM is generally regarded as a waste product of catecholamine metabolism resulting from DA autooxidation, and accumulates with age in nigral neurons (Swan, 1963; Graham, 1978). Interestingly, this accumulation takes place in the cell bodies, and not the terminals of SNc neurons, although the latter portion of the cell is more exposed to DA. The presence of NM in particular SNc cells has been interpreted as a risk factor, in that it is seen as indicative of local OS, whatever the reasons for this stress. It should also be noted that Gibb (1992) found that the ventral portion of the SNc (that is, the region more vulnerable to cell death in IP) was less melanized than the dorsal tier.

Whether NM affects cellular function is disputed. It has been suggested that its accumulation can reach a point where cellular function is compromised, even fatally; its concentration has been negatively correlated with cytoplasmic mRNA levels and nucleolar volume (Mann et al., 1977; see also Mann and Yates, 1983; Marsden, 1983). Further, NM has been described as a "double-edged sword" with reference to free radical reactions: it sequesters redox-active metal ions (such as iron), but may also promote their participation in free radical-generating processes (Lindquist et al., 1987; Pilas et al., 1988; Miller et al., 1990; Swartz et al., 1992; Youdim et al., 1994). It may also represent a depot for other cellular toxins, such as MPP^+ (see, for example, D'Amato et al., 1987). That NM is not an essential part of neurodegeneration in IP is clear from the fact that nonmelanized nuclei, such as the nucleus basalis Meynert and the subcortical structures discussed above, are also affected (Braak et al., 1995, 1996); this does not, however, exclude a role for NM in the cells where it is found.

What further complicates the issue are questions concerning the chemical structure of NM. These large amorphous molecules long proved resistant to analysis by standard chemical methods; a number of approaches provided information about its chemical composition, but the application of nuclear magnetic resonance (NMR) spectroscopy was necessary to determine structural details (Aime et al., 1996). NM is known to include covalently bound DA and cysteine, and it has been proposed that the implied presence of reactive catecholamine intermediates indicates a role for covalent bonding in the pathology of IP (Graham, 1978; Carstam et al., 1991). Further, 5-S-cysteinyl recovered from the cerebrospinal fluid has been reported to be an early indicator of the disease, its levels correlated with the extent of nigral degeneration (Carlsson and Fornstedt, 1991).

Advanced glycation end-products (AGEs) are of group of substances formed by a cascade of reactions following the reaction of protein amino groups (especially the N-terminal and side chain amino groups of lysine and arginine) with monosaccharide molecules (reviewed in Buccala and Cerami, 1992). Accumulation of such products is a feature of the ageing human brain, and is promoted by OS and the presence of transition metals, including iron; in turn, oxidation processes in AGE synthesis generate highly reactive dicarbonyl products and oxygen free radicals, thus creating a vicious circle. Due to this property and their promotion of protein cross-linking, AGEs have been implicated in degenerative processes associated with ageing, both in the periphery and the CNS (Buccala and Cerami, 1992). In particular, Münch et al. (1997) have investigated the role of such accumulation in neuronal damage in AD, where they appear to accelerate the polymerization of β-amyloid peptide, the major component of AD plaques. Castellani et al. (1996) identified immunohistochemically two AGEs (pentosidine and pyrraline), as well as heme oxygenase-1, a putative marker for OS, in the IP SN and in the neocortex in DLBD; the heme oxygenase-1 reactivity was localized by immunoelectron microscopy to the filaments of cortical LBs. The involvement of AGEs in the neurodegeneration in IP thus merits further investigation, especially as it provides a conceptual link between the OS presumed to exist in IP and the appearance of LBs, suggesting a susceptibility of the cytoskeleton to OS.

Summary. There thus exists a great deal of evidence for the OS in the SNc of the IP brain:

— evidence of oxidative damage to proteins, lipids, DNA
— reduced glutathione levels
— increased SOD levels
— increased iron levels in the absence of increased ferritin levels
— increased aluminium levels
— increased levels of heme oxygenase-1 and of two AGE species.

That such stress is the cause, and not simply the result of the neurodegenerative process in IP, however, remains to be demonstrated. The causal relationships amongst the various phenomena identified also needs to be explored. Perhaps a first step in this direction is the finding that reduced GSH levels are possibly an early characteristic of the disease. The demonstration of a similar defect in the periphery would be of considerable assistance, not least in establishing whether a compromised OS defensive system is a lifetime defect and a general physiological feature of these patients, or whether this loss is manifested at a later stage of life, for whatever reason. In conclusion, OS as a decisive factor in the picture of IP has yet to be unequivocally demonstrated, but a role for such stress seems likely.

Mitochondrial problems

Closely associated with OS models is the idea that abnormalities of the mitochondrial electron transport system, and in particular of complex I (see

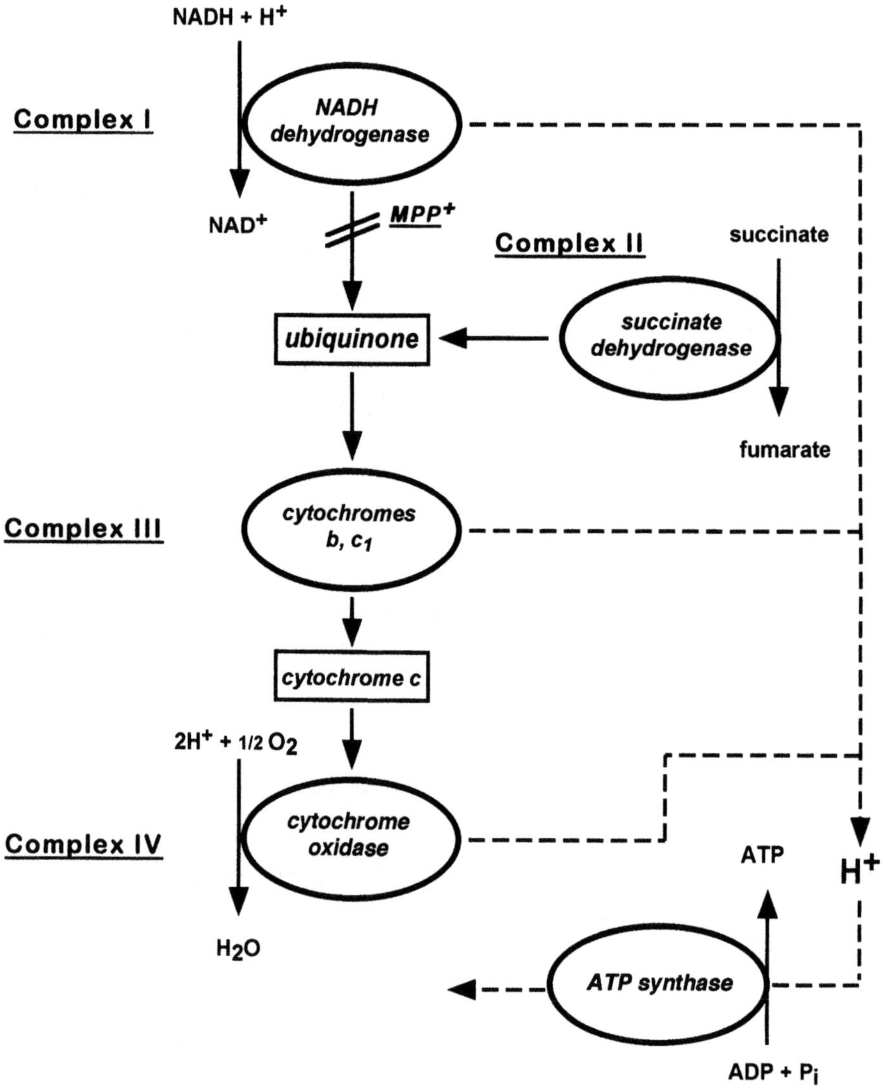

Fig. 1. Schematic representation of the mitochondrial electron transport chain

Fig. 1), are involved in the pathophysiology of IP. As already discussed, this hypothesis was suggested by the impairment of complex I activity by MPTP, probably following the accumulation of MPP^+ by the mitochondrion. Most commonly reported in IP is a decline of about 30% in SNc complex I activity, as well as in the striatum, skeletal muscle and platelet, without clear structural changes or mitochondrial DNA deficits (reviews: Reichmann et al., 1993; Schapira, 1994). The other components of the transport chain are usually reported to be normal in the SNc, although impairment of complex IV has also been reported (Itoh et al., 1997). This pattern of dysfunction is unique to IP, and has been interpreted as a phenomenon which predisposes the ageing human brain to neurodegeneration in these regions. Hattori et al. (1991) found in an immunohistochemical investigation that the decline in complex I reactivity occurred without evidence of autolytic cell damage, indicating that the mitochondrial changes are not secondary to cell death. Perhaps significantly, ILBD is reported to be associated with an SNc complex I activity intermediate between that of normal controls and IP patients (Schapira et al., 1990).

A progressive decrease in mitochondrial electron transport activity has been detected in the normal ageing primate brain (DiMonte et al., 1992), and is well known in peripheral tissues (for example, Yen et al., 1989). This may be linked with the increased evidence of damage to mitochondrial DNA with ageing, which is particularly susceptible to free radical-induced damage (Wallace, 1992a). Beal et al. (1993) suggested that these phenomena, taken together, would provide an explanation for the late onset and progressive nature of neurodegenerative disorders; a defect in energy metabolism might not become significant until combined with further changes associated with normal ageing. The regional specificity of such disorders might then be associated with differential isozyme expression in the affected regions; tissue-specific expression of NADH dehydrogenase (complex I) and cytochrome-c oxidase (complex IV) have been reported (Wallace, 1992b), and higher than normal expression of particular cytochrome-*c* oxidase subunits in particular brain regions has been correlated with susceptibility to damage in AD (Chandrasekaran et al., 1992). An interplay between genotypes for mitochondrial NADH dehydrogenase and CYP2D6 in IP has recently been suggested (Kosel et al., 1996). It may also be important that α-ketoglutarate dehydrogenase (α-KGDH) activity is also reduced in IP (Mizuno et al., 1994); the combination of the two defects may be greater than a simple additive effect. Further, decreased α-KGDH activity would be expected to lead to an intracellular accumulation of glutamate, with excitotoxic effects for an already weakened nerve cell.

However, it must be emphasized that the relationship between such an impairment and cellular damage remains to be established. Several inhibitors of complex I, including paraquat, 6-OHDA and MPP^+, lead to the generation of superoxide and hydroxyl free radicals and H_2O_2, as well as to an impairment of ATP production (Jenner and Olanow, 1996b). Amongst the possible endogenous inhibitors of the complex are the NO radical and the β-carboline TaClo. The latter is an interesting compound: a potent complex I inhibitor which induces a progressive parkinsonian-like response when injected into the SN, it can be synthesized endogenously from tryptamine and either chloral, a commonly used hypnotic, or a metabolite of trichloroethylene, a widely used industrial solvent (Bringmann et al., 1995). Chronic administration of L-DOPA or DA also inhibits complex I activity in rats, an effect prevented by reduced glutathione, ascorbate, SOD or catalase, but only partly suppressed by MAO inhibitors, suggesting the involvement of autooxidation in the process (Przedborski et al., 1993).

Summary. The complex I deficit (along with reduced GSH levels) is perhaps the best established biochemical characteristic of the SNc in IP, but its position in the etiological process needs further investigation. The resulting energy deficit would of itself be damaging to the neuron, but the generation of free radicals caused by reduced activity at complex I could also be important. That MPTP possibly exerts its principle effect at this point is no proof that a similar action by a toxin underlies IP. Indeed, that the decrease is not so marked in IBLD suggests that mitochondrial insufficiency is a later event in the IP process. It is also important to note that the impairment of complex I activity in the brain is largely restricted to the nigrostriatal system; but it is also evident in several peripheral tissues, offering the possibility of a biomarker for IP.

Disturbance of intracellular calcium homeostasis

The calcium hypothesis proposes that an uncontrolled rise in intracellular calcium levels leads to uncontrolled stimulation of a number of Ca^{2+}-dependent enzymes, leading to metabolic dysfunction and ultimately necrotic cell death (reviews: Siesjö, 1990; Gerlach et al., 1996). Such a rise could result from the opening of voltage-dependent Ca^{2+} channels (possibly as a consequence of a mitochondrial energy failure), inappropriate excitatory stimulation or via free radical-induced membranal leakiness. This hypothesis thus proposes a "downstream" consequence of a pathological process; correspondingly, a neuroprotective effect might be exerted by agents which corrected this influx, without altering the underlying pathology.

Amongst the processes of possible significance would be the activation of the calpains I and II, leading to cytoskeletal alterations, which may be significant for the role of LBs in IP. On the other hand, the activation of protein kinase C (PKC) and NOS would lead to increased NO production, damage to the cell membrane by phospholipase A_2, and the release of arachidonic acid, whose extracellular metabolism would further expose the neuron to free radical attack. A vicious circle would thus be initiated, leading ultimately to neuronal death (Gerlach et al., 1996).

Such changes have been implicated in the early stages of other neurodegenerative disorders, including AD (Landfield et al., 1991). We are aware of only one report that links MPTP with elevated cytosolic free Ca^{2+} levels (Kass et al., 1988). Although measured Ca^{2+} levels in the IP brain are normal (Riederer et al., 1989), a significant loss of the Ca^{2+}-binding protein calbindin *D28k* has been reported in the SN (Chan-Palay et al., 1991; also in AD: Iacopino and Christakos, 1990). Further, it has been noted that calbindin-positive neurons in the SN appear less susceptible to the neurodegenerative changes seen in both IP and MPTP-induced parkinsonism (Yamada et al., 1990; German et al., 1992; Hirsch et al., 1992), leading to its being seen as a marker of cellular resistance. Calbindin expression declines with age in both the human and rat brainstem (Iacopino and Christakos, 1990), but the situation in dopaminergic SN neurons has not been specifically addressed. Hirsch et al. (1997) concluded that the at-risk population of neurons in the SNc were characterized by the presence of NM, a higher concentration of the free-radical detoxifying glutathione peroxidase and the absence of the major cellular Ca^{2+}-binding protein D28k — taken together, indicative of an increased susceptibility to OS and disturbed Ca^{2+} homeostasis. However, not all calbindin D28k-containing nuclei are spared from degeneration in IP, an example of a vulnerable D28k-positive region being the nucleus basalis Meynert (Ichimaya et al., 1989; see also discussion in Gibb, 1992).

An increase in the concentration of immunoreactive m-calpain (which requires millimolar Ca^{2+} concentrations for activation; the normal intracellular level is about 0.1μM) in the SN and locus ceruleus of IP patients, an increased density of m-calpain-positive fibres in these nuclei, and the co-presence of ubiquitin, have been interpreted as being suggestive of degeneration in such populations (Mouatt-Prigent et al., 1996; Hirsch et al., 1997). Interestingly, calpain immunoreactivity was also associated with the LB. A diminished concentration of dihydropyridine binding sites (that is, of voltage-dependent Ca^{2+} channels) has been reported in the IP nigrostriatal system (Nishino et al., 1986), perhaps related to the loss of these neurons. Interestingly, similar findings have been documented for the temporal cortex in AD (Pigott et al., 1991) and the striatum of Huntington's chorea patients (Sen et al., 1993); that is, at the sites of specific cell loss for these disorders.

Summary. Disturbances of Ca^{2+} homeostasis are not seen as the primary cause of IP, but rather as a possible participant in the process ending in neuronal death. Such mechanisms have, for instance, been implicated in excitotoxic cell damage. Sustained elevation of cytosolic Ca^{2+} levels is a response to several toxins, including the cytotoxic quinones and hydroperoxides; contradictory roles for Ca^{2+} have also been proposed in apoptosis (Fawthrop, et al., 1991; Franklin and Johnson, 1992). Manipulation of intracellular Ca^{2+} status might therefore be of benefit in IP therapy, if difficult to implement in a specific manner.

Excitotoxicity

"Excitotoxicity" occurs when excessive release and accumulation of glutamate lead to the persistent activation of NMDA or AMPA receptors, the disturbance of intracellular Ca^{2+} homeostasis and ultimately to the death of the hyperstimulated neuron (reviews: Olney, 1978, 1989; Coyle and Puttfarcken, 1993). Such receptors are activated not only by their natural ligand glutamate, but also by a number of endo- and exotoxins, such as β-N-methylamino-L-alanine (BMMA; contained in the seed of the sago palm, and possibly involved in the Guam parkinsonian syndrome; Spencer et al., 1987). The role of glutamate receptors in acute ischemic neurodegeneration in the CNS was first proposed by Olney in 1978; excitotoxic damage is also associated with a number of idiopathic neurodegenerative diseases, including ALS and HD (Beal et al., 1993).

Impaired energy metabolism at the mitochondrial level may lead to disruption of normal membrane function, including loss of the voltage-dependent Mg^{2+} block of NMDA receptors, allowing persistent activation of these receptors even in the presence of normal glutamate levels. Experiments with rats involving mitochondrial chain inhibitors, including aminooxyacetic acid (AOAA), MPP^+, malonic acid and 3-nitropropionic acid (3-NPA), have indicated that they lead to CNS degeneration via an excitotoxic mechanism, resembling that seen in human IP and HD (Ikonomidou and Turski, 1996). The effect can be ameliorated in these models by treatment with either glutamate receptor antagonists or with agents which support energy metabolism.

Although a role for striatal glutamatergic receptors in the expression of parkinsonian symptoms (especially rigidity) is suspected (see, for example, Klockgether and Turski, 1993), evidence for the involvement of excitotoxicity in IP is not so clear. No clearcut differences with respect to glutamatergic parameters between the normal and IP brain have been found (Rinne et al., 1988; Difazio et al., 1992). Direct injection of kainate/NMDA receptor agonists into the striatum selectively destroys intrinsic striatal neurons, sparing afferent dopaminergic fibres from the nigra, thus modelling the situation in HD better than IP (Irwin and Langston, 1995). Stronger evidence for excitotoxic involvement in IP remains to be presented, although a role in the action of MPTP has been proposed (initially by Turski et al., 1991; reviewed in Fornai et al., 1996); NMDA antagonists and decortication have been reported to block MPTP-induced nigral neurodegeneration in both rats and primates, but not in mice, perhaps by antagonizing toxin-induced elevation of glutamate and aspartate levels in the striatum (reviewed in Langston, 1995).

Summary. A role for excitotoxicity in IP has thus far not been substantiated, but, in view of the significance of this phenomenon for CNS neurodegeneration in general, the question will, correctly, be further pursued.

Toxin theories of IP

The above processes all represent plausible explanations of the processes leading to neurodegeneration in IP, but the question remains as to what "triggers" the fatal sequence of events. A number of agents have been suggested to activate this trigger, although none has established itself as the ultimate "cause" of the disease, either in general or in a defined subset of patients. The involvement of an external toxin was suggested by epidemiological studies which found a greater risk for IP associated with exposure to industrial or rural environments (!) and with the use of agricultural chemicals, amongst a range of other seemingly unconnected variables (Spencer and Butterfield, 1995; De Michele et al., 1996). Various nonidiopathic parkinsonian syndromes are connected with, for instance, manganese poisoning. The discovery of MPTP toxicity led to the search for the "natural" IP toxin responsible for the onset of the disease, but to date a positive identification has not been made.

It has also been hypothesized that the problem in IP may be a "two hit" phenomenon (Langston, 1996; Kuhn and Müller, 1997); due to altered handling of a xenobiotic by the responsible cytochrome P450 (CYP450) isozymes, a normally inoffensive molecule becomes a threat to normal cellular function. Xenobiotica are usually handled in a two stage process: an "activation" phase (phase I), followed by a "conjugation" phase (phase II). Most work has focussed on variants of the CYP450D6 (debrisoquine 4-hydroxylase) enzyme, partially responsible for the metabolism of nicotine (IP is negatively correlated with smoking; Baron, 1986) and of MPTP and MPP^+. Barbeau et al. (1985) reported an increased frequency of subjects with reduced activity of this enzyme in IP, but subsequent work has been contradictory, possibly due to problems of patient treatment (CYP450D6 metabolizes a range of pharmaceuticals, including antidepressants). Attempts to link IP with particular variants or with so-called "poor metabolizers" have also proved inconclusive (see review by Landi et al., 1996). Instead, Langston (1996) has proposed that it is possible that problems must arise in several enzyme systems for a neurotoxic effect to emerge; for instance, they identified a variant of the (phase I) CYP450D6 isozyme which was associated with only a weak increase in the risk for IP; when combined, however, with a particular variant of the phase II enzyme GSTM I, the risk for IP was greatly enhanced — indeed, 11–14 times above that of the normal population.

Such a model proposes the interaction between genetic and environmental factors which most workers assume to be at play in IP; not only must the genetic predisposition be present, but the patient must at some stage be exposed to a toxin which reveals this weakness. The concept is valuable in that it may be possible to identify which individuals are at particular risk of developing the disease, without an absolute requirement for identifying a specific toxin. The search for such a toxin has correctly been compared to looking for a "needle in a haystack" (Kuhn and Müller, 1997), especially in light of the fact that we are aware of toxin-related diseases where symptoms are manifested only years after exposure (see Koller and Montgomery, 1997).

It is also possible that a toxin might be synthesized in the CNS itself — for instance, DA quinones or 6-OHDA from DA, or an MPTP-like compound from 5-HT — but the evidence for the synthesis of such molecules in the CNS remains controversial. A number of candidates with similar cytotoxic properties are nonetheless under investigation, including several isoquinoline and β-carboline species which share similar cytotoxic properties, such as selective uptake by dopaminergic neurons, impairment of mitochondrial function and a potential for inducing the degeneration of SNc cells, and which are known to be synthesized in the CNS (Collins and Neafsey, 1985; Dostert et al., 1988; Irwin and Langston, 1996). Such molecules have described as "weak" neurotoxins, in that they do not have the immediate disastrous effects of MPTP, but the effect of long-term exposure is still an open question. Members of the β-carboline family, in particular, bear a strong resemblance to MPTP/MPP^+, and have been isolated from IP patients (Matsubara et al., 1995). Kuhn et al. (1995) reported a rise in the plasma levels of the β-carbolines harman and norharman in IP patients, but the reason for the rises was not clear. An interesting member of the β-carboline family is TaClo, discussed briefly above (review: Bringmann et al., 1995). It has not only been suggested that this compound might be synthesized in the brain from tryptamine and an exogenous organic substance, it is also the only toxin apart from iron(III) to produce a progressive parkinsonian syndrome in the rat, demonstrated both in their motor behaviour and by assessment of striatal DA levels (Grote et al., 1995; Sontag et al., 1995). It is believed that TaClo must be N-methylated in situ before it exhibits toxicity. Both TaClo and its N-methylated version inhibit the mitochondrial complex at lower concentrations than does MPP^+; the N-methylated compound is more potent, and also inhibits complex II (Janetzky et al., 1995).

Summary. The idea that a toxin induces IP has long been entertained, but the broad geographical and environmental distribution of the disorder implies that either the toxin is universally distributed, and a genetic weakness with respect to this toxin plays the critical role in the disease, or that there are a number of distinct toxins which are capable of producing the damage seen in IP. Whatever the case, the fact is that no direct connection between either an endogenous or an exogenous toxin with IP has yet emerged; the search, however, continues.

Disturbance of normal cell maintenance

The regeneration, maintenance of function and development of neurons are influenced by a group of proteins known as neurotrophic factors. The possibility that cell death in IP is attributable to an age-related or pathologic decline in these protective growth factors has also been proposed (reviews: Hefti, 1994; Kupsch et al., 1995). The major problem with interpreting results in this area is the fact that many investigations employ cultured neurons or examine neuronal survival in rats; whether such models may be compared with aged human neuronal populations is uncertain.

Amongst the candidate factors, glial-derived neurotrophic factor (GDNF), a member of the family of transforming growth factors-β, has thus far received the most attention. A neuroprotective effect against the neurodegenerative effects of MPTP and other nigrostriatal injury has been demonstrated in rodents: intrastriatally administered GDNF was shown by Tomac et al. (1995) to rescue tyrosine hydroxylase-positive cells in the SN of MPTP-treated mice, even when applied a week after the toxin; it was more effective applied in the striatum than directly in the SN, but did not induce sprouting in the striatum, so that it in the normal animal may be a target-derived neurotrophic factor. GDNF mRNA is not expressed in the adult rat striatum, but is found in a range of adult human CNS structures, including the striatum, cortex and spinal cord (Springer et al., 1994). Even more spectacular was the demonstration of a functional recovery from the effects of MPTP in monkeys treated with GDNF (Gash et al., 1996). The clinical application of GDNF to the treatment of IP is currently under cautious consideration (Olson, 1997).

No reduction in GDNF or any other neurotrophin has thus far been identified in IP. On the other hand, Nagatsu's laboratory has identified an increased level of transforming growth factors-α and -β-1, of interleukins-1β,

-2 and -6 and of tumor necrosis factor-α in the nigrostriatal regions and ventricular CSF of Parkinson's patients, accompanied by increases in the levels of the *bcl-2* protein, thought to be a repressor of programmed cell death (apoptosis), and of the soluble *sFas* protein, believed to play a role in the regulation of apoptosis; these latter rises were correlated with each other and with the levels of IL-1β and -6 (Mogi et al., 1996). Although the significance of these findings for the etiology of IP is unclear, the topographic and disease specificity of the changes is intriguing, and point to a possible involvement of neuroimmune mechanisms in IP (see also Kuhn and Müller, 1995).

Basic fibroblast growth factor (FGF-2) and brain-derived neurotrophic factor (BDNF) have trophic effects on mesencephalic dopaminergic neurons, and, like GDNF, promote cell survival and structural integration in embryonic mesencephalic neuronal grafts (Beck, 1994); nigral FGF-2 levels are reported to decline in IP (Tooyama et al., 1993). Although likely candidates, no role for these molecules in IP has been determined.

Finally, immunohistochemical studies have found that the insulin receptor is lacking in IP in the SN, amongst other brain regions (Moroo et al., 1994). Whether altered glucose metabolism plays a role in IP has not been clarified, but has been suggested to play a role in AD (Blum-Degen et al., 1995).

Summary. There is at the moment no evidence that either a normal or pathological decline in neurotrophic factors underlies the problems of IP. The only report of this nature concerns a decline in hippocampal BDNF mRNA levels in AD (Phillips et al., 1991). Nevertheless, the application of members of this class of compounds, whose broad spectrum of action is only beginning to be understood, in therapy, in order to rescue neurons which would otherwise die, merits further attention. The criticism could also be made that the research has thus far concentrated on the effects of neurotrophins on acute, as opposed to chronic, dopaminergic neuronal injuries; their clinical potential might therefore be reduced with respect to the recovery of older neurons in IP, although they may be of major significance in supporting neuronal graft therapies. The high molecular weight of the trophic factors will, in any case, render their delivery to the CNS difficult, so that stimulation of native production might prove to be the better strategy. Finally, an involvement of (auto)immune mechanisms in IP remains speculative, but provides yet another plausible avenue for investigation.

Genetic factors

In the absence of a clearly identifiable cause for IP, it is assumed that some degree of genetic predisposition to the disease must exist (reviews: Vieregge, 1994; Golbe, 1995; Duvoisin, 1996). However, despite intensive investigation of a number of candidate genes, the specification of this involvement remains problematic. Early twin concordance studies failed to indicate a clear genetic link; this might be attributable to a long preclinical phase for IP, which would obscure familial associations of the disorder, as does a lack of consistency with respect to the diagnosis of IP. More recent PET studies have, in fact, suggested a higher concordance than previously suspected, with the identification of abnormalities of dopaminergic transmission in nonsymptomatic siblings (see reviews and Johnson et al., 1990).

Calne (1994) noted that in each of the major degenerative disorders (IP, AD, HD) there are a minority of patients (~15%) whose family history for the disease over several generations suggests the involvement of a genetic factor. The limited penetrance of this factor, however, underscores the crucial nature of an environmental trigger, whether a toxin, infection or of another nature. The first family history positive for IP was described by Mjönes in 1949, in which 79 of 194 members (41%) suffered from the disorder. This and a number of more recent studies have detected a possible autosomal dominant inheritance pattern with reduced penetrance (Golbe et al., 1993; Lazzarini et al., 1994; Waters and Miller, 1994), although early studies found evidence for a multifactorial etiology (for example, Kondo and Kurland, 1973). Calne et al. (1987), however, urged caution in the interpretation of such studies, as the effects of sharing a common environmental background are usually not sufficiently controlled for, so that an environmental effect might be misinterpreted.

In a recent review, Gasser (1997) argued that genetic factors probably play a greater role in IP than previously assumed; the assumption of a simple monogenetic Mendelian inheritance is, however, unlikely. More likely is a polygenetic, low penetrance explanation, modulated by one or more extragenetic factors. Amongst the polymorphic genes for which an allelic association with IP has been reported are those for CYP450D6 (on chromosome 22, close to genes for a DA receptor, COMT and PDGF-B; Gasser, 1997), MAO-A (Hotamisligil et al., 1994) and MAO-B (in a non-coding sequence; Kurth et al., 1993). Amongst the factors with which a genetic link with familial IP has been excluded are tyrosine hydroxylase, BDNF, amyloid precursor protein, glutathione peroxidase, superoxide dismutase and CYP2D6 (Gasser et al., 1994). Attempts to establish a link with mitochondrial genes have also been inconclusive (Maraganore et al., 1991).

Golbe (1995), however, has argued that the inheritance patterns observed in familial IP are inconsistent with multifactorial hypotheses, leading to the question of whether familial and "sporadic" IP should be regarded as distinct pathological entities, a suspicion reinforced by the somewhat atypical pathological picture presented in some histories of familial IP (Duvoisin, 1996).

The most exciting report in this connection was the recent identification by Polymeropoulos et al. (1997) of an allele for the gene for α-synuclein carried by the affected members of four unrelated families (one Italian and three Greek) positive for familial IP; it was not found in 100 controls or 52 sporadic IP patients in southern Italy, or in two other Greek families with a history of IP. The function of α-synuclein is not known. It is believed to be a presynaptic protein involved in neural plasticity, and is expressed in the brain regions susceptible to damage in IP; a fragment of this protein has been detected in both AD plaques (Nussbaum and Polymeropoulos, 1997) and Lewy bodies (Heintz and Zoghbi, 1997; Spillantini et al., 1997). It is expected that the discovery may explain a small percentage of IP patients, but the first positive identification of a gene linked to the disease is nonetheless encouraging, and may also afford insights into the etiology of sporadic IP.

Summary. It would appear unlikely that genetic factors play no role in IP, despite the elusiveness of a precise connection. Even the toxin theories of IP generally assume a genetic component in order to explain the selectivity of the disease. The identification of α-synuclein as the significant factor in the familial IP of four southern European families is exciting, but further work will be necessary to determine whether this or, indeed, any other single mutation will be able to explain the greater part of the incidence of IP.

What is the cause of IP?

A palette of "explanations" are thus available for the etiology of IP, and not all possibilities have been discussed here:

- oxidative stress (the free radical hypothesis): possibly due to failure of antioxidant defences;
- mitochondrial defects: especially an impairment of the activity of complex I;
- an endogenous or exogenous toxin;
- age- or pathology-related changes in normal cell maintenance, such as the neurotrophin system;
- excitotoxic injury of specific neuronal populations;
- disturbance of intracellular calcium homeostasis;
- a genetic cause for the disease, or any of the above processes superimposed on a genetically determined predisposition.

A number of pathological processes and phenomena have been identified in PD, but the causal relationships amongst these phenomena remain unclear. It is all too easy to seize upon individual features of the disease in order to construct an "etiology"; but in the end, the event which sets the degenerative process in train must be identified, if the process is to be properly understood and controlled. The search for the underlying "cause" is hampered, as in many diseases of the brain, by the fact that the disorder does not appear to occur spontaneously in other species — indeed, we have been reminded in recent years of the significant differences between man and other higher primates with respect to the organization of the striatal and other dopaminergic systems; the remarkably restricted expression of one of the central PD-related substances, neuromelanin, further complicates this problem.

All proffered models would explain cell loss in the SNc, all would allow for a "latent", preclinical period in the disease. The essential missing pieces of the puzzle are the nature and speed of the degeneration involved, and the significance of the LB (and PB) for the course of the disease. Are the cells "sick" for an extended period before finally succumbing to their fate, or are they individually and rapidly extinguished as they become exposed to the cause of death? Is the LB predictive of future cell death, and, if so, on what sort of time-scale? And the LB seems more the result of an active process than cellular "garbage" — what does this process indicate, and what sets it in train? And what significance attaches to the absence of tau protein in these structures?

Another vital question concerns the rate of decline in the SNc; which metabolic process(es) determine its dynamic, and is the rate of degeneration constant throughout the disease — or, as suggested by Lee et al. (1994), does the pathological degeneration occur at the start of the clinical phase, rendering later intervention somewhat ambitious? Does neurodegeneration commence at the nerve terminal, as in MPTP-induced parkinsonism, or at the cell body? The various possibilities are presented schematically in Fig. 2.

Any explanation of neurodegeneration in IP will be constrained by limitations in our understanding of the processes underlying normal neuronal maintenance and death; how does neurodegeneration in IP differ from that of normal ageing (if indeed it does)? The cells most at risk in IP appear to be distinct from those SNc neurons whose loss in the course of normal ageing is greatest, but, as already noted, whether different processes are at play has not been established. The nature of neuronal death in IP is undetermined; there are, however, indications that it ensues via a combination of apoptosis, or "programmed cell death", and necrotic degeneration (for reviews of apoptosis in neurodegeneration, see Riederer et al., 1996; Przuntek et al., 1997). One of the hallmarks of apoptosis is the fact that it is an active cellular process, so that only a highly specified population of cells in a particular region is "selected" for death. This contrasts with necrosis, where a nonspecific mixture of cell-types in a tissue die together. It is proposed that excessive apoptotic cell death might ensue if, as the result of pathologically altered intracellular conditions, a signal were produced which initiates the process, or the threshold of response to such a signal was lowered, or perhaps as the result of reduced production of an antiapoptotic signal which had previously protected the neuron from destruction. The nature of this signal has not been determined in any of the neurodegenerative diseases in which apoptotic cell death is suspected (AD, ALS, IP), but it has been suggested that it might be elicited by OS, mitochondrial defects, increased Ca^{2+} levels, excitotoxic agents, a decline in neurotrophic factors and other signs of cellular pathology (Przuntek, et al., 1997). The initiation of cell death would presumably be the result of a complex interaction of cellular and extracellular factors relevant to the survival and function of the cell. In ALS, it was found that apoptotic cell death was associated with a mutation in the gene for the Cu/Zn-SOD enzyme, and that cellular death could be hindered by introduction of the normal enzyme (Rosen et al., 1993; Greenlund et al., 1995). Also of potential relevance to IP is the fact that receptor-stimulated apoptosis appears associated with the family of tumor necrosis factor receptors (Przuntek et al., 1997), to which the GDNF receptor belongs. Schapira (1997) noted that mitochondria appear to play an important role in the final common pathway leading to apoptotic cell death.

Confirmation of a role for apoptosis in IP has not yet emerged, but there are some suggestions in this direction (Mochizuki et al., 1996; Anglade et al.,

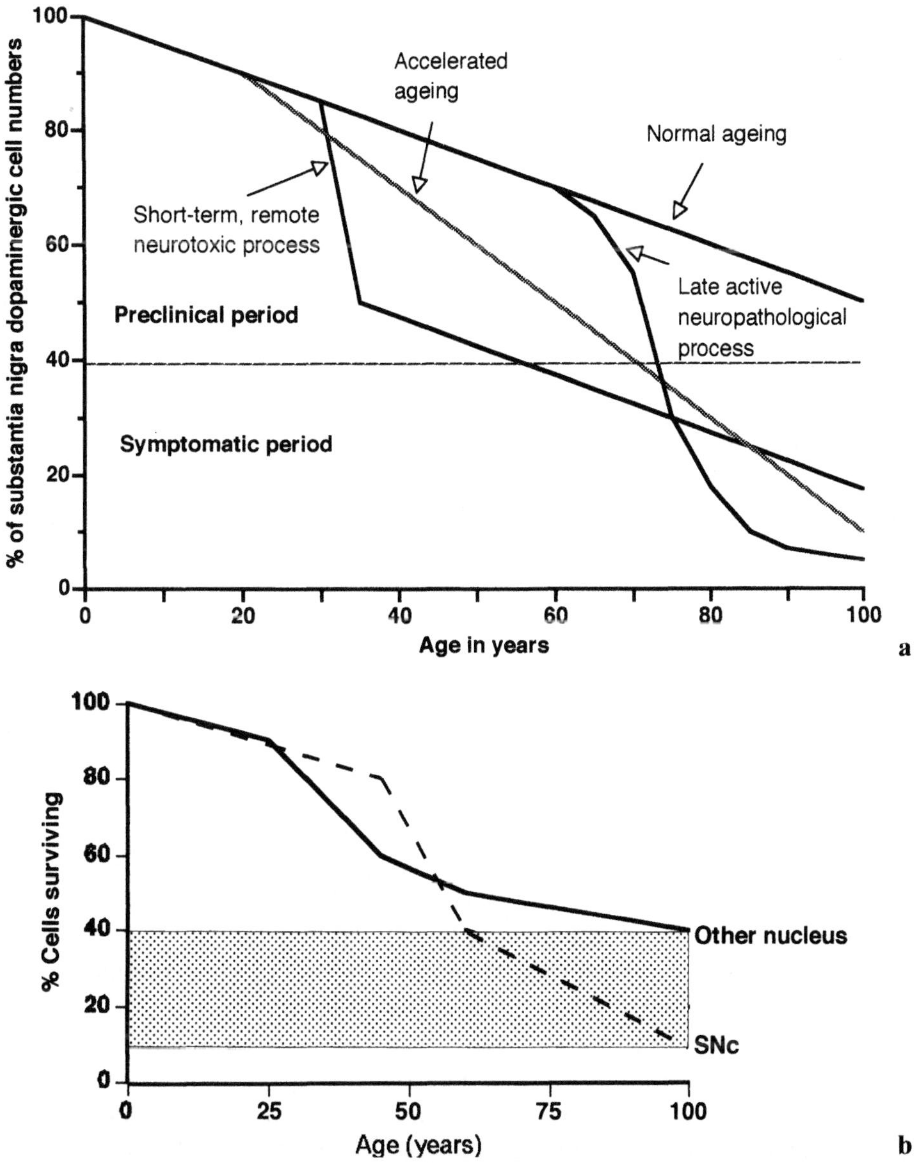

Fig. 2. Hypothetical models for neural degeneration in Parkinson's disease. (**a**) Different models of the rate of progression of SNc neurodegeneration (from Gerlach and Riederer, 1997; based on McGeer et al., 1988). (**b**) Model illustrating differential rates of neurodegeneration in different nuclei of the IP brain: The shaded region represents the period in which SNc neuronal loss is great enough to produce the classical IP symptomology, but where sufficient cells remain to make neuroprotection a viable strategy. The degree of neuronal loss in other nuclei which produces clinically detectable symptoms will, of course, be different, as will the window of opportunity for a neuroprotective strategy

1997). Schapira (1996) notes that there is evidence of active gliosis at the time of SN neurodegeneration, indicative of an active process, and not of the inflammation normally associated with necrosis. Further, MPP^+ concentrations which inhibit mitochondrial ATP synthesis in PC12 dopaminergic cells by between 15 and 50% can dose-dependently induce apoptotic cell death; beyond this point, necrotic cell death dominates (Schapira, 1996). L-DOPA can also induce apoptosis, an effect inhibited by antioxidants, although not by agents inhibiting its conversion to DA. Tompkins et al. (1997) recently reported a higher detection rate for apoptotic cells in the SN of patients with LB-associated disorders (PD, PD with AD, DLBD) than in patients with AD alone or no degenerative disease. DA and DA-melanin have been shown to be capable of initiating apoptotic changes in PC12 cells (Offen et al., 1997), although this has been criticized on the basis of the nonphysiological concentration of DA employed. It is assumed that, in normal circumstances, the brain is able to resist DA genotoxicity. Nevertheless, it is interesting that DNA repair has been found to be defective in IP, a situation which would favor pro-apoptotic processes (Robbins et al., 1985). It may thus be possible that low level exposure to a toxin initially stimulates apoptotic losses in the SNc, paving the way at the same time to progression to necrotic neurodegeneration. The apoptosis/necrosis question becomes critical in attempting either to define the etiology of IP or in developing a rational neuroprotective approach for those afflicted: should we be looking for a "signal" which instructs the pigmented cells of the SNc to enter their terminal phase, or for the absence of a signal which postpones the onset of apoptosis?

Many basic questions therefore remain unanswered, and two facts must be borne in mind when seeking their answers. The first is that the "triggering event" may not necessarily be the factor which drives the later progression of the disease. That is, a short-term insult may be sufficient to unleash a chain of events leading to presentation of the disease, but the process which ultimately causes the decline of the SNc may progress independently of this event. This early incident may be an infection, a toxin, a genetic factor — in any case, its elucidation promises to be difficult. And there may be several factors, each of which can independently elicit the disease, making intervention at this level even more complicated.

It is more likely that a neuroprotective therapy will be more effectively directed at the subsequent process — an attempt to halt the decline in function which leads to clinical manifestation of IP. In this case, the dynamic of the neurodegenerative process is of paramount importance, because it determines the window of opportunity in which a neuroprotective strategy would be meaningful. Synaptic plasticity allows the brain to tolerate a certain degree of neuronal loss: post mortem investigations indicate that in the case of the SNc, a remarkable 60% nigrostriatal neuronal loss (corresponding to 80% of striatal DA content) can be lost before clinical IP is manifested (Riederer and Wuketich, 1976). However, more recent PET investigations suggest that even a loss of 20% is clinically detectable (Morrish et al., 1996). Important is, that the disease process can probably not be reversed by a drug-based approach; lost neurons cannot be replaced, and it is not clear to what extent the surviving neurons can be encouraged to sprout and compensate functionally for their lost comrades. The aim of neuroprotection is to halt the disease process at the point a which it is intercepted, and must consequently be instituted as early as possible — preferably before the appearance of clinical signs, or, at the latest, soon after their appearance. PET studies are not a practical solution for mass screenings of potential IP sufferers, so that the development of alternative tools for the early identification of IP patients — whether via peripheral biochemical markers or by assessment of the nonmotor phenomena discussed above — is essential. The question must also be addressed as to whether "sick" neurons transmit their problem to their "healthy" neighbours. This has clear consequences for the choice of therapeutic strategy; the transplantation of healthy neurons, for instance, would be of no long-term benefit in this case, unless their future was secured by the concomitant institution of a neuropreservative therapy.

Secondly, as illustrated in Fig. 3, the hypotheses presented are not mutually exclusive, but rather interact with each in such a complicated manner that a precise outcome would be difficult to predict. The models pertain to different aspects of what must be seen as a "cellular syndrome", rather than as alternatives; what needs to be determined are the precise relationships between these aspects, and at which point in this tangled web of biological events it would be most advantageous and practical to intervene. Contrary to what often seems to be assumed, these biological events need not be successive stages of a temporal chain of events, but may be branches of a cascade of disasters emanating from one or more "primary" problems. For example, the mitochondrion model could lead to excessive OS as well as to cellular energy deficits, which would have an impact on cellular defence mechanisms, such as glutathione — but these may already have been compromised, thus rendering the mitochondrial defect critical. The great value of the MPTP model of IP is that it illustrates the fact that a linear causal chain of events may be overly simplistic: IP at the SNc cellular level may be a general metabolic crisis whose interconnections are resistant to unravelling. The alternative biochemical models of IP essentially offer insights into what may be happening in the dying cell, but not "why" — the identification of the earliest hallmarks of the disease needs to be established with greater clarity before this question can be answered. This naturally complicates the search for an effective neuroprotective strategy, as it is probable that a range of biochemical processes will require simultaneous taming in order to bring the disease process under control.

But it would seem in any case, given the apparently sporadic nature of IP, that a multifactorial etiology is to be expected. This could take any of several forms:

- a basic problem — a genetic predisposition or the result of an insult — allows a later "trigger" problem to be effective;
- two or more distinct events must be present simultaneously in order to elicit the disease;
- a problem which is normally handled effectively by the cell becomes critical due to lowered cellular defences, perhaps due to of ageing of the cell.

If a vulnerability is assumed, it may be possible to identify a biological marker which allows the determination of those at risk; candidates at the moment include mutations in CYP450 and mitochondrial protein genes.

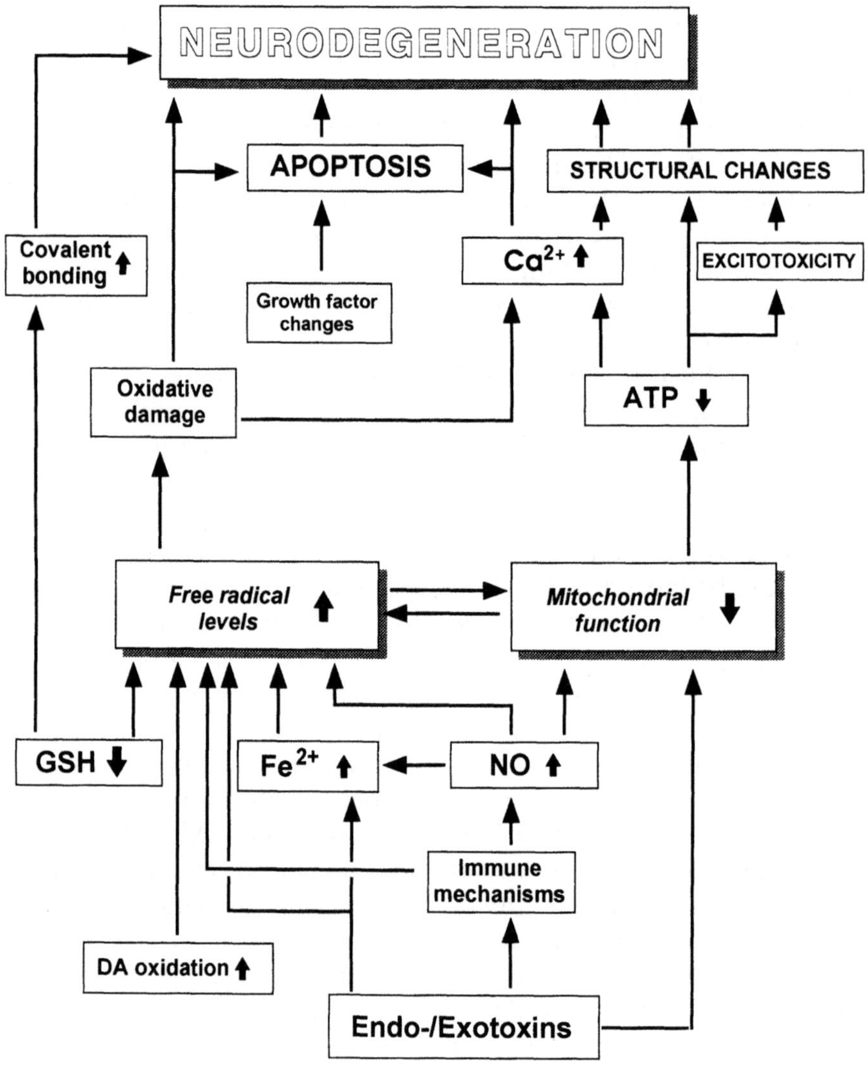

Fig. 3. Cellular mechanisms proposed to be involved in the neurodegenerative process in IP. As depicted, cellular processes, including the cascades of events leading to neurodegeneration, interact in complex fashions; no single etiological cause of IP can be viewed in isolation

An aspect of IP which is not explained by any of the models is the relevance of extranigral pathology in IP, recently reviewed in detail by Braak and coworkers (1995, 1997). IP must be regarded not only as a multisystemic disease at the biochemical level, but also at the anatomical and functional levels. It has long been recognized that damage in different areas of the brain might be responsible for the motor and psychic symptoms of the disease (including dementia). The allocortical loop, which exercises limbic influence on the frontal association cortex, is also marked by specific changes. Braak et al. (1995) have noted that the anterior cingulate proneocortex is severely affected in IP, with large numbers of LBs, especially in the multiform layer.

As this layer projects to the amygdala, this phenomenon is of more than academic interest, especially, as discussed above, the psychic (depression, anxiety) and autonomic manifestations (back pain, seborrheic dermatitis, constipation) of IP possibly antedate the motor complaint. The amygdala itself is a site of major lesions, as are often its projection targets; a major reduction in the volume of medial temporal structures (including the hippocampus) has been identified in morphological investigations (Double et al., 1996). In a comparison of AD and IP, Braak's laboratory noted the curious specificity of the lesions in AD for afferent cortical structures, while in IP efferent subcortical structures appear targeted, with sparing of neighbouring cell types. On the other hand, it is not to be forgotten, that the progression of the degeneration is not equally rapid for all the cell groups involved, and that patients with particular types of IP deteriorate more rapidly than others: for example, late age of onset and the akinesia-rigor form of IP are both associated with rapid decline of the patient.

The sensitivity of particular cell groups to degeneration in IP and in AD, frequently associated with IP, might be summarized as follows:

IP DA > NA ≫ 5-HT > ACh
AD ACh > 5-HT ≫ NA > DA

This differential, but not absolutely selective sensitivity of the different transmitter systems is important when attempting to elucidate their causes. Further, it is clear that combinations of the two disorders can and do occur, and that glutamatergic disturbances are also described for both; then we could propose, instead of a strict delineation of the two disorders, a spectrum of degeneration, thus:

$$\text{AD} \Leftrightarrow \text{AD} + \text{IP} \Leftrightarrow \text{DLBD} \Leftrightarrow \text{IP} + \text{AD} \Leftrightarrow \text{IP}$$

The overlap in the presentation of classical neuropathological markers also illustrates the continuity of the neurodegenerative spectrum. What such a model suggests, however, is that as much attention must be paid to the features which the various neurodegenerative features share in common, as to their distinctive characteristics; such a view also draws our attention to the multisystem nature of these diseases, rather than devoting our full attention to one particular brain region.

The temporal relationship between changes in the SN and extranigral regions also remains to be determined; the latter are not directly connected with the SN, so that the antero- or retrograde transmission of a problem from a primary SN lesion is unlikely. It should also be noted here, that the only biopsy of an MPTP toxicity sufferer found no evidence of extranigral pathology, or of major LB activity (Davis et al., 1979); however, more work is needed to confirm this finding. It is also important to remember that not all symptoms respond to L-DOPA therapy, such as freezing and postural instability, particularly in the latter stages of the disease (Fahn, 1992); the interpretation of these phenomena is disputed, but implicate nondopaminergic mechanisms in the disease.

The question then arises of whether we are looking at a distinct entity when we use the term "IP". We have already delineated IP — perhaps better designated as "LB-associated parkinsonism" — from parkinsonian syndromes of known etiological origin, and suggested that familial IP may differ from sporadic IP. But it has also been proposed that IP be seen as part of a continuum of neurodegeneration, with itself at one end, AD at the other, and a range of variations in between (Calne and Eisen, 1989). Such a view is supported by recent findings of LBs in a variety of neurodegenerative diseases, and, in particular, by the overlap between ILBD and AD (Smith and Prayson, 1996). "Lewy body disease" would then be a general term for a family of disorders sharing the LB as part of their pathology, and whose specific clinical manifestations would be related to the location of these structures. This proposal does not solve the problem of how LBs arise, but perhaps directs attention to a common feature which has not yet been incorporated into an etiological model of any of these disorders.

In summary, IP must be regarded as a multisystemic disorder in which multiple aspects of neuronal function are compromised; further, its demarcation from other neurodegenerative diseases becomes more difficult when the range of symptoms presented is taken into consideration. No model of IP which restricts itself to explaining the death of cells in the SNc will account for all aspects of the disorder, a fact often overlooked, although consideration of IP as a broader, "spectrum" disease might bring us closer to understanding it.

Conclusions

Current therapies aim to control the symptoms of IP, and are not capable of actually curing or halting the disease. This will require knowledge of what happens early in IP, in the presymptomatic phase, and how it relates to events in the symptomatic phase. If ILBD is correctly regarded as representing an a- or presymptomatic form of IP, the earliest and most specific signs of IP appear to be the decline in SNc GSH levels and thereafter a decline in mitochondrial complex I activity. Beyond this, however, the answers are less clear: are these changes causative or symptomatic, and, in either case, why do they arise? Do the changes in various parts of the IP brain arise independently, or is there a link we have not yet detected?

All models of degeneration in IP assume a long "latent" phase and a comparatively rapid decline thereafter. On the basis of this conception of the dynamic of the disease, even a highly effective neuroprotective strategy will have a narrow therapeutic window within which it could significantly alter the course of the disease. Intervention should actually commence before the appearance of symptoms, in order to have a large hope of success, but it has thus far not been possible, despite application of a range of biochemical and neuroimaging approaches, to devise a practical means of diagnosing IP before the appearance of clinical symptoms.

The need for precision in understanding the preclinical course of IP is dictated by this tight time frame. Most strategies which prove neuroprotective in the laboratory involve administration of the rescuing agent within days of the toxin — not a practicable solution for IP patients. And this is separate from the question of time: most laboratory work is necessarily focussed on acute effects, and the relationship with the chronic and progressive nature of IP remains unclear. All this in turn assumes that the course of IP is determined by a process which we are capable of steering.

It is often tempting to seize one aspect of the syndrome as the central aspect and to build an etiological theory around it, rather than to regard it as one of several pieces of the puzzle; or, even worse, to confuse a model of the disease with the disease itself. The fact that a distinction between "IP" and "parkinsonism" is considered necessary indicates that a variety of reasons can elicit the same symptoms, despite being distinct neurological entities. The same caution applies to the MPTP model: while it reproduces the most significant clinical features of IP, there are indications that at the neural level significant differences may exist between it and true IP; MPTP toxicity could conceivably model the middle of the causal chain in IP, without giving hints to the endogenous trigger of the disease:

"Trigger" for IP → [chain of events] → complex I deficiency → [chain of events] → IP
↑
$MPTP/MPP^+$

Many questions remain open: for instance, why do we find NM only in man, and, of all places, in one of the most ancient departments of the brain? Perhaps our upright gait or our manual dexterity impose challenges for the work capacity of our SN, challenges which have led to the evolution of unique coping strategies, such as NM, but also with a residual risk of oxidative and metabolic stress, which in the end, for some people, ends in its failure — like an exhausted muscle or heart.

There is currently no indication that a model is in the offing which could explain all aspects of IP. It is more reasonable to assume that IP is the result of a multifactorial process, and that a cascade of interacting and overlapping biochemical mechanisms determine the course of the disease. Unfortunately, this will render the development of neuroprotective strategies difficult, so that less specific strategies may prove to be more effective. But when the key to IP is found, given the above proposed spectrum of neurodegenerative disorders, it is probable that the answers will be of broader relevance than for IP alone. Conversely, developments in the investigation of AD, HD, and other disorders are likely to produce clues concerning the nature of IP.

In conclusion, there exists a great body of evidence to suggest that oxidative stress is an important factor in the demise of nigral neurons in IP, but what is responsible for this challenge, and why the normal cellular defence mechanisms are no longer capable of dealing with it, remain mysteries. We have sufficient explanations of how nigral cells might die, but not why particular cells die. This is almost certainly the result of our limited understanding of normal neuronal life and death, and of the unique qualities of the human SNc. Similarly, the fact that IP is clearly, on the basis of the pathophysiological data, a multisystemic disease has not yet received the attention it deserves. Despite the significant growth in our understanding of the processes involved in IP, there is a long way to go before we can speak of an "answer".

References

Adams JD, Odunze IN (1991) Biochemical mechanisms of 1-methyl-4-phenyl-1,2,3,6-tetrahydropyridine toxicity. Biochem Pharmacol 41: 1099–1105

Aime S, Fasano M, Bergamasco B, Lopiano L, Quattrocolo G (1996) Nuclear magnetic resonance spectroscopy characterization and iron content determination of human mesencephalic neuromelanin. In: Battistin L, Scarlato G, Caraceni T, Ruggieri S (eds) Advances in neurology, vol 69, Parkinson's disease. Lippincott-Raven, Philadelphia, pp 263–270

Alam ZI, Jenner A, Daniel SE, Lees AJ, Cairns N, Marsden CD, Jenner P, Halliwell B (1997) Oxidative DNA damage in the parkinsonian brain: a selective increase in 8-hydroxyguanine in substantia nigra? J Neurochem 69: 1196–1203

Ambani LM, Van Woert MH, Murphy S (1975) Brain peroxides and catalase in Parkinson's disease. Arch Neurol 32: 114–118

Anglade P, Agid Y, Hirsch EC, Vyas S (1997) Apoptosis in dopaminergic neurons of the human substantia nigra during normal aging. Histol Histopathol 12: 603–610

Bancher C, Lassmann H, Budka H, Budka H, Jellinger K, Grundke-Iqbal I, Iqbal K, Wiche G, Seitelberger F, Wisniewski HM (1989) An antigenic profile of Lewy bodies: immunocytochemical indication for protein phosphorylation and ubiquitination. J Neuropathol Exp Neurol 48: 81–93

Barbeau A, Cloutier T, Roy M, Plasse L, Paris S, Poirier J (1985) Ecogenetics of Parkinson's disease: 4-hydroxylation of debrisoquine. Lancet II: 1213–1216

Baron JA (1986) Cigarette smoking and Parkinson's disease. Neurology 36: 1490–1496

Beal MF, Hyman BT, Koroshetz W (1993) Do defects in mitochondrial energy metabolism underlie the pathology of neurodegenerative diseases? Trends Neurosci 16: 125–131

Beck KD (1994) Functions of brain-derived neurotrophic factor, insulin-like growth factor-I and basic fibroblast growth factor in the development and maintenance of dopaminergic neurons. Prog Neurobiol 44: 497–516

Beckman JS, Beckman TW, Chen J, Marshall PA, Freeman PA (1990) Apparent hydroxyl radical production by peroxynitrite: implications for endothelial injury from nitric oxide and superoxide. Proc Natl Acad Sci USA 87: 1620–1624

Ben-Shachar D, Youdim MBH (1991) Intranigral iron injection induces behavioral and biochemical "Parkinsonism" in rats. J Neurochem 57: 2133–2135

Ben-Shachar D, Eshel G, Finberg JPM, Youdim MBH (1991) The iron chelator desferrioxamine (desferal) retards 6-hydroxydopamine-induced degeneration of nigrostriatal neurons. J Neurochem 56: 1441–1444

Birkmayer W, Riederer P (1985) Die Parkinson-Krankheit: Biochemie, Klinik, Therapie 2nd edn. Springer, Wien New York, pp 60–101

Blum-Degen D, Frölich L, Hoyer S, Riederer P (1995) Altered regulation of brain glucose metabolism as a cause of neurodegenerative disorders? J Neural Transm [Suppl] 46: 139–147

Blunt SB, Jenner P, Marsden CD (1993) Suppressive effect of l-dopa on dopamine cells remaining in the ventral tegmental area of rats previously exposed to the neurotoxin 6-hydroxydopamine. Mov Disord 8: 129–133

Braak H, Braak E, Yilmazer D, de Vos RAI, Jansen ENH, Bohl J, Jellinger K (1994) Amygdala pathology in Parkinson's disease. Acta Neuropathol 88: 493–500

Braak H, Braak E, Yilmazer D, Schultz C, de Vos RAI, Jansen ENH (1995) Nigral and extranigral pathology in Parkinson's disease. J Neural Transm [Suppl] 46: 15–32

Braak, H, Braak E, de Vos RAI, Jansen EHN, Bohl J (1997) Extranigrale Pathologie der Parkinson-Krankheit — limbisches System und vegetative Kerne. In: Fischer PA (ed) Parkinson-Krankheit. Entwicklungen in Diagnostik und Therapie. Schattauer, Stuttgart, pp 25–36

Bringmann G, God R, Feineis D, Wesemann W, Riederer P, Rausch WD, Reichmann H, Sontag KH (1995) The TaClo concept: 1-trichloromethyl-1,2,3,4-tetrahydro-β-carboline (TaClo), a new toxin for dopaminergic neurons. J Neural Transm [Suppl] 46: 235–244

Brookes DJ (1991) Detection of preclinical Parkinson's disease with PET. Neurology 41 (Suppl 2): 24–27

Buccala R, Cerami A (1992) Advanced glycosylation: chemistry, biology and implications for diabetes and aging. Adv Pharmacol 23: 1–34

Calne DB (1994) Is idiopathic parkinsonism the consequence of an event or a process? Neurology 44: 5–10

Calne DB, Eisen AA (1989) The relationship between Alzheimer's disease, Parkinson's disease and motoneuron disease. Can J Neurol Sci 16: 547–550

Calne S, Shoenberg BS, Martin W, Uitti RJ, Spencer P, Calne DB (1987) Familial Parkinson's disease: possible role of environmental factors. Can J Neurol Sci 14: 303–315

Carlsson A, Fornstedt B (1991) Catechol metabolites in the cerebrospinal fluid as possible markers in the early diagnosis of Parkinson's disease. Neurology 40 (Suppl 2): 50–52

Carstam R, Brinck C, Hindemith-Augustsson A, Rorsman H, Rosengren E (1991) The neuromelanin of the human substantia nigra. Biochim Biophys Acta 1097: 152–160

Castellani R, Smith MA, Richey PL, Perry G (1996) Glycoxidation and oxidative stress in Parkinson disease and diffuse Lewy body disease. Brain Res 737: 195–200

Chandrasekaran K, Rapoport SI, Brady DR, Stoll J (1992) Localization of cytochrome oxidase (COX) activity and COX mRNA in the hippocampus and entorhinal cortex of the monkey brain: correlation with specific neuronal pathways. Brain Res 579: 333–336

Chan-Palay V, Zetzsche T, Hochli M (1991) Parvalbumin neurons in the hippocampus in senile dementia of the Alzheimer type, Parkinson's disease and multi-infarct dementia. Dementia 2: 297–313

Cleeter MWJ, Cooper JM, Schapira AHV (1992) Irreversible inhibition of mitochondrial complex I by 1-methyl-4-phenylpyridinium: evidence for free radical involvement. J Neurochem 58: 786–789

Cleeter MWJ, Cooper JM, Darley-Usmar VM, Moncada S, Schapira AHV (1994) Reversible inhibition of cytochrome c oxidase, the terminal enzyme of the mitochondrial respiratory chain, by nitric oxide: implications for neurodegenerative disorders. Acta Biochem Biophys 288: 481–487

Collins MA, Neafsey EJ (1985) β-Carboline analogues of N-methyl-4-phenyl-1,2,3,6-tetrahydropyridine (MPTP): endogenous factors underlying idiopathic parkinsonism. Neurosci Lett 55: 179–184

Connor JR, Snyder BS, Arosio P, Loeffler DA, LeWitt P (1995) A quantitative analysis of isoferritins in select regions of aged, parkinsonian and Alzheimer's diseased brains. J Neurochem 65: 717–724

Coyle JT, Puttfarcken P (1993) Oxidative stress, glutamate and neurodegenerative disorders. Science 262: 689–695

D'Amato RJ, Alexander GM, Schwartzman RJ, Kitt CA, Price DL, Snyder SH (1987) Evidence for neuromelanin involvement in MPTP-induced neurotoxicity. Nature 327: 324–326

Damier P, Hirsch E, Javoy-Agid F, Zhang P, Agid Y (1993) Glutathione peroxidase, glial cells and Parkinson's disease. Neuroscience 52: 1–6

Davis GC, Williams AC, Markey SP, Ebert MH, Caine ED, Reichert CM, Kopin IJ (1979) Chronic Parkinsonism secondary to intravenous injection of meperidine analogues. Psychiat Res 1: 249–254

Dawson TM, Dawson VL, Snyder SH (1992) A novel neuronal messenger molecule in brain: the free radical, nitric oxide. Ann Neurol 32: 297–311

DeLeve LD, Kaplowitz N (1991) Glutathione metabolism and its role in hepatotoxicity. Pharmacol Ther 52: 287–305

De Michele G, Filla A, Volpe G, Gogliettino A, Ambrosio G, Campanella G (1996) Etiology of Parkinson's disease. The role of environment and heredity. In: Battistin L, Scarlato G, Caraceni T, Ruggieri S (eds) Advances in neurology, vol 69, Parkinson's disease. Lippincott-Raven, Philadelphia, pp 19–24

Dexter DT, Wells FR, Agid FJ (1987) Increased nigral iron content in postmortem parkinsonian brain. Lancet II: 1219–1220

Dexter DT, Cater CJ, Wells FR, Javoy-Agid F, Agid Y, Lees A, Jenner P, Marsden CD (1989) Basal lipid peroxidation in substantial nigra is increased in Parkinson's disease. J Neurochem 52: 381–389

Dexter DT, Carayon A, Vidailhet M, Ruberg M, Agid F, Agid Y, Lees AJ, Wells FR, Jenner P, Marsden CD (1990) Decreased ferritin levels in brain in Parkinson's disease. J Neurochem 55: 16–20

Dexter DT, Carayon A, Javoy-Agid F (1991) Alterations in the levels of iron, ferritin and other trace row diroudopically metals in Parkinson's disease and other neurodegenerative diseases affecting the basal ganglia. Brain 114: 1953–1975

Dexter DT, Ward RJ, Wells FR, Daniel SE, Lees AJ, Peters IJ, Jenner P, Marsden CD (1992) α-Tocopherol levels in brain are not altered in Parkinson's disease. Ann Neurol 32: 591–593

Dexter DT, Holley AE, Flitter WD Slater TF, Wells FR, Daniel SE, Lees AJ, Jenner P, Marsden CD (1994a) Increased levels of lipid hydroperoxides in the parkinsonian substantia nigra: an HPLC and ESR study. Mov Disord 9: 92–97

Dexter DT, Sian J, Rose S, Hindmarsh JS, MannVM, Cooper JM, Wells FR, Daniel SE, Lees AJ, Schapira AHV, et al (1994b) Indices of oxidative stress and mitochondrial function in individuals with incidental Lewy body disease. Ann Neurol 35: 38–44

Difazio MC, Hollingsworth Z, Young AB, Penney JBJ (1992) Glutamate receptors in the substantia nigra of Parkinson's disease brains. Neurology 42: 402–406

DiMonte DA (1991) Mitochondrial DNA and Parkinson's disease. Neurology 41 (Suppl 2): 38–42

Dostert P, Strolin-Benedetti M, Dordain G (1988) Dopamine-derived alkaloids in alcoholism and in Parkinson's and Huntington's disease. J Neural Transm 74: 61–74

Doty RL, Deems DA, Stellar S (1988) Olfactory dysfunction in parkinsonism: a general deficit unrelated to neurologic signs, disease stage, or disease duration. Neurology 38: 1237–1244

Double KL, Halliday GM, McRitchie DA, Reid WGJ, Hely MA, Morris JGL (1996) Regional brain atrophy in idiopathic Parkinson's disease and diffuse Lewy body disease. Dementia 7: 304–313

Double KL, Maywald M, Schmittel M, Riederer P, Gerlach M (1997) In vitro studies of ferritin iron release and neurotoxicity. J Neurochem 70: 2492–2499

Duffy PE, Tennyson VM (1965) Phase and electron microscopic observations of Lewy bodies and melanin granules in the substantia nigra and locus coeruleus in Parkinson's disease. J Neuropathol Exp Neurol 24: 398–414

Duvoisin RC (1996) Recent advances in the genetics of Parkinson's disease. In: Battistin L, Scarlato G, Caraceni T, Ruggieri S (eds) Advances in neurology, vol 69, Parkinson's disease. Lippincott-Raven, Philadelphia, pp 33–40

Fahn S (1992) Adverse effects of levodopa. In: Olanow CW, Lieberman AN (eds) The scientific basis for the treatment of Parkinson's disease. Parthenon, Carnforth (UK), pp 89–112

Faucheaux BA, Hirsch EC, Villares J, Selimi F, Mouatt-Prigent A, Javoy-Agid F, Agid Y (1993) Distribution of ^{125}I-ferrotransferrin binding sites in the mesencephalon of control subjects and patients with Parkinson's disease. J Neurochem 60: 2238–2241

Faucheaux BA, Nillesse N, Damier P, Spik G, Mouatt-Prigent A, Pierce A, Leveugle B, Kubis N, Hauw JJ, Agid Y, Hirsch EC (1995) Expression of lactoferrin receptors is increased in the mesencephalon of patients with Parkinson's disease. Proc Natl Acad Sci USA 92: 9303–9307

Fawthrop DJ, Boobis AR, Davies DS (1991) Mechanisms of cell death. Arch Toxicol 65: 437–444

Fearnley JM, Lees A (1991) Ageing and Parkinson's disease: substantia nigra regional selectivity. Brain 114: 2283–2301

Fornai F, Vaglini F, Maggio R, Bonuccelli U, Corsini GU (1996) Excitatory amino acids and MPTP toxicity. In: Battistin L, Scarlato G, Caraceni T, Ruggieri S (eds) Advances in neurology, vol 69, Parkinson's disease. Lippincott-Raven, Philadelphia, pp 167–176

Forno LS (1995) Pathological considerations in the etiology of Parkinson's disease. In: Ellenberg JH, Koller WC, Langston JW (eds) Etiology of Parkinson's disease. Marcel Dekker, New York, pp 65–95

Forno LS, Langston JW (1993) Lewy bodies and aging: relation to Alzheimer's and Parkinson's diseases. Neurodegeneration 2: 19–24

Forno LS, DeLanney LE, Irwin I, Langson JW (1995) Ultrastructure of eosinophilic inclusion bodies in the amygdala-parahippocampal region of aged squirrel monkeys treated with 1-methyl-4-phenyl-1,2,3,6-tetrahydropyridine, a dopaminergic toxin. Neurosci Lett 184: 44–47

Forno LS, DeLanney LE, Irwin I, Langson JW (1996) Electron microscopy of Lewy bodies in the amygdala-parahippocampal region: comparison with inclusion bodies in the MPTP-treated squirrel monkey. In: Streifler MB, Korczyn AD, Melamed E, Youdim MBH (eds) Advances in neurology, vol 53, Parkinson's disease: anatomy, pathology and therapy. Raven Press, New York, pp 217–228

Franklin JL, Johnson EM (1992) Suppression of programmed neuronal cell death by sustained elevation of cytosolic calcium. Trends Neurosci 15: 501–508

Gash DM, Zhang Z, Ovadia A, Cass WA, Yi A, Simmerman L, Russell D, Martin D, Lapchak PA, Collins F, Hoffer BJ, Gerhardt GA (1996) Functional recovery in parkinsonian monkeys treated with GDNF. Nature 380: 252–255

Gasser T (1997) Stand der neurogenetischen Forschung bezüglich der Parkinson-Erkrankung. In: Fischer PA (ed) Parkinson-Krankheit. Entwicklungen in Diagnostik und Therapie. Schattauer, Stuttgart, pp 59–71

Gasser T, Wszolek ZK, Trofatter J, Ozelius L, Uitti RJ, Lee CS, Gusella J, Pfeiffer RF, Calne DB, Breakefield XO (1994) Genetic linkage studies in autosomal dominant parkinsonism: evaluation of seven candidate genes. Ann Neurol 36: 387–396

Gerlach M, Riederer P (1993) The pathophysiological basis of Parkinson's disease. In: Szelenyi (ed) Inhibitors of monoamine oxidase. Birkhäuser, Basel, pp 25–50

Gerlach M, Riederer P (1996) Animal models of Parkinson's disease: an empirical comparison with the phenomenology of the disease in man. J Neural Transm 103: 987–1041

Gerlach M, Riederer P, Przuntek H, Youdim MBH (1991) MPTP mechanisms of neurotoxicity and their implications for Parkinson's disease. Eur J Pharmacol [Mol Pharmacol Sect] 208: 273–286

Gerlach M, Ben-Shachar D, Riederer P, Youdim MBH (1994) Altered brain metabolism of iron as a cause of neurodegenerative diseases? J Neurochem 63: 793–807

Gerlach M, Riederer P, Youdim MBH (1996) Molecular mechanisms for neurodegeneration: synergism between reactive oxygen species, calcium and excitotoxic amino acids. In: Battistin L, Scarlato G, Caraceni T, Ruggieri S (eds) Advances in neurology, vol 69, Parkinson's disease. Lippincott-Raven, Philadelphia, pp 177–194

German DC, Manaye KF, Sonsalla PK, Brooks BA (1992) Midbrain dopaminergic cell loss in Parkinson's disease and MPTP-induced parkinsonism: sparing of calbindin D28k-containing cells. Ann NY Acad Sci 648: 42–62

Gibb WRG (1989) The diagnostic relevance of Lewy bodies and other inclusions in Parkinson's disease. In: Przuntek H, Riederer P (eds) Early diagnosis and preventative therapy in Parkinson's disease. Springer, Wien New York, pp 171–180

Gibb WRG (1992) Melanin, tyrosine hydroxylase, calbindin and substance P in the human midbrain and substantia nigra in relation to nigrostriatal projections and differential neuronal susceptibility in Parkinson's disease. Brain Res 581: 283–291

Gibb WRG, Lees AJ (1989) The significance of the Lewy body in the diagnosis of idiopathic Parkinson's disease. Neuropathol Appl Neurobiol 15: 27–44

Gibb WRG, Esiri MM, Lees AJ (1985) Clinical and pathologic features of diffuse cortical Lewy body disease (Lewy body dementia). Brain 110: 1131–1153

Gibb WRG, Scott T, Lees AJ (1991) Neuronal inclusions of Parkinson's disease. Mov Disord 6: 2–11

Götz ME, Künig G, Riederer P, Youdim MBH (1994) Oxidative stress. Free radical production in neural degeneration. Pharmac Ther 63: 37–122

Golbe LI (1995) Genetics of Parkinson's disease. In: Ellenberg JH, Koller WC, Langston JW (eds) Etiology of Parkinson's disease. Marcel Dekker, New York, pp 115–140

Golbe LI, Lazzarini AM, Schwarz KO, Mark MH, Dickson DW, Duvoison RC (1993) Autosomal dominant parkinsonism with benign course and typical Lewy-body pathology. Neurology 43: 2222–2227

Goldman JE, Yen SH, Chiu FC, Peress NS (1983) Lewy bodies of Parkinson's disease contain neurofilament antigen. Science 221: 1082–1084

Good P, Olanow CW, Perl DP (1992) Neuromelanin-containing neurons of the substantia nigra accumulate iron and aluminum in Parkinson's disease. A LAMMA study. Brain Res 593: 343–346

Graham DG (1978) Oxidative pathways for catecholamines in the genesis of neuromelanin and cytotoxic quinones. Mol Pharmacol 14: 633–643

Greenlund LJ, Deckwerth TL, Johnson EMJ (1995) Superoxide dismutase delays neuronal aptosis: a role for reactive oxygen species in programmed neuronal death. Neuron 14: 303–315

Grote C, Clement HW, Wesemann W, Bringmann G, Feineis D, Riederer P, Sontag KH (1995) Biochemical lesions of the nigrostriatal system by TaClo (1-trichloromethyl-1,2,3,4-tetrahydro-β-carboline) and derivatives. J Neural Transm [Suppl] 46: 275–281

Gutteridge JM, Quinlan GJ, Clark I, Halliwell B (1985) Aluminum salts accelerate peroxidation of membrane lipids stimulated by iron salts. Biochim Biophys Acta 835: 441–447

Halliday GM, Blumberg PC, Cotton RGH, Blessing WW, Geffen LB (1990) Loss of brainstem serotonin- and substance P-containing neurons in Parkinson's disease. Brain Res 510: 104–107

Halliwell B (1992) Reactive oxygen species and the central nervous system. J Neurochem 59: 1609–1623

Hansen LA, Galasko D (1992) Lewy body disease. Curr Opin Neurol Neurosurg 5: 889–894

Hattori N, Tanaka M, Ozawa T, Mizuno Y (1991) Immunohistochemical studies on complexes I, II, III and IV of mitochondria in Parkinson's disease. Ann Neurol 30: 563–571

Hefti F (1994) Neurotrophic factor therapy for central nervous system degenerative disease. J Neurobiol 25: 1418–1435

Hefti F, Melamed E, Bhawan J, Wurtman R (1981) Long term administration of L-dopa does not damage dopaminergic neurons in the mouse. Neurology 31: 1194–1195

Heintz N, Zoghbi H (1997) α-synuclein — a link between Parkinson and Alzheimer diseases? Nat Genet 16: 325–327

Hill W, Lee VMY, Hurtig H, Murray JM, Trojanowski JQ (1991) Epitopes located in spatially separated domains of each neurofilament subunit are present in Parkinson's disease Lewy bodies. J Comp Neurol 309: 150–160
Hirsch EC, Graybiel AM, Agid Y (1988) Melanized dopaminergic neurons are differentially affected in Parkinson's disease. Nature 334: 345–348
Hirsch EC, Brandel JP, Galle P, Javoy-Agid F, Agid Y (1991) Iron and aluminum increase in the substantia nigra of patients with Parkinson's disease: an X-ray microanalysis. J Neurochem 56: 446–451
Hirsch EC, Mouatt A, Thomasset M, Javoy-Agid F, Agid Y, Graybiel AM (1992) Expression of calbindin D_{28K}-like immunoreactivity in catecholaminergic cell groups in the human midbrain. Normal distribution and distribution in Parkinson's disease. Neurodegeneration 1: 83–93
Hirsch EC, Faucheux B, Damier P, Mouatt-Prigent A, Agid Y (1997) Neuronal vulnerability in Parkinson's disease. J Neural Transm [Suppl] 50: 79–88
Hotamisligil GS, Girmen AS, Fink JS, Tivol E, Shalish C, Trofatter J, Baenziger J, Diamond S, Markham C, Sullivan J, et al (1994) Hereditary variations in the monoamine oxidase as a risk factor for Parkinson's disease. Mov Disord 9: 305–310
Hunot S, Boissièrre F, Faucheux B, Brugg B, Mouatt-Prigend A, Agid Y, Hirsch EC (1996) Nitric oxide synthase and neuronal vulnerability in Parkinson's disease. Neuroscience 72: 355–363
Iacopino AM, Christakos S (1990) Specific reduction of calcium-binding protein (28-kilodalton calbindin-D) gene expression in aging and neurodegenerative diseases. Proc Natl Acad Sci USA 87: 4078–4082
Ichimaya Y, Emson PC, Mountjoy CQ, Lawson DEM, Iizuka P (1989) Calbindin D_{28k}-immunoreactive cholinergic neurons in the nucleus basalis of Meynert in Alzheimer-type dementia. Brain Res 499: 402–406
Ikonomidou C, Turski L (1996) Neurodegenerative disorders: clues from glutamate and energy metabolism. Crit Rev Neurobiol 10: 239–263
Irwin I, Langston JW (1995) Endogenous toxins as potential etiologic agents in Parkinson's disease. In: Ellenberg JH, Koller WC, Langston JW (eds) Etiology of Parkinson's disease. Marcel Dekker, New York, pp 153–201
Itoh K, Weis S, Mehraein P, Muller-Hocker J (1997) Defects of cytochrome c oxidase in the substantia nigra of Parkinson's disease: an immunohistochemical and morphometric study. Mov Disord 12: 9–16
Janetzky B, God R, Bringmann G, Reichmann H (1995) 1-Trichloromethyl-1,2,3,4-tetrahydro-β-carboline, a new inhibitor of complex I. J Neural Transm [Suppl] 46: 265–273
Jellinger KA (1990) New developments in the pathology of Parkinson's disease. In: Streifler MB, Korczyn AD, Melamed E, Youdim MBH (eds) Advances in neurology, vol 53, Parkinson's disease: anatomy, pathology and therapy. Raven Press, New York, pp 1–16
Jellinger KA (1991) Pathology of Parkinson's disease. Changes other than the nigrostriatal pathway. Mol Chem Neuropathol 14: 153–197
Jellinger KA (1995) Neurodegenerative disorders with extrapyramidal features. J Neural Transm [Suppl] 46: 33–58
Jellinger P, Paulus W, Grundke-Iqbal I, Riederer P, Youdim MBH (1990) Brain iron and ferritin in Parkinson's disease and Alzheimer's diseases. J Neural Transm [PD Sect] 2: 327–340
Jellinger K, Youdim MBH, Ben-Shachar D, Stachelberger H, Riederer P, Rumpelmair G, Kienzl E (1992) Iron-melanin complex in substantia nigra of parkinsonian brains: an X-ray microanalysis. J Neurochem 59: 1168–1171
Jellinger K, Linert L, Kienzl E, Youdim MBH (1995) Chemical evidence for 6-hydroxydopamine to be an endogenous toxic factor in the pathogenesis of Parkinson's disease. J Neural Transm 46: 297–314

Jenner P, Olanow CW (1996a) Oxidative stress and the pathogenesis of Parkinson's disease. Neurology 47 [Suppl 3]: S161–S170
Jenner P, Olanow CW (1996b) Pathological evidence for oxidative stress in Parkinson's disease and related degenerative disorders. In: Olanow CW, Jenner P, Youdim M (eds) Neurodegeneration and neuroprotection in Parkinson's disease. Academic Press, London, pp 24–45
Johnson WG, Hodge SE, Duvoisin RC (1990) Twin studies and the genetics of Parkinson's disease — a reappraisal. Mov Disord 5: 187–194
Kass GEN, Wright JM, Nicotera P, Orrenius S (1988) The mechanism of 1-methyl-4-phenyl-1,2,3,6-tetrahydropyridine toxicity: role of intracellular calcium. Arch Biochem Biophys 260: 789–797
Kish SJ, Morito CH, Hornykiewicz O (1985) Glutathione peroxidase activity in Parkinson's disease brain. Neurosci Lett 58: 343–346
Klockgether T, Turski L (1993) Toward an understanding of the role of glutamate in experimental parkinsonism: agonist-sensitive sites in the basal ganglia. Ann Neurol 34: 585–593
Koller WC, Montgomery EB (1997) Issues in the early diagnosis of Parkinson's disease. Neurology 49 [Suppl 1]: S10–S25
Kondo K, Kurland RT (1973) Parkinson's disease, genetic analysis and evidence of a multifactorial etiology. Mayo Clin Proc 48: 465–474
Kosaka K (1978) Lewy bodies in the cerebral cortex: report of three cases. Acta Neuropathol (Berl) 42: 127–134
Kosel S, Lucking SB, Egensperger R, Mehraein P, Graeber MB (1996) Mitochondrial NADH dehydrogenase and CYP2D6 genotypes in Lewy-body parkinsonism. J Neurosci Res 44: 174–183
Kuhn W, Müller T (1995) Neuroimmune mechanisms in Parkinson's disease. J Neural Transm [Suppl] 46: 229–234
Kuhn W, Müller T (1997) Hypersusceptibilität gegen Xenobiotika. Die potentielle Bedeutung ökogenetischer Faktoren für die Ätiologie des Morbus Parkinson. In: Fischer PA (ed) Parkinson-Krankheit. Entwicklungen in Diagnostik und Therapie. Schattauer, Stuttgart, pp 87–93
Kuhn W, Müller T, Große H, Rommelspacher H (1995) Plasma harman and norharman in Parkinson's disease. J Neural Transm [Suppl] 46: 291–295
Kupsch A, Oertel WH, Earl CD, Sautter J (1995) Neuronal transplantation and neurotrophic factors in the treatment of Parkinson's disease — update February 1995. J Neural Transm [Suppl] 46: 193–207
Kurth JH, Kurth MC, Poduslo SE, Schwankhaus JD (1993) Association of a monoamine oxidase B allele with Parkinson's disease. Neurology 33: 368–372
Landfield PW, Applegate MD, Schwitzer-Osborne SE, Naylor CE (1991) Phosphate/calcium alterations in the first stages of Alzheimer's disease: Implications for etiology and pathogenesis. J Neurol Sci 106: 221–229
Landi MT, Ceroni M, Martignoni E, Bertazzi PA, Caporaso NE, Nappi G (1996) Gene-environment action in Parkinson's disease. The case of CYP2D6 polymorphism. In: Battistin L, Scarlato G, Caraceni T, Ruggieri S (eds) Advances in neurology, vol 69, Parkinson's disease. Lippincott-Raven, Philadelphia, pp 61–72
Langston JW (1996) The etiology of Parkinson's disease with emphasis on the MPTP story. Neurology 47 [Suppl 3]: S153–S160
Langston JW, Ballard P, Tetrud JW, Irwin I (1983) Chronic parkinsonism in humans due to a product of meperidine-analog synthesis. Science 219: 970–980
Lazzarini AM, Myers RH, Zimmerman TR, Mark MH, Golbe JI, Sage JI, Johnson WG, Duvoisin RC (1994) A clinical genetic study of Parkinson's disease: evidence for dominant transmission. Neurology 44: 499–506
Lee CS, Schulzer M, Mak E, Snow BJ, Tsui JK, Calne S, Hammerstad J, Calne B (1994) Clinical observations on the rate of progression of idiopathic parkinsonism. Brain 117: 501–507

Leigh P, Probst A, Gale G, Dale GE, Power DP, Brion JP, Dodson A, Anderton BH (1989) New aspects of the pathology of neurodegenerative disorders as revealed by ubiquitin antibodies. Acta Neuropathol (Berl) 79: 61–72
Leveugle B, Faucheux BA, Bouras C, Nillesse N, Spik G, Hirsch EC, Agid Y, Hof PR (1996) Immunohistochemical analysis of the iron binding protein lactotransferrin in the mesencephalon of Parkinson's disease cases. Acta Neuropathol 91: 566–572
Lewy FH (1912) Paralysis agitans. I. Pathologische Anatomie. In: Lewandowsky M (ed) Handbuch der Neurologie, vol III. Springer, Berlin, pp 920–933
Lindquist NG, Larsson BS, Lyden-Sokolowski A (1987) Neuromelanin and its possible protective and destructive properties. Pigment Cell Res: 133–136
Mann DMA, Yates PO (1983) Possible role of neuromelanin in the pathogenesis of Parkinson's disease. Mech Age Dev 21: 193–203
Mann DMA, Yates PO, Barton CM (1977) Neuromelanin and RNA in cells of substantia nigra. J Neuropathol Exp Neurol 36: 379–383
Mann VM, Cooper JM, Daniel SE, Srai K, Jenner P, Marsden CD, Schapira AH (1994) Complex I, iron and ferritin in Parkinson's disease substantia nigra. Ann Neurol 36: 876–881
Maraganore DM, Harding AE, Marsden CD (1991) A clinical and genetic study of familial Parkinson's disease. Mov Disord 6: 205–211
Marklund S, Adolfsson R, Gottfries C, Winblad B (1985) Superoxide dismutase isoenzymes in normal brains and in brains from patients with dementia of Alzheimer type. J Neurol Sci 67: 319–325
Marsden CD (1983) Neuromelanin and Parkinson's disease. J Neural Transm [Suppl] 19: 121–141
Martilla RJ, Rinne UK (1981) Epidemiology of Parkinson's disease: an overview. J Neural Transm 51: 135–148
Martilla RJ, Lorentz H, Rinne UK (1988) Oxygen toxicity protecting enzymes in Parkinson's disease: increase of superoxide dismutase-like activity in the substantial nigra and basal nucleus. J Neurol Sci 86: 321–331
Matsubara K, Koyabashi S, Koyabashi Y, Yamashita K, Koide H, Hatta M, Iwamoto K, Tanaka O, Kimura K (1995) β-Carbolinium cations, endogenous MPP^+ analogs, in the lumbar cerebrospinal fluid of patients with Parkinson's disease. Neurology 45: 2240–2245
McCall T, Vallance P (1991) Nitric oxide takes center stage with newly defined roles. Trends Pharmacol Sci 13: 1–6
McGeer PL, Itagaki S, Akiyama K, McGeer EG (1988) Rate of cell death in parkinsonism indicates active neuropathological process. Ann Neurol 24: 574–576
Miller DM, Buettner GR, Aust SD (1990) Transition metals as catalysts of auto-oxidation reactions. Free Radical Biol Med 8: 95–108
Mizuno Y, Matuda S, Yoshino H, Mori H, Hattori N, Ikebe SI (1994) An immunohistochemical study on α-ketoglutarate dehydrogenase complex in Parkinson's disease. Ann Neurol 35: 204–210
Mjönes H (1949) Paralysis agitans: a clinical and genetic study. Acta Psychiatr Neurol 54: 1–95
Mochizuki H, Goto K, Mori H, Mizuno Y (1996) Histochemical detection of apoptosis in Parkinson's disease. J Neurol Sci 137: 120–123
Mogi M, Harada M, Kondo T, Mizuno Y, Narabayashi H, Riederer P, Nagatsu T (1996) The soluble form of Fas molecule is elevated in parkinsonian brain tissues. Neurosci Lett 220: 195–198
Moroo I, Yamada T, Makino H, Tooyama I, McGeer PL, McGeer EG, Hirayama K (1994) Loss of insukin receptor immunoreactivity from the substantia nigra pars compacta neurons in Parkinson's disease. Acta Neuropathol 87: 343–348
Morrish PK, Sawle GV, Brooks PJ (1996) An [^{18}F]dopa PET and clinical study of the rate of progression in Parkinson's disease. Brain 119: 585–591

Mouant-Prigent A, Karlsson JO, Agid Y, Hirsch EC (1996) Increased m-calpain expression in the mesencephalon of patients with Parkinson's disease but not in other neurodegenerative disorders involving the mesencephalon: a role in cell death? Neuroscience 73: 979–987

Münch G, Thome J, Foley P, Schinzel R, Riederer P (1997) Advanced glycation end products in ageing and disease. Brain Res Rev 23: 134–143

Mufson EJ, Brandabur MM (1994) Sparing of NADPH-diaphorase striatal neurons in Parkinson's and Alzheimer's diseases. Neuroreport 5: 705–708

Nishino N, Noguchi-Kuno SA, Sugiyama T, Tanaka C (1986) [^{3}H]Nitrendipine binding sites are decreased in the substantia nigra and striatum of the brain from patients with Parkinson's disease. Brain Res 377: 186–189

Nussbaum RL, Polymeropoulos MH (1997) Genetics of Parkinson's disease. Hum Mol Genet 6: 1687–1691

Oestreicher E, Sengstock GJ, Riederer P, Olanow CW, Dunn AJ, Arendash G (1994) Degeneration of nigrostriatal dopaminergic neurons increases iron in within the substantia nigra: a histochemical and neurochemical study. Brain Res 660: 8–18

Offen D, Ziv I, Barzilai A, Gorodin S, Glater E, Hochman A, Melamed E (1997) Dopamine-melanin induces apoptosis in PC12 cells: possible implications for etiology of Parkinson's disease. Neurochem Int 31: 207–216

Olanow CW (1997) Attempts to obtain neuroprotection in Parkinson's disease. Neurology 49 [Suppl 1]: S26–S33

Olney JW (1978) Neurotoxicity of excitatory amino acids. In: McGeer EG, Olney JW (eds) Kainic acid as a tool in neurobiology. Raven, New York, pp 95–121

Olney JW (1989) Excitatory amino acids and neuropsychiatric disorders. Biol Psychiatry 26: 505–525

Olson L (1997) The coming of age of the GDNF family and its receptors: gene delivery in a rat Parkinson model may have clinical implications. Trends Neurosci 20: 277–279

Paulus W, Jellinger K (1991) The neuropathologic basis of different clinical subgroups of Parkinson's disease. J Neuropathol Exp Neurol 50: 743–755

Pearce RKB, Owen A, Daniel S, Jenner P, Marsden CD (1997) Alterations in the distribution of glutathione in the substantia nigra in Parkinson's disease. J Neural Transm 104: 661–677

Perry TL, Goden DV, Hansen S (1982) Parkinson's disease: a disorder due to nigral glutathione deficiency. Neurosci Lett 33: 305–310

Perry TL, Young VW, Ito M, Foulks JG, Wall RA, Godin DV, Clavier RM (1984) Nigrostriatal dopaminergic neurons remain undamaged in rats given high doses of L-dopa and carbidopa chronically. J Neurochem 43: 990–993

Perry RH, Irving D, Tomlinson BE (1990) Lewy body prevalence in the aging brain: relationship to neuropsychiatric disorders, Alzheimer-type pathology and catecholaminergic nuclei. J Neurol Sci 100: 223–233 (published erratum in J Neural Sci (1991) 102:121)

Phillips HS, Hains JM, Armanini M, Laramee GR, Johnson SA, Winslow JW (1991) BDNF mRNA is decreased in the hippocampus of individuals with Alzheimer's disease. Neuron 7: 695–702

Pigott MA, Candy JM, Perry RH (1991) [^{3}H]Nitrendipine binding in temporal cortex in Alzheimer's and Huntington's diseases. Brain Res 565: 42–47

Pilas B, Sarna T, Kalyanaraman B, Swartz RM (1988) The effect of melanin on iron associated decomposition of hydrogen peroxide. Free Radical Biol Med 4: 285–293

Pileblad E, Magnusson T, Fornstedt B (1996) Reduction of brain glutathione by L-buthionine sulfoximine potentiates the dopamine-depleting action of 6-hydroxydopamine in rat striatum. J Neurochem 52: 978–980

Poewe W, Gerstenbrand F, Ransmayr G, Plorer S (1983) Premorbid personality of Parkinsonian patients. J Neural Transm [Suppl] 19: 215–224

Polymeropoulos MH, Lowedern C, Leroy E, Ide SE, Dehija A, Dutra A, Pike B, Root H, Rubenstein J, Boyer R, et al (1997) Mutation in the α-synuclein gene identified in families with Parkinson's disease. Science 276: 2045–2047

Przedborski S, Kostic V, Jackson-Lewis V, Naini AB, Simonetti S, Fahn S, Carlson E, Epstein CJ, Cadet JL (1992) Transgenic mice with increased Cu/Zn-superoxide dismutase activity are resistant to MPTP-induced neurotoxicity. J Neurosci 12: 1658–1667

Przedborski S, Jackson-Lewis V, Muthane U, Jiang H, Ferreira M, Naini AB, Fahn S (1993) Chronic levodopa administration alters cerebral mitochondrial respiratory chain activity. Ann Neurol 34: 715–723

Przuntek H, Müller T, Kuhn W, Hoffmann V (1997) Ist Apoptose, ein zentraler Mechanismus der Neurodegeneration, durch Selegiline beeinflußbar? In: Fischer PA (ed) Parkinson-Krankheit. Entwicklungen in Diagnostik und Therapie. Schattauer, Stuttgart, pp 259–274

Radi R, Beckman JS, Bush KM, Freeman BA (1991) Peroxynitrite-induced membrane peroxidation: the cytotoxic potential of superoxide and nitric oxide. Arch Biochem Biophys 288: 481–487

Reichmann H, Lestienne P, Jellinger K, Riederer P (1993) Parkinson's disease and the electron transport chain in post mortem brain. In: Narabayashi H, Nagatsu T, Yanagisawa N, Mizuno Y (eds) Advances in neurology, vol 60, Parkinson's disease: from basic research to treatment. Raven, New York, pp 297–299

Reif DW, Simmons RD (1990) Nitric oxide mediates iron release from ferritin. Arch Biochem Biophys 283: 537–541

Riederer P, Youdim MBH (eds) (1993) Iron in central nervous system disorders. Springer, Wien New York

Riederer P, Wuketich S (1976) Time course of nigrostriatal degeneration in Parkinson's disease: a detailed study of influential factors in human brain amine analogues. J Neural Transm [P-D Sect] 38: 277–301

Riederer P, Sofic E, Rausch WD, Schmidt B, Reynolds GP, Jellinger K, Youdim MBH (1989) Transition metals, ferritin, glutathione and ascorbic acid in Parkinsonian brains. J Neurochem 52: 515–520

Rinne JO, Halonen T, Riekinnen PJ, Rinne UK (1988) Free amino acids in the brain of patients with Parkinson's disease. Neurosci Lett 94: 182–186

Robbins JH, Otsuka F, Nee LE (1985) Parkinson's disease and Alzheimer's disease: hypersensitivity to x-rays in cultured cell lines. J Neurol Neurosurg Psychiatry 48: 916–923

Román GC, Zhang ZX, Ellenberg JH (1995) The neuroepidemiology of Parkinson's disease. In: Ellenberg JH, Koller WC, Langston JW (eds) Etiology of Parkinson's disease. Marcel Dekker, New York, pp 203–243

Rosen DR, Siddique T, Patterson D, Figlewicz DA, Sapp P, Hentati A, Donaldson D, Goto J, O'Regan JP, Deng HX, et al (1993) Mutations in Cu/Zn superoxide dismutase gene are associated with familial amyotrophic lateral sclerosis. Nature 362: 59–62 (published erratum: Nature 364: 362)

Rubanyi GM, Ho EH, Cantor EH, Lumma WC, Botelho LH (1991) Cytoprotective function of nitric oxide: inactivation of superoxide radicals produced by human leukocytes. Biochem Biophys Res Commun 181: 1392–1397

Saggu H, Cooksey J, Dexter D, Wells FR, Lees A, Jenner P, Marsen CD (1989) A selective increase in particulate superoxide dismutase activity in Parkinsonian substantia nigra. J Neurochem 53: 692–697

Sanchez-Ramos JR, Övervik E, Ames BN (1994) A marker of oxyradical-mediated DNA damage (8-hydroxy-2′-deoxyguanosine) is increased in nigro-striatum of Parkinson's disease brain. Neurodegeneration 3: 197–204

Sawle GV, Wroe SJ, Lees AJ, Brooks DJ, Frackowiak RS (1992) The identification of presymptomatic parkinsonism: clinical and [^{18}F]dopa positron emission tomography studies in an Irish kindred. Ann Neurol 32: 609–617

Schapira AHV (1994) Evidence for mitochondrial dysfunction in Parkinson's disease — a critical appraisal. Mov Disord 9: 125–13

Schapira AHV (1996) Neurotoxicity and the mechanisms of cell death in Parkinson's disease. In: Battistin L, Scarlato G, Caraceni T, Ruggieri S (eds) Advances in neurology, vol 69, Parkinson's disease. Lippincott-Raven, Philadelphia, pp 161–165

Schapira AHV (1997) Mitochondrial disorders. Curr Opin Neurol 10: 43–47

Schapira AHV, Mann VM, Cooper JM, Dexter D, Daniel SE, Jenner P, Clark JB, Marsden CD (1990) Anatomic and disease specificity of NADH CoQ1 reductase (complex I) deficiency in Parkinson's disease. J Neurochem 55: 2142–2145

Scherman D, Desnos C, Darchen F, Javoy-Agid F, Agid Y (1989) Striatal dopamine deficiency in Parkinson's disease: role of aging. Ann Neurol 26: 551–557

Seaton TA, Marsden CD, Jenner P (1996) Mitochondrial respiratory enzyme function and superoxide dismutase activity following brain glutathione depletion in the rat. Biochem Pharmacol 13: 1657–1663

Sen AP, Boksa P, Quirion R (1993) Brain calcium channel-related dihydropyridine and phenylalkylamine binding sites in Alzheimer's, Parkinson's and Huntington's diseases. Brain Res 611: 216–221

Sengstock GJ, Olanow CW, Dunn AJ, Arendash GW (1992) Iron induces degeneration of nigrostriatal neurons. Brain Res Bull 28: 645–649

Sengstock GJ, Olanow CW, Dunn AJ, Barone S, Arendash GW (1994) Progressive changes in striatal dopaminergic markers, nigral volume and rotational behavior following iron infusion into rat substantia nigra. Exp Neurol 130: 82–94

Sian J, Dexter DT, Lees AJ, Daniel S, Jenner P, Marsden CD (1994) Glutathione-related enzymes in brain in Parkinson's disease. Ann Neurol 36: 356–361

Siesjö BK (1990) Calcium in the brain under physiological and pathological conditions. Eur Neurol 30: 3–9

Singer TP, Castagnoli N, Ramsay RR, Trevor AJ (1987) Biochemical events in the development of parkinsonism induced by of 1-methyl-4-phenyl-1,2,3,6-tetrahydropyridine. J Neurochem 49: 1–8

Smith TA, Prayson RA (1996) Lewy body disease. South Med J 89: 1174–1180

Snow BJ (1996) Fluorodopa PET scanning in Parkinson's disease. In: Battistin L, Scarlato G, Caraceni T, Ruggieri S (eds) Advances in neurology, vol 69, Parkinson's disease. Lippincott-Raven, Philadelphia, pp 449–457

Sofic E, Riederer P, Heinsen H, Beckman H, Reynolds GP, Hebenstreit G, Youdim MBH (1988) Increased iron(III) and total iron content in post mortem substantia nigra of Parkinsonian brain. J Neural Transm 74: 199–205

Sofic E, Lange KW, Jellinger K, Riederer P (1992) Reduced and oxidized glutathione in the substantia nigra of patients with Parkinson's disease. Neurosci Lett 142: 128–130

Sontag KH, Heim C, Sontag TA, God R, Reichmann H, Wesemann W, Rausch WD, Riederer, Bringmann G (1995) Long-term behavioural effects of TaClo (1-trichloromethyl-1,2,3,4-tetrahydro-β-carboline) after subchronic treatment in rats. J Neural Transm [Suppl] 46: 283–289

Spencer PS, Butterfield PG (1995) Environmental agents and Parkinson's disease. In: Ellenberg JH, Koller WC, Langston JW (eds) Etiology of Parkinson's disease. Marcel Dekker, New York, pp 319–365

Spencer PS, Nunn PB, Hugon J, Ludolph AC, Ross SM, Roy DN, Robertson RC (1987) Guam amyotrophic lateral sclerosis-parkinsonism-dementia linked to a plant excitant neurotoxin. Science 237: 517–522

Spencer JPE, Jenner A, Aruoma OI, Evans PJ, Kaur H, Dexter DT, Jenner P, Lees AJ, Marsden DC, Halliwell B (1994) Intense oxidative DNA damage promoted by l-dopa and its metabolites: implications for neurodegenerative disease. FEBS Lett 353: 246–250

Spencer-Smith T, Parker WD, Bennett JP (1994) L-Dopa increases nigral production of hydroxyl radicals in vivo: potential l-dopa toxicity? Neuroreport 5: 1009–1011

Spillantini MG, Schmidt ML, Lee VM, Trojanowski JQ, Jakes R, Goedert M (1997) α-synuclein in Lewy bodies (letter). Nature 388: 839–840
Spina MB, Cohen G (1989) Dopamine turnover and glutathione oxidation: implications for Parkinson's disease. Proc Natl Acad Sci USA 88: 1398–1400
Springer JE, Mu X, Bergmann LW, Trojanowsky Q (1994) Expression of GDNF mRNA in rat and human nervous tissue. Exp Neurol 127: 167–170
Swan GA (1963) Chemical structure of melanins. Ann NY Acad Sci 100: 1005
Swartz HM, Sarna T, Zecca L (1992) Modulation by neuromelanin of the availability and reactivity of metal ions. Ann Neurol 32 [Suppl]: S69–S75
Taussig D, Plante-Bordeneuve V (1997) Les syndromes parkinsoniens familiaux "atypiques". Maladies de Parkinson ou entites autonomes? Presse Med 26: 290–296
Temlet JA, Landsberg JP, Watt F, Grime GW (1994) Increased iron in the substantia nigra compacta of the MPTP-lesioned hemiparkinsonian African green monkey: evidence from proton microprobe elemental microanalysis. J Neurochem 62: 134–146
Tipton KF, Singer TP (1993) Advances in our understanding of the mechanisms of the neurotoxicity of MPTP and related compounds. J Neurochem 61: 1191–1206
Toffa S, Kunikowska GM, Zeng BY, Jenner P, Marsden CD (1997) Chronic glutathione depletion in rat brain does not cause nigrostriatal pathway degeneration. J Neural Transm [PD Sect] 104: 67–75
Tomac A, Lindquist E, Lin LFH, Ögren SO, Young D, Hoffer BJ, Olsen L (1995) Protection and repair of the nigrostriatal dopaminergic system by GDNF in vivo. Nature 373: 335–339
Tompkins MM, Basgall EJ, Zamrini E, Hill WD (1997) Apoptotic-like changes in Lewy-body-associated disorders and normal aging in substantia nigral neurons. Am J Pathol 150: 119–131
Tooyama I, Kawamata T, Walker D, Yamada I, Hanai K, Kimura H, Iwane M, Igarashi K, McGeer EG, McGeer PL (1993) Loss of basic fibroblast growth factor in substantia nigra neurons in Parkinson's disease. Neurology 43: 372–376
Turski L, Bressler K, Rettig KJ, Löschmann PA, Wachtel H (1991) Protection of substantia nigra from MPP^+ neurotoxicity by N-methyl-d-aspartate antagonists. Nature 349: 414–418
Vieregge P (1994) Genetic factors in the etiology of Parkinson's disease. J Neural Transm 8: 1–37
Walinshaw G, Waters CM (1995) Induction of apoptosis in catecholaminergic PC12 cells by L-dopa: imolications for the treatment of Parkinson's disease. J Clin Invest 95: 2458–2464
Wallace DC (1992a) Mitochondrial genetics: a paradigm for aging and degenerative diseases? Science 256: 628–632
Wallace DC (1992b) Diseases of the mitochondrial DNA Ann Rev Biochem 61: 1175–1212
Waters CH, Miller CA (1994) Autosomal dominant Lewy body parkinsonism in a four-generation family. Ann Neurol 35: 59–64
Wesemann W, Blaschke S, Solbach M, Grote C, Clement HW, Riederer P (1994) Intranigral injected iron progressively reduces striatal dopamine metabolism. J Neural Transm [PD Sect] 8: 209–214
Wüllner U, Löschmann PA, Schulz JB, Schmid A, Dringen R, Eblen F, Turski L, Klockgether T (1996) Glutathione depletion potentiates MPTP and MPP^+ toxicity in nigral dopaminergic neurones. Neuroreport 7: 921–923
Yamada T, McGeer PL, Baimbridge KG, McGeer EG (1990) Relative sparing in Parkinson's disease of substantia nigra neurons containing calbindin D28K. Brain Res 526: 303–307
Yen TC, Chen YS, King KL, Yeh SH, Wei YH (1989) Liver mitochondrial functions decline with age. Biochem Biophys Res Commun 165: 994–1003

Yoritaki A, Hattori N, Uchida K, Tanaka M, Stadtman ER, Mizuno Y (1996) Immunohistochemical detection of 4-hydroxynonenal protein adducts in Parkinson's disease. Proc Natl Acad Sci USA 93: 2696–2713

Yoshida E, Mokuno K, Aoki SI, Takahashi A, Riku S, Murayama T, Yanagi T, Kato K (1994) Cerebrospinal fluid levels of superoxide dismutases in neurological diseases detected by sensitive enzyme immunoassays. J Neurol Sci 124: 25–31

Youdim MBH, Ben-Shachar D, Riederer P (1994) The enigma of neuromelanin in Parkinson's disease substantia nigra. J Neural Transm [Suppl] 43: 113–132

Authors' address: Prof. P. Riederer, Clinical Neurochemistry, Department of Psychiatry, University of Würzburg, Füchsleinstrasse 15, D-97080 Wüzburg, Federal Republic of Germany

Clinical efficacy of budipine in Parkinson's disease

H. Przuntek and **T. Müller**

Department of Neurology, Ruhr University of Bochum,
Federal Republic of Germany

Summary. The lipophilic t-butyl analog of 1-alkyl-4,4-diphenyl piperidine, budipine, possesses a polyvalent spectrum of mechanisms of action. Budipine experimentally increased the brain content of norepinephrine, serotonine, dopamine and histamine in reserpine treated rats. Budipine did not alter the receptor affinity of these neurotransmitters but antagonizes the effect of NMDA at its receptor binding site in vitro. Budipine reduced MPP^+ toxicity in the nigrostriatal system of mice. This complex pharmacologic profile is not comparable to the one of convenient antiparkinsonian drugs. In clinical trials budipine reduced tremor, akinesia and rigidity. Budipine induced a relevant additional positve effect in patients with an optimal dopaminergic therapy based on levodopa and dopamine agonists, such as bromocriptine. Current available data suggest that the need for levodopa application in early stages of the disease may be postponed by budipine and that the long-term application of budipine may induce a levodopa-sparing effect.

Introduction

The primary objective of symptomatic drug treatment in Parkinson's disease is the compensation of the striatal dopamine deficit by application of levodopa. Dopamine agonists, mainly stimulating postsynaptic dopaminergic receptor binding sites, may also be used for this purpose. Antagonizing of the anticholinergic and glutamate pathway may also improve dopaminergic transmission (Jellinger, 1999). Neurotransmitters, such as acetylcholine, norepinephrine, histamine, GABA, glutamate or serotonine, are also involved in the complex symptomatology, progression and intensity of idiopathic Parkinson's disease (Bonnet et al., 1987; Braak et al., 1995; Jellinger 1999). This supports the view, that idiopathic Parkinson's disease does not represent one disease entity.

Therefore we propose the development of a novel classification of different subtypes with respect to the variety of the expression of additional clinical features and extrapyramidal symptomatology in the course of the disease based on the known and postulated pathophysiologic and symptomatic pathogenetic forms of the Parkinsonian syndrome (Fig. 1). We assume,

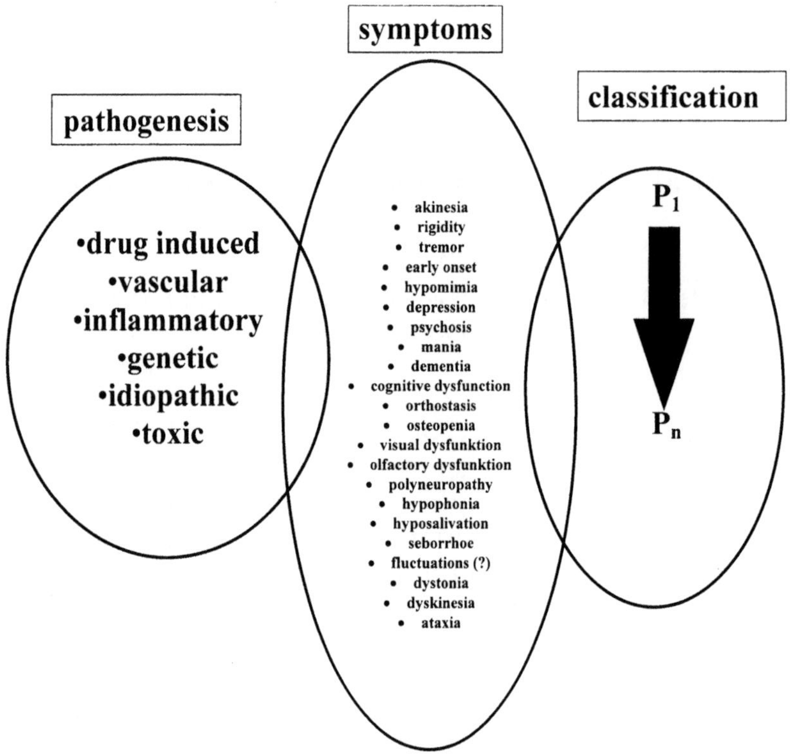

Fig. 1. Model for the development of a novel subclassification of the Parkinsonian syndrome

that reduced neuronal and glial energy metabolism due to different kinds of neurodegenerative processes via e.g. apoptosis, growth factors and/or inflammatory mediators in these subtypes causes neuronal cell death to a various extent. This induces alterations of other neurotransmitter systems beside dopamine and the motor loops and subsequently influences the clinical phenomenology of the disease (Fig. 2).

Up to now most antiparkinsonian drugs mainly substitute the nigrostriatal dopamine deficit and subsequently improve motor symptoms in Parkinson's disease. They scarcely therapeutically influence the clinical symptomatology resulting from the non-dopaminergic neurodegeneration in Parkinson's disease.

In consequence of the above mentioned hypothesis new therapeutic strategies — with e.g. drugs with a broad spectrum of modes of action — are necessary to improve the complex changes beyond the dopaminergic system in Parkinsonian subjects. The lipophilic t-butyl analog of 1-alkyl-4,4-diphenyl piperidine budipine may represent such a novel possibility of treating Parkinson's disease. Budipine compensates and balances most of the altered neurotransmitter systems within their physiologic ranges due to its polyvalent modes of action and does not overstimulate them, which may cause e.g. dyskinesias, fluctuations and dystonia in case of the dopaminergic system (Hacknack et al., 1979; Russ et al., 1986; Jakisch et al., 1993).

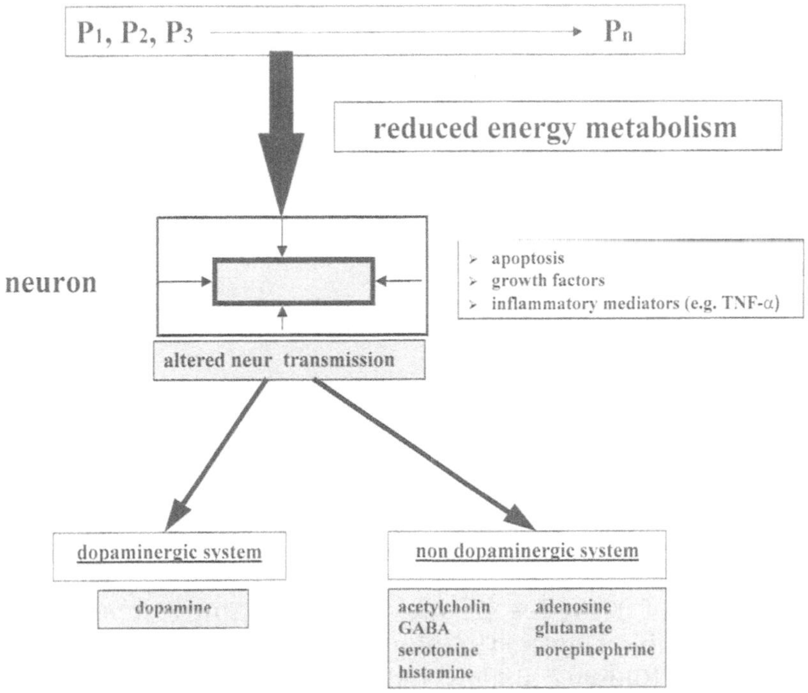

Fig. 2. Hypothetic model of the neurodegenerative process and the influence on neurotansmitter metabolism in idiopathic Parkinson's disease

Pharmacological profile of budipine

Budipine increased the content of norepinephrine, dopamine, serotonine, histamine in the brain of rats, treated with reserpin (Menge and Brand, 1982; Przuntek and Stasch, 1984). Budipine did not alter the receptor affinity of dopamine, serotonine, norepinephrine, GABA and endorphine according to in vitro trials (Przuntek and Stasch, 1984). Budipine suppresses dopamine reuptake, increases dopamine release from presynaptic vesicles and reversibly inhibits monoamine oxidase-B (Przuntek et al., 1985; Stasch et al., 1988; Niznik et al., 1990; Jakisch et al., 1993). Thus budipine may develop an indirect sympathomimetic effect due to its effect on the metabolism of norepinephrine (Jakisch et al., 1993) and inhibit serotonin reuptake (Menge and Brand, 1982; Stasch et al., 1988). Jakisch and al. demonstrated that release of ^{3}H labeled acetylcholine in response to the excitatory amino acid N-methyl-D-aspartate (NMDA) was suppressed by budipine in sections of the caudate nucleus of the rabbit. Budipine blocked the NMDA receptor binding site for the NMDA agonist [^{3}H]MK-801 in a concentration-dependent manner (Jakisch et al., 1993; Jakisch et al., 1994). Budipine showed affinity for the PCP and sigma$_1$ binding sites of the NMDA receptor in potential therapeutic concentrations (Kornhuber et al., 1995). Budipine has been suggested to be neuroprotective partially due to its NMDA antagonism (Kornhuber et al., 1995). In accordance with this hypothesis earlier trials demonstrated the reduction of MPP^+ toxicity in the nigrostriatal system of mice (Przuntek and Russ, 1985; Przuntek et al., 1985; Russ et al., 1986; Mihatsch et al., 1988). The antimuscarinergic action of budipine contributes to its ability to reduce tremor (Menge and Brand, 1982; Menge and Brand, 1985; Jakisch et al., 1993). This complex pharmacologic profile suggests, that budipine might therapeutically influence altered nondopaminergic systems involved in the pathogenesis of Parkinson's disease (Bonnet et al., 1987; Braak et al., 1995).

Clinical trials

Additional application of budipine reduced the levodopa dosage due to a retrospective analysis of a one year lasting therapeutic trial in 18 Parkinson patients. 16 subjects needed 22.3% less levodopa after 10 weeks with an additional improvement of their Webster score, 5 patients were only treated with 65.6% of their original levodopa dosage in the further course of this trial after one year (Siegfried and Fischer, 1985). Budipine — administered with an oral daily dosage of 20 mg for 12 weeks — significantly improved the total score of the Columbia University Rating Scale (CURS) in 29 patients in the By 701/14A trial. Budipine mainly reduced tremor and at least improved bradykinesia (Jellinger and Bliesath, 1987).

The efficacy of budipine on tremor was quantitatively evaluated with long-term electromyographic recordings in an open-label trial. Tremor intensity was reduced by 25%, tremor incidence by 34%. Moreover a 20% improvement of the scores in the Unified Parkinson's Disease Rating Scale occurred in patients treated with budipine (Spieker et al., 1995; 1999).

Efficacy and safety of adjuvant daily budipine application was evaluated in patients optimally treated with levodopa and bromocriptine in a multicenter, double-blind, placebo-controlled, randomized trial (FKF/004; participants, see footnote[1]). 99 patients received either budipine with a dose of up to 60 mg daily or placebo. 84 completed the trial. No significant differences appeared between both groups on entry into the trial. Subjects on budipine showed first improvements after 4 weeks during the 16 weeks lasting trial. CURS scores of budipine treated patients were significantly lower compared to those of subjects on placebo at the endpoint of the trial after 16 weeks. Improvement of CURS score did not depend on disease severity and levodopa dosage. Beneficial effects of budipine did not relate on patients' age and gender or the application of selegiline. Akinesia (6.8 ± 5.1 on budipine versus 2.7 ± 3.8 on placebo; $p = 0.0001$), rigidity (3.1 ± 2.3 on budipine versus 1.1 ± 2.5 on placebo; $p = 0.0001$; one-sided explorative Mann-Whitney-U test) and tremor (2.8 ± 2.7 on budipine versus 1.2 ± 3.0 on placebo; $p = 0.0017$) significantly

[1] Budipine Study Group: H. Przuntek, Bochum; B. Rathay, Karlsruhe; S. Spieker, Tübingen; P. H. Kraus, Bochum; U. Büttner, Munich; E. Teshmar, Saarlouis; D. Volc, Vienna; W. Kuhn, Bochum; G. Fuchs, Wolfach; F. Ulm, Kassel; D. Müller, Hamburg; S. Bittkau, Karlstadt, L. Lachenmayer, Hamburg; J. Glaß, Neubrandenburg; H. Haller, Herbolzheim; H. Bliesath, Constance.

improved in patients on budipine compared to those on placebo. The drop-out rate was 20.4% in the budipine — and 12% in the placebo group.

Efficacy and safety of budipine at a daily dosage of 60mg was evaluated in untreated, so-called "de-novo" Parkinsonian patients in early stages of the disease in a 6 months lasting trial (participants, see footnote[2]). Budipine was slowly weekly increased in 10mg steps up to a maximal daily dose of 60mg. Primary objectives were (a) the timepoint for the need of additional levodopa application and (b) the changes of the total scores for motor symptoms (III) of the Unified Parkinson's Disease Rating Scale (UPDRS) after 3 months of treatment. The number of patients requiring levodopa was significantly lower in subjects treated with budipine (24.6%, $p < 0.001$; Fischer's exact test, two-sided) compared those with placebo (55.4%) after 6 months. The UPDRS (III) motor score of patients with budipine dropped by 2.47 ± 7.09 (mean $\pm$ standard deviation) compared to 1.03 ± 6.41 (mean $\pm$ standard deviation) of subjects with placebo after 3 months. However this difference was not statistically significant ($p = 0.1514$; two-sided Mann-Whitney-U test). The drop-out rate was 26.1% in the budipine — and 31.1% in the placebo group.

An open-label prospective 4-year safety trial analysed occurence of side effects, laboratory data and subjective assessment of efficacy and tolerance of budipine by physicians and patients. 2962 Parkinsonian patients between 50 and 85 years with inadequate response to previous combination therapy strategies with other antiparkinson drugs and subjects with a good response to budipine in previous controlled studies were enrolled in this trial. They received daily oral dosages between 10 and 100mg of budipine. 2346 subjects continuously received budipine after enrollment, 1314 for more than 1 year. 274 patients discontinued budipine treatment because of the occurrence of adverse events mainly within the first 3 months of the study. The drug was excellently or well tolerated according to 74% of the physicians and 70.2% of the patients. Most common side effects of budipine were nausea, dryness of the mouth, dizziness, nervousness, unrest, gastrointestinal symptoms and fatigue.

Discussion

In vitro and animal trials demonstrated the complex pharmacological profile of budipine, which influences both dopaminergic and non dopaminergic systems involved in the pathogenesis of Parkinson's disease and its complex symptomatology due to the alterations of neurotransmitters such as

[2] Budipine Study Group: H. Przuntek, Bochum; D. Müller, Hamburg; B. Rathay, Karlsruhe; W. Kuhn, Bochum; E. Teshmar, Saarlouis; B. Tettenborn, Mainz; F. Ulm, Kassel; W. Poewe, Innsbruck; S. Bittkau, Karlstadt; B. Conrad, Munich; G. Fuchs, Wolfach; W. Weinrich, Hannover; J. Glaß, Neubrandenburg; K.D. Neher, Ravensburg; L. Lachenmayer, Hamburg; U. Hobert, Würzburg; H. Haller, Herbolzheim; T. Klockgether, Tubingen; F. von Baumgarten, Marktheidenfeld; P.H. Kraus, Bochum; S. Petrovici, Cologne; E. Schneider, Hamburg; J. Sgonina, Hamburg.

acetylcholine, serotonine, norepinephrine, histamine, GABA, glutamate and endorphine. Budipine significantly improved main motor symptoms of Parkinson's disease according to the results of clinical trials. Budipine is effective in all stages of Parkinson's disease. If budipine is well tolerated, it may be administered alone or in combination with other antiparkinsonian drugs. Budipine induced a relevant additional positive effect in patients with optimal therapy based on levodopa in combination with a dopamine agonist. Further studies are neccessary for the evaluation of a putative beneficial effect of budipine on the occurrence and intensity of dyskinesias and motor fluctuations, because budipine like amantadine is a NMDA antagonist (Klockgether et al., 1993; Kornhuber et al., 1995). NMDA receptor hyperfunction may contribute to the pathogenesis of levodopa associated motor complications such as motor fluctuations or dyskinesias (Chase et al., 1996), which may be reduced by blocking of NMDA receptors (Verhagen Metmann et al., 1998). The antimuscarinergic mechanism of action of budipine may mainly be responsible for the tremorlytic potency of budipine, but may also induce above mentioned side effects in clinical trials and the 4-year safety study. Therefore budipine should not be applicated in subjects with severe dementia in the course of neurodegenerative disorders, recurrent confusion, glaucoma, prostatic hypertrophy with urinary retention, cardiac or renal failure. Current available data suggest that budipine may postpone the timepoint of additional necessary levodopa treatment in early stages of Parkinson's disease and may posess a levodopa sparing effect on long-term administration. However, this preliminary results must be confirmed by placebo-controlled, randomized, double-blind multicenter trials. Budipine may also be neuroprotective due to its levodopa sparing effect in view of the putative toxicity of levodopa (Fahn, 1996; 1997). Whether budipine may slow the progression of the disease or not, will have to be demonstrated in future clinical long-term trials.

References

Bonnet AM, Loria Y, Saint Hilaire MH, Lhermitte F, Agid Y (1987) Does long-term aggravation of Parkinson's disease result from nondopaminergic lesions? Neurology 37: 1539–1542

Braak H, Braak E, Yilmazer D, de Vos RAI, Jansen ENH (1995) Nigral and extranigral pathology in Parkinson's disease. J Neural Transm [Suppl] 46: 15–32

Chase TN, Engber TM, Mouradian MM (1996) Contribution of dopaminergic and glutaminergic mechanisms to the pathogenesis of motor response complications. Adv Neurol 69: 497–501

Fahn S (1996) Is Levodopa toxic? Neurology 47 [Suppl 3]: 184–195

Fahn S (1997) Levodopa-induced neurotoxicity: does it represent a problem for the treatment of Parkinson's disease? CNS Drugs 8(5): 376–393

Hacknack G, Menge HG, Eistetter K, Krüger U, Schäefer H, Klosa J (1979) 4,4-Diphenylpiperidine- ein neuer Substanztyp für die Therapie des Morbus Parkinson. Österr Chemie Z 4: 61–62

Jackisch R, Huang HY, Reimann W, Limberger N (1993) Effects of the antiparkinsonian drug budipine on neurotransmitter release in central nervous system tissue in vitro. J Pharmacol Exp Ther 264: 889–898

Jackisch R, Kruchen A, Sauermann W, Hertting G, Feuerstein TJ (1994) The antiparkinsonian drugs budipine and biperiden are use-dependent (uncompetitive) NMDA receptor antagonists. Eur J Pharmacol 264: 207–211

Jellinger K (1999) Post mortem studies in Parkinson's disease — is it possible to detect brain areas for specific symptoms? J Neural Transm [Suppl] 56: 1–29

Jellinger K, Bliesath H (1987) Adjuvant treatment of Parksinson's disease with budipine: a double-blind trial versus placebo. J Neurol 234: 280–282

Klockgether T, Jacobsen P, Loschmann PA, Turski L (1993) The antiparkinsonian agent budipine is an N-methyl-D-aspartate antagonist. J Neural Transm [PD-Sect] 5: 101–106

Kornhuber J, Herr B, Thome J, Riederer P (1995) The antiparkinsonian drug budipine binds to NMDA and sigma receptors in post mortem brain tissue. J Neural Transm [Suppl] 46: 131–137

Menge H, Brand U (1982) Zusammenfassende Darstellung der Pharmakologie von Budipin, einem neuen 4,4 Diphenylpiperidinderivat für die Parkinson-Therapie. Drug Res 32: 85–98

Menge HG, Brand U (1985) Pharmacologic bases of antiparkinsonian therapy. In: Gerstenbrand F, Poewe W, Stern G. (eds) Clinical experiences with budipine in Parkinson therapy. Springer, Berlin Heidelberg New York Tokyo, pp 82–92

Mihatsch W, Russ H, Przuntek H (1988) Intracerebroventricular administration of 1-methyl-4-phenylpyridinium ion in mice: effects of simultaneously administered nomifensine, deprenyl, and 1-t-butyl-4,4-diphenylpiperidine. J Neural Transm 71: 177–188

Niznik HB, Tyndale RF, Sallee FR, Gonzales FJ, Hardwick JP, Inaba T, Kalow W (1990) The dopamine transporter and cytochrome P450IID1 (debrisoquine 4-hydroxylase) in brain: resolution and identification of two distinct (^{3}H)GBR-12935 binding proteins. Arch Biochem Biophys 276: 424–432

Przuntek H, Russ H (1985) Budipine and the MPTP binding site. Lancet 2: 35–36

Przuntek H, Stasch JP (1984) Biochemical and pharmacologic aspects of the mechanism of action of budipine. In: Gerstenbrand F, Poewe W, Stern G. (eds) Clinical experiences with budipine in Parkinson therapy. Springer, Berlin Heidelberg New York Tokyo, pp 106–111

Przuntek H, Russ H, Henning K, Pindur U (1985). The protective effect of 1-tert.butyl-4,4-diphenylpiperidine against the nigrostriatal neurodegeneration caused by 1-methyl-4-phenyl-1,2,3,6-tetrahydropyridine. Life Sci 37: 1195–1200

Russ H, Pindur U, Przuntek H (1986) The interaction of 1-alkyl-4,4-diphenylpiperidines with the 1-methyl-4-phenyl-1,2,3,6-tetrahydropyridine receptor binding site. J Neural Transm 65: 157–166

Schaefer H, Hackmack G, Eistetter K, Kruger U, Menge HG, Klosa J (1984) Synthesis, physical-chemical properties and pharmacologically-oriented screening studies on budipine and related 4,4-diphenylpiperidines. Arzneimittelforschung 34: 233–240

Siegfried J, Fischer R (1985) Budipine: A new chemical substance in the treatment of Parkinson's disease. In: Gerstenbrand F, Poewe W, Stern G (eds) Clinical experiences with budipine in Parkinson therapy. Springer, Berlin Heidelberg New York Tokyo, pp 153–157

Spieker S, Löschmann P, Jentges C, Boose A, Klockgether T, Dichgans J (1995) Tremorlytic activity of budipine: a quantitative study with long-term tremor recordings. Clin Neuropharmacol 18(3): 266–272

Spieker S, Breit S, Klockgether T, Dichgans J (1999) Tremorlytic activity of budipine in Parkinson's disease. J Neural Transm [Suppl] 56: 165–172

Stasch JP, Russ H, Schacht U, Witteler M, Neuser D, Gerlach M, Leven M, Kuhn W, Jutzi P, Przuntek H (1988) 4,4-Diphenylpiperidine derivates and their sila analogues.

A comparative study of their interaction with neural receptor binding sites and synaptosomal monoamine uptake. Arzneimittelforschung 38: 1075–1078
Verhagen Metman L, Del Dotto P, van den Munckhof BS, Fang J, Mouradian MM, Chase TN (1998) Amantadine as treatment for dyskinesias and motor fluctuations in Parkinson's disease. Neurology 50: 1323–1326

Authors' address: Professor Dr. H. Przuntek, Department of Neurology, Ruhr University of Bochum, Gudrunstrasse 56, D-44791 Bochum, Federal Republic of Germany

Multiple mechanisms of action: the pharmacological profile of budipine

M. Eltze

Department of Pharmacology, Byk Gulden, Konstanz, Federal Republic of Germany

Summary. Four major components of the mechanism of action have been identified for the antiparkinsonian drug budipine up to now. 1) The primary action of budipine is an indirect dopaminergic effect as shown by facilitation of dopamine (DA) release, inhibition of monoamine oxidase type B (MAO-B) and of DA (re)up-take and stimulation of aromatic L-amino acid decarboxylase (AADC), which in sum might be responsible for enhancing the endogenous dopaminergic activity. 2) Radioligand and functional studies at the N-methyl-D-aspartate (NMDA) type glutamate receptor characterize budipine as a low-affinity, uncompetitive antagonist with fast kinetics and moderate voltage-dependency at the phencyclidine (PCP) binding site, comparable to that observed with amantadine, thereby counteracting an increased excitatory glutamatergic activity. 3) The antimuscarinic action of budipine, verified by functional and binding studies at native muscarinic M_1-M_3 and human recombinant m1-m5 receptor subtypes in vitro, is up to 125-fold weaker than that of biperiden and corresponds to its approximately 100-fold lower potency to cause experimentally-induced peripheral antimuscarinic effects and explains only part of its high potency, which equals biperiden, to suppress cholinergically evoked tremor. 4) An additional inhibition of striatal gamma-aminobutyric acid (GABA) release by budipine may be beneficial to suppress an increased striatal GABAergic output activity. The contribution of other observed effects to the therapeutic action of budipine, i.e. weak stimulation of noradrenaline and serotonin release, binding to brain $sigma_1$ receptors and blockade of histamine H_1 receptors, is not yet clear. By means of these multiple mechanisms, budipine might correct the imbalance of striatal output pathways by restoring DA levels in the striatum, and positively influence the secondary changes in other transmitter systems (glutamate, acetylcholine, GABA) observed in Parkinson's disease.

Introduction

Budipine (1-tert-butyl-4,4-diphenylpiperidine), synthetized by Hackmack et al. (1979) and extensively investigated in vitro and in vivo for its pharmacological properties and the possible therapeutic use in Parkinson's disease (PD)

by Menge's group (Brand and Menge, 1980; Eltze, 1980; Menge et al., 1982; Menge and Brand 1985; Schaefer et al., 1984) and Przuntek's group (Gerlach et al., 1983; Mihatsch et al., 1988; Przuntek et al., 1985; Przuntek and Stasch, 1985; Russ et al., 1983, 1986), has been introduced in the therapy of PD for more than 10 years. The clinical use of budipine, generally administered in a daily dose of 30–60 mg, is especially recommended in view of its impressive beneficial effects on tremor (Iizuka and Fischer, 1986; Jellinger and Bliesath, 1987; Oppel, 1985; Spieker et al., 1995; Wang, 1985), its advantage in the combination therapy with L-DOPA (Poewe et al., 1985; Siegfried and Fischer, 1983, 1985) leading to improved therapeutic responses at lower L-DOPA doses, and its relatively low propensity to evoke side effects (Siegfried and Fischer, 1985). The high oral activity and a half-time of 31 h in humans (Zech et al., 1985), renders budipine most suitable for the management of PD patients with fluctuating responses to L-DOPA. From its ability to improve nearly all of the cardinal symptoms of this disease, i.e. tremor, bradykinesia and muscular rigidity, a complex mechanism of action comprising effects on multiple transmitter systems can be expected for this drug.

The mechanism of action of budipine, as it has been tentatively defined by initial pharmacological observations, previously suggested various neurotransmitter systems related to PD, particularly those of dopamine (DA), serotonin (5-HT), noradrenaline (NA) and acetylcholine (Menge et al., 1982; Menge and Brand, 1985). Budipine was found to be effective in attenuating the decrease in whole brain transmitter contents of DA, NA, 5-HT and histamine following reserpine treatment of rats (Gerlach et al., 1983), and showed dopamimetic-like properties in vivo, such as potentiation of apomorphine- and L-DOPA-induced behavioral effects, antagonism of neuroleptic-induced cataleptic, rigid and akinetic states, without stimulating DA receptors directly (Menge et al., 1982; Menge and Brand, 1985; Przuntek and Stasch, 1985; Stasch et al., 1988). Furthermore, budipine acted as a reversible inhibitor of the DA metabolizing enzyme monoamine oxidase type B (MAO-B) in vitro (Menge et al., 1982). The antimuscarinic effect of budipine has been found to be relatively weak (Menge et al., 1982; Menge and Brand, 1985; Keim et al., 1998) and hence, although responsible for the 100-fold weaker anticholinergic side effects as compared to biperiden, could only partly explain its high potency to inhibit cholinergically induced tremor in animals and its clinical success in suppressing tremor in Parkinson patients (Jellinger and Bliesath, 1987; Spieker et al., 1995). Moreover, although several in vivo data (e.g. a mild central stimulation) indicated interactions with the noradrenergic and serotonergic system (Menge et al., 1982; Menge and Brand, 1985), which in part may be transmitter release and/or inhibition of re-uptake (Eltze, 1980; Przegalinski et al., 1982), these observations could hardly explain the antiparkinsonian effects of budipine.

Meanwhile, the complexity of budipine's action on the dopaminergic systems was demonstrated in more detail. Budipine is capable of enhancing the release of striatal DA (Jackisch et al., 1993). Furthermore, MAO-B inhibition (Kuhn et al., 1987) and/or DA-(re)uptake inhibition (Jackisch et al., 1993), two properties that would increase the levels of striatal DA, have been suggested to be the underlying mechanism(s) for the protective action by budipine on the neurotoxic effect of 1-methyl-4-phenyl-1,2,3,6-tetrahydropyridine (MPTP) and its active metabolite, 1-methyl-4-phenylpyridinium ion (MPP^+), on the monoamine deficits that resemble experimental PD in animals (Mihatsch et al., 1988; Przuntek et al., 1985; Russ et al., 1986). Additionally, budipine markedly increases the activity of the DA synthetizing enzyme aromatic L-amino acid decarboxylase (AADC) in rat striatum and substantia nigra (Fisher et al., 1997; Starr et al., 1997).

Receptor binding, transmitter release and electrophysiological experiments in vitro as well as seizure studies in animals characterize budipine as a low-affinity, uncompetitive (use-dependent) antagonist with fast channel blocking and unblocking kinetics at the N-methyl-D-aspartate (NMDA) type of glutamate receptors interacting with the phencyclidine (PCP) binding site (Jackisch et al., 1994; Klockgether et al., 1993, 1996; Parsons et al., 1998). Additionally, an interaction with the $sigma_1$ binding site has been reported with budipine at a concentration even lower than that to block the PCP binding site of the NMDA receptor (Kornhuber et al., 1995). Finally, it is possible that some aspects of the therapeutic properties of budipine may be accounted for by its ability to act as a weak releaser of NA (Menge et al., 1982; Jackisch et al., 1993), as a weak 5-HT uptake inhibitor and/or releaser (Eltze, 1980; Menge et al., 1982; Mihatsch et al., 1988; Przegalinski et al., 1982), and to block histamine H_1 receptors (Menge et al., 1982).

With this background, the present review deals with current status of the mechanism of action underlying the clinical efficacy of budipine.

Interaction of budipine with the dopaminergic system

It is known that the progressive loss of DA-containing cells in the pars compacta of the substantia nigra and disturbance of nigrostriatal dopaminergic functions causes severe motor deficits, such as paucity and slowness of movements, increased muscle tone and tremor, that occur in Parkinson's disease, and that the DA precursor L-DOPA and DA receptor agonists used alone or in combination are effective symptomatic treatments of PD. Whether budipine is a direct DA receptor agonist, or more likely, must be regarded to act indirectly for enhancing dopaminergic transmission, is discussed below.

Indirect dopaminergic effects

Budipine neither stimulates dopamine D_1 and D_2 receptors directly (binding affinities $IC_{50} > 10^{-4}$M; Przuntek and Stasch, 1985; Stasch et al., 1988) nor does it inhibit DOPA decarboxylase ($IC_{50} > 3 \times 10^{-3}$M) (Menge et al., 1982). Nevertheless, several studies point to indirect dopamimetic properties of the drug. Although budipine did not cause stereotypies in mice and rats of the kind observed after administration of the DA receptor agonist, apomorphine, a potentiation of the behavioral effects of subacute doses of apomorphine and

L-DOPA was observed in rats (Menge et al., 1982). The property that budipine antagonizes tremor induced by LON-954 in mice (Brand and Menge, 1980; Menge et al., 1982; Menge and Brand, 1985), a compound known to induce a deficit in DA (Coward et al., 1977), and reverses reserpine- and various neuroleptic-induced rigid, cataleptic and akinetic states in mice and rats suggests that it might facilitate dopaminergic transmission (Menge et al., 1982). However, it has been argued that this effect does not rely exclusively on a dopaminergic action because also antimuscarinic drugs as well as NMDA receptor antagonists attenuate catalepsy (Schmidt and Bubser, 1989). Yet, the antimuscarinic biperiden is ineffective against rigidity induced by reserpine in mice and rats, whereas budipine in turn behaved as potent as bromocriptine and amantadine (Menge and Brand, 1985). In a turning model in rats after stereotactic lesions by unilateral 6-hydroxydopamine injection into the substantia nigra, direct DA receptor agonists cause contralateral rotations by activating supersensitive postsynaptic DA receptors, whereas compounds such as amphetamine enhance the release of DA and induce ipsilateral rotations. The failure of budipine (0.78–12.5 mg/kg i.p.) to evoke either contralateral or ipsilateral rotations, not even following postsynaptic DA receptor sensitization after a 3-day treatment with haloperidol, suggests that, at least under these experimental conditions, budipine appears to possess neither direct nor indirect dopaminergic activities (Menge et al., 1982). In another study in rats bearing unilateral 6-hydroxydopamine-evoked lesions in substantia nigra, budipine (3.1–12.5 mg/kg i.p.) and the NMDA receptor antagonist, 3[(±)-2-carboxypiperazine-4-yl]-propyl-1-phosphonic acid (CPP, 0.1–0.4 mg/kg i.p.), increased the number of contralateral rotations induced by apomorphine (Klockgether et al., 1996). However, in a model with increasing dopamine D_2 receptor binding capacity in rat striatum by chronic administration of the neuroleptic haloperidol, which, for instance, can be prevented by subsequent administration of L-DOPA (Burt et al., 1977), it was found that budipine (2×10 mg/kg s.c. per day, for two weeks) prevented the increase in dopamine D_2 receptor binding capacity during chronic haloperidol administration (Hennies, 1983, unpublished).

Effects on dopamine levels

Previous studies in rats did not reveal any change in the content of DA measured in rat striatum, thalamus and subthalamus after a single oral dose of 50 mg/kg budipine. Likewise, the DA turnover was not significantly altered by budipine, neither after a single dose of 50 mg/kg p.o. nor at 8 mg/kg p.o. on five consecutive days (Menge et al., 1982). Also in a recent study, budipine after a single dose (10 mg/kg i.v.) did not facilitate the release of striatal DA and its metabolites DOPAC, HVA and 3-methoxytyramine measured in vivo by brain microdialysis in naive rats (Klockgether et al., 1996). In mice, however, repeated administration of budipine (daily dose 35.5 mg/kg i.p., for six days) caused a slight but persistent increase in the levels of DA and its metabolites

DOPAC and HVA in the caudate nucleus 5 weeks after the last administration (Mihatsch et al., 1988).

Great interest was attracted by cases of poisoning with 1-methyl-4-phenyl-1,2,3,6-tetrahydropyridine (MPTP) which selectively destroys the dopaminergic neurones of the nigrostriatal pathway and produces motor disorders that closely resemble idiopathic parkinsonism (Ballard et al., 1985). MPTP enters the brain where it is captured by acidic organelles in astrocytes; these cells contain MAO-B, which oxidizes MPTP to the 1-methyl-4-phenylpyridinium ion (MPP^+), which is taken up into dopaminergic neurones through the DA uptake system, where it develops its toxic effects by inhibiting mitochondrial respiration. Principally, there exist two major possibilities to block the neurotoxic effect of MPTP. Firstly, by preventing the formation of MPP^+ by inhibition of MAO-B and secondly, by use of DA uptake inhibitors which block the transport of MPP^+ by the catecholamine uptake carrier into the dopaminergic neuron (Ricaurte et al., 1985).

In rats treated with MPTP (1.3 mg/kg i.p./day) for one month, a permanent decrease in the content of striatal DA and 5-HT and their respective metabolites DOPAC, HVA and HIAA could be observed, the effect of which was prevented by co-administration of budipine (4 mg/kg i.p./day) (Przuntek et al., 1985). In a similar study in mice treated directly by i.cv. injection of MPP^+ (total dose 5.4 mg/kg), budipine (35.5 mg/kg i.p./day) administered 30 min before the neurotoxin over 8 days attenuated the decrease in DA content in caudate nucleus and clearly counteracted the decrease in its metabolites DOPAC and HVA, but had no effects upon the changes of 5-HT and 5-HIAA concentrations measured five weeks after the combined drug administration (Mihatsch et al., 1988). The mechanism(s) which might explain the effects of budipine to prevent the decrease in striatal dopamine levels after MPTP in rats and MPP^+ in mice could be inhibition of MAO-B in astrocytes, which capture MPTP from blood and oxidizes MPTP to MPP^+, and /or inhibition of MPP^+ uptake into dopaminergic neurones where it inhibits mitochondrial respiration resulting in ATP depletion and cell death.

Effects of DA uptake and release

The possibility that budipine exerts indirect dopamimetic properties by inhibiting the reuptake of DA has been suggested several times (Menge et al., 1982; Mihatsch et al., 1988; Niznik et al., 1990; Przuntek et al., 1985). Direct measurement of the high-affinity uptake of DA into synaptosomes prepared from rabbit caudate nucleus showed budipine to be effective ($IC_{50} = 1.1 \times 10^{-5}$ M) (Jackisch et al., 1993); however, budipine was less potent than the DA uptake inhibitor, nomifensine ($IC_{50} = 4.8 \times 10^{-8}$ M), or the DA receptor antagonist, haloperidol ($IC_{50} = 2.2 \times 10^{-6}$ M) (Reimann et al., 1979).

The presynaptic effects of budipine on spontaneous and electrical stimulation-induced [^{3}H]DA release in that brain region primarily involved in the pathology of PD, the striatum, was studied in superfused slices of nucleus caudatus from rabbits (Jackisch et al., 1993). Budipine (10^{-6}–10^{-5} M) strongly increased the spontaneous [^{3}H]outflow, which, in the presence and even in the absence of nomifensine, was shown to consist partly of unmetabolized DA.

Indirect dopamimetic properties of budipine could also be detected in the presence of the DA-uptake inhibitor, nomifensine, when both the evoked and the spontaneous release of acetylcholine in rabbit caudate nucleus were significantly reduced by budipine (10^{-5} M), suggesting the presence of increased endogenous DA levels at cholinergic nerve terminals (Hertting et al., 1980), in this case evoked indirectly by budipine. This conclusion is also supported by the finding that the inhibitory effect of budipine was attenuated by the dopamine D_2 receptor antagonist domperidone. This observation indicates that budipine exhibits indirect dopamimetic properties by releasing DA which in turn leads to a reduced release of acetylcholine in this model (Jackisch et al., 1993).

Effects on DA biosynthesis

The biosynthesis of DA involves two enzymes, namely tyrosine hydroxylase and aromatic L-amino acid decarboxylase (AADC), the former classically regarded as being the rate-limiting enzymatic step, although it has recently been found that the activity of AADC can also be modulated by DA (Zhu et al., 1992). Furthermore, it has been reported that the NMDA channel blocker, MK-801, increased brain AADC activity, suggesting that endogenous glutamate also regulates DA biosynthesis (Hadjiconstantinou et al., 1995) in a way, that the inhibition of glutamatergic transmission in the basal ganglia accelerates the enzymatic conversion of L-DOPA to DA at presynaptic sites (Starr, 1995). In a recent investigation, an entirely new aspect of the mechanism of action was found for budipine (Fisher et al., 1997; Starr et al., 1997): In rats treated with various ligands for glutamate receptors, budipine (12 mg/kg i.p.) increased the activity of AADC in corpus striatum and substantia nigra 2–4 fold, similar to other NMDA receptor antagonists like MK-801 (0.1–1 mg/kg i.p.), memantine (20 mg/kg i.p.), amantadine (40 mg/kg i.p.) and dextromethorphan (40 mg/kg i.p.), whereas the glycine site antagonist, (+)-HA 966, elevated enzyme activity only in corpus striatum. In contrast, the NMDA polyamine site antagonist, eliprodil, the glutamate binding site antagonist, CGP 40116, and the AMPA receptor antagonist, 1,2,3,4-tetrahydro-6-nitro-2,3-dioxo[f]quinoxaline-7-sulfonamide (NBQX), proved to be ineffective. These findings show that glutamate antagonism by ion-channel blockade of the NMDA receptor appears to be the site of action by which budipine may increase AADC activity (Fisher et al., 1997; Starr et al., 1997).

These observations make it most likely that an indirect dopaminergic action contributes to the antiparkinsonian action of budipine. The interactions of budipine with the dopaminergic system are listed in Table 1.

Table 1. Interaction of budipine with the dopaminergic system

In vitro studies	References
No affinity at dopamine D_1 and D_2 receptors: $IC_{50} > 10^{-4}$ M	11
No inhibition of DOPA decarboxylase: $IC_{50} > 3 \times 10^{-3}$ M	7
Inhibition of MAO-B in rat liver: $IC_{50} = 5 \times 10^{-6}$ M	7
Inhibition of MPTP binding (MAO-B) in rat brain: $IC_{50} = 2 \times 10^{-6}$ M	13
Inhibition of high-affinity synaptosomal DA-uptake: $IC_{50} = 1.1 \times 10^{-5}$ M	6
Enhancement of spontaneous release of DA in rabbit nucleus caudatus (at 10^{-6}–10^{-5} M)	6
In vivo and ex vivo studies	
No apomorphine-like stereotypies in mice and rats	7
Potentiation of behavioral effects of subacute doses of L-DOPA and apomorphine in rats	7
Inhibition of catalepsy evoked by tetrabenazine, perphenazine and haloperidol	7
Inhibition of reserpine- and tetrabenazine-induced rigidity in mice and rats	8
Inhibition of haloperidol- and perphenazine-induced akinesia in mice and rats	8
Inhibition of tremor due to LON-954-induced dopaminergic-cholinergic imbalance in mice	1, 7, 8
Increase in levels of DA and its metabolites in mouse caudate nucleus after repeated administration	9
Inhibition of haloperidol-induced increase in dopamine D_2 receptor binding capacity in rat striatum	5
Prevention of reserpine-induced decrease in content of DA in rat brain	4, 11
Prevention of MPTP-induced decrease in contents of DA, 5-HT and their metabolites in rat nucleus caudatus	10
Prevention of MPP^+-induced decrease in content of DA and its metabolites in mouse nucleus caudatus	9
Inhibition of MAO-B but not of MAO-A in rat brain	2
Increase in DA synthesis via stimulation of AADC in substantia nigra and striatum of rats	3, 12

References: *1* Brand and Menge (1980); *2* Demisch (1986) unpublished results; *3* Fisher et al. (1997); *4* Gerlach et al. (1983); *5* Hennies (1983) unpublished results; *6* Jackisch et al. (1993); *7* Menge et al. (1982); *8* Menge and Brand (1985); *9* Mihatsch et al. (1988); *10* Przuntek et al. (1985); *11* Przuntek and Stasch (1985); *12* Starr et al. (1997); *13* Stasch et al. (1988)

Interaction of budipine with NMDA receptors

In recent years, an increasing body of evidence has drawn attention to the importance of the excitatory amino acid glutamate in the pathophysiology of PD (for recent reviews, see McBain and Mayer, 1994; Starr, 1995). Accordingly, glutamate might play a double role: as a neurotransmitter it intervenes in the alterations of basal ganglia neurotransmission which results from the nigrostriatal damage; as a neurotoxin it contributes to the neurodegenerative process in the substantia nigra. Glutamate activates three major types of ionotropic receptors, namely, N-methyl-D-aspartate (NMDA), α-amino-2,3-dihydro-5-methyl-3-oxo-4-isoxazolepropanoic acid (AMPA) and kainic acid (KA) receptors, and several types of metabotropic receptors. From these glutamate receptor subtypes, NMDA-sensitive receptors are coupled to high-conductance cationic channels permeable to K^+, Na^+ and Ca^{2+}.

Although the principal neurochemical defect in PD is the loss of striatal DA due to degeneration of dopaminergic nigral neurones, the loss of DA-mediated nigrostriatal inhibition upsets the delicate neurotransmitter balance in striatum and leads to secondary postsynaptic changes in other transmitter systems, which also contribute to the movement difficulties experienced in parkinsonian patients. In addition to the pathological hyperactivity which develops in striatal cholinergic interneurones, allowing muscarinic antagonists, such as biperiden, to be used to treat milder stages of PD, several glutamatergic pathways lying within the basal ganglia also increase their firing rates following DA depletion (Bergman et al., 1990; Brotchie et al., 1991). Presently, there are two major hypotheses suggesting a link between increased glutamatergic excitation and PD. The first hypothesis suggests that DA loss leads to increased glutamatergic excitation in the subthalamus nucleus and the basal ganglia output nuclei, and that overexcitation within these nuclei is an essential step in the evolution of the parkinsonian syndrome. The second hypothesis suggests that the neurotoxic action of glutamate contributes to the selective degeneration of the substantia nigra in PD. Therefore, it has been proposed that glutamate receptor blockers, like for instance NMDA receptor antagonists, might be useful adjuvants in the therapy of PD (Carlsson and Carlsson, 1990; Klockgether and Turski, 1989). In support of this hypothesis it was found that some antiparkinsonian agents, such as memantine and amantadine exhibit NMDA receptor antagonistic properties (Lupp et al., 1992), whereas injection of NMDA, but not of non-NMDA receptor agonists, into the rostral striatum may elicit experimental parkinsonism in animals (Klockgether and Turski, 1993). Thus, recent evidence suggests that PD is also a glutamate hyperactivity disorder and hence treatable with glutamate antagonists that facilitate motor recovery with L-DOPA (Starr, 1995).

In vitro studies

The possibility of whether budipine can act as a NMDA receptor antagonist was studied in rabbit caudate nucleus slices previously labeled with [^{3}H]choline for release studies in response to NMDA receptor stimulation (Jackisch et al., 1994). In this model, budipine uncompetitively inhibited the NMDA-evoked acetylcholine release with a functional K_i value of 4.6×10^{-6}M, thereby being equipotent to memantine ($K_i = 3.25 \times 10^{-6}$M), but more potent than amantadine ($K_i = 3 \times 10^{-5}$M) (Lupp et al., 1992). In radioligand binding studies, budipine displaced [^{3}H]MK-801 specifically bound to the NMDA receptor-linked ion channel in the same tissue, although with lower affinity (IC_{50} = 3.8 × 10^{-5}M), the possible reason for which is discussed below together with its interaction with $sigma_1$ binding sites.

Similarly, in a further study the affinity of budipine in comparison with amantadine and memantine was determined at the PCP binding site of the NMDA receptor in postmortem human brain tissue (Kornhuber et al., 1995). By the ability to displace [^{3}H]MK-801 binding in this tissue, a K_i value of 1.17 × 10^{-5}M was determined for budipine, whereas respective values of 1.05 × 10^{-5}M and 5 × 10^{-7}M were obtained for amantadine and memantine.

In a recent study, the potency, kinetics and voltage-dependency of the blockade by budipine of the NMDA receptor was investigated in cultured rat striatal neurones (Parsons et al., 1998). In the concentration range of 10^{-5}–10^{-4}M, budipine very rapidly, and in concentration- and moderate voltage-dependent manner, inhibited inward Ca^{2+} current response to NMDA with an IC_{50} value of 6 × 10^{-5}M, its potency being 24-fold weaker than that of memantine, but equipotent to amantadine in this experimental set-up.

The low affinity and fast kinetics of NMDA receptor blockade by budipine may have therapeutic implications. The low incidence of psychotomimetic side effects under budipine treatment (Bien and Wörz, 1985; Jellinger and Bliesath, 1987) is comparable to amantadine and is clearly lower than with other uncompetitive NMDA receptor antagonists such as MK-801 and PCP (Kornhuber and Weller, 1995). It has been suggested that the low affinity (pK_i ~ 5.0) and the associated fast and reversible voltage-dependent channel blocking kinetics of budipine and amantadine are responsible for their better tolerability (Kornhuber et al., 1995). These properties may allow budipine to block prolonged pathological low level activation of the NMDA receptors while leaving their transient physiological activation relatively intact. In contrast to the compounds with higher affinity (pK_i > 7.0) at the NMDA receptor, like MK-801 and PCP which evoke psychosis, low-affinity ligands at the PCP binding site of the NMDA receptor exhibit faster channel kinetics with rapid unblocking rates, which might be advantageous to attenuate the pathological effects of the sustained, but relatively small increases in extracellular glutamate to micromolar concentrations but, like Mg^{2+}, leave the channel following its physiological activation, where short-lived high millimolar concentrations of glutamate are observed (Kornhuber and Weller, 1997; Porter and Greenamyre, 1995).

Quantitative receptor autoradiography with [^{3}H]MK-801 in rat brain has shown that budipine displaces the ligand from NMDA receptor ion channel in frontal cortex, striatum, entorhinal cortex, hippocampal CA1, dentate gyrus and cerebellum in the concentration range of 4 × 10^{-5}–10^{-4}M (Porter and Greenamyre, 1995). Interestingly, there was a general trend that drugs known to be better tolerated in humans, namely budipine, amantadine and memantine, had lower affinities than compounds known to evoke severe behavioral or psychotomimetic effects, namely MK-801, 1-[1-(2-thienyl)cyclohexyl]piperidine (TCP), PCP and ketamine. Moreover, the first group of compounds including budipine, had 2 to 4-fold higher affinities at NMDA receptors in cerebellum than in forebrain, whereas the reverse was true for the second group of compounds. Thus it is feasible that the reasons for the more favourable behavioral profile of the first group of compounds including budipine, might be related not only to their rapid kinetic properties but also to subtype specifity due to heterogeneity of the NMDA receptor (Kornhuber and Weller, 1997; Porter and Greenamyre, 1995).

Based mainly on the hypothesis that oxidative stress is an additional important pathological factor in PD, NMDA receptor antagonists have also been regarded as neuroprotective agents by preventing or retarding the damage of dopaminergic neurons, and thereby slowing the disease progression (Ebadi et al., 1996; Kornhuber and Weller, 1997). Accordingly, the therapeutic effect of some uncompetitive NMDA receptor antagonists in PD may be supported by their possible antioxidant properties (Lupp et al., 1996). In an NADPH/Fe^{2+}-stimulated lipid peroxidation system in rat liver, budipine and MK-801 inhibited lipid peroxidation, whereas the competitive NMDA antagonist, 2-amino-5-phosphonopentanoate (D-AP5), and all agonists, such as NMDA, glutamate, aspartate and glycine, were without effect. Thus, uncompetitive NMDA antagonists like budipine and MK-801 appear to act as radical scavangers (Lupp et al., 1996), and inhibition of free radical production could theoretically have a beneficial effect on PD (Di Paola and Utti, 1996; Greenamyre and O'Brian, 1991).

In vivo effects

Budipine displaced specific [^{3}H]TCP binding at the NMDA receptor with an IC_{50} of 3.6×10^{-5}, whereas no affinity at AMPA and KA receptors ($pK_i <$ 3.0) was detected in rat cortical membranes (Klockgether et al., 1993, 1996), and since TCP is a ligand of NMDA receptor-gated ion channel, budipine expectedly acts as an NMDA receptor antagonist in vivo. Accordingly, budipine (ED_{50} = 10.2 mg/kg i.v.) and the NMDA receptor antagonist, 3-[(±)-2-carboxypiperazin-4-yl]-propyl-1-phosphonic acid (CPP; ED_{50} = 4.4 mg/kg i.v.), but not biperiden ($ED_{50} >$ 50 mg/kg i.v.), dose-dependently increased the threshold of NMDA to induce seizures in mice, while budipine (up to 50 mg/kg i.v) had no effect on seizures induced by non-NMDA receptor stimulation, i.e. via AMPA receptors or KA receptors (Klockgether et al., 1993). Similar to specific NMDA receptor antagonists such as CPP, budipine (0.1–1 mg/kg i.v.) depressed polysynaptic spinal reflexes in mice induced by percutaneous tibial nerve stimulation with a maximal effect 30 min after application, but not monosynaptic (Hofmann) reflexes known to be sensitive to AMPA receptor blockade by 6-cyano-7-nitroquinoxaline-2,3-dione (CNQX)(Turski et al., 1990). The suggestion that an NMDA receptor antagonistic property of budipine might contribute to its antiparkinson action is also supported by the following observation: In a 6-hydroxydopamine rotational model of PD in mice, the action of budipine corresponds to that of the NMDA receptor antagonist CPP (Klockgether et al., 1996). In particular, while budipine and CPP did not induce ipsilateral or contralateral rotations when given alone, both drugs typically potentiate the action of dopaminergic drugs, like apomorphine (Löschmann et al., 1991). The interactions of

Table 2. Interaction of budipine with NMDA receptors

In vitro studies	References
Affinity at NMDA receptors in rabbit nucleus caudatus, rat cortex and human brain: $pK_i = 4.5–5.0$	1, 2, 3, 4, 5, 6
Affinity at AMPA and kainic acid (KA) receptors: $pK_i < 3.0$	2
Uncompetitive inhibition of NMDA-evoked acetylcholine release in rabbit nucleus caudatus	4
Use-dependent inhibition with fast onset and offset rates of NMDA-evoked Ca^{2+} inward current in rat striatal neurones	5
In vivo studies	
Selective inhibition of seizures evoked by NMDA but not by AMPA or KA receptor stimulation in mice	2
Selective suppression of polysynaptic flexor reflexes (NMDA-receptor mediated) but not of monosynaptic reflexes (AMPA-receptor-mediated) in mice	2

References: *1* Klockgether et al. (1993); *2* Klockgether et al. (1996); *3* Kornhuber et al. (1993); *4* Jackisch et al. (1994); *5* Parsons et al. (1998); *6* Porter and Greenamyre (1995)

budipine with NMDA receptors in vitro and in vivo are summarized in Table 2.

Interaction of budipine with muscarinic receptors

In PD, the degeneration of dopaminergic neurones in the substantia nigra causes a loss of inhibitory signals onto postsynaptic cholinergic neurones which results in abnormal high acetylcholine release in the striatum. Thus, the dopamine deficiency in the striatum in patients with PD leads to secondary changes in the activity of non-dopaminergic transmitter systems lying downstream to the dopaminergic nigrostriatal area. Traditional theories of the pathophysiology of PD emphasize the role of enhanced cholinergic activity in the striatum and the most compelling evidence for the participation of a cholinergic mechanism in PD is the antiparkinsonian action of relatively strong antimuscarinic drugs, such as biperiden and trihexyphenidyl.

Budipine possesses a high anti-tremor potency in animal experiments comparable to that of biperiden, whereas its peripheral antimuscarinic activity has been found to be 100-fold weaker than that of biperiden (Menge et al., 1982). This is evident from a number of in-vivo studies, such as inhibition of cholinergically-induced tremor by tremorine, oxotremorine and physostigmine in mice and rats, in which budipine was shown to be more than or at least equally as potent as biperiden. However, there exists a marked difference between the two drugs. The anti-tremor potency of budipine is long lasting and can be detected even after a pretreatment time of 5h after both i.v. and p.o. administration. In contrast, the anti-tremor effect of biperiden differs markedly between i.v. and p.o. administration, is short lasting and considerably weaker after p.o. administration (Menge et al., 1982).

Functional experiments on isolated rabbit vas deferens supplied with muscarinic M_1 receptors, on rat left atrium supplied with M_2 receptors and on guinea-pig ileum supplied with M_3 receptors have characterized budipine as a competitive antagonist with affinity constants (pA_2 values) of 7.17, 6.60 and 6.68, respectively (Keim et al., 1998; Menge et al., 1982). In binding studies at human recombinant muscarinic m1-m5 receptors expressed in CHO cells using [^{3}H]N-methylscopolamine as radioligand, budipine had affinity constants (pK_i values) of 7.45, 7.02, 6.30, 6.78 and 6.73, respectively (Keim et al., 1998). However, compared with the affinities of biperiden at native muscarinic receptors (M_1 = 8.80, M_2 = 7.57, M_3 = 8.14) and human recombinant muscarinic receptors (m1 = 9.32, m2 = 8.20, m3 = 8.41, m4 = 8.62, m5 = 8.20), the affinity of budipine to muscarinic receptor subtypes is up to 125-fold lower, which corresponds to its approximately 100-fold lower potency to cause peripheral antimuscarinic effects such as mydriasis in mice, or inhibition of carbachol- and acetylcholine-evoked salivation and blood pressure decrease in rats (Menge et al., 1982; Menge and Brand, 1985). Therefore, the high potency of budipine to inhibit cholinergically induced tremor cannot be explained solely by its weak antimuscarinic activity. Since budipine has a 5-fold higher affinity than biperiden ($IC_{50} = 3.6 \times 10^{-5}$M versus 1.7×10^{-4}M) for the PCP binding site within the NMDA receptor complex, and only budipine (ED_{50} = 10.2 mg/kg i.v.) but not biperiden ($ED_{50} > 50$ mg/kg i.v) attenuates seizures evoked by NMDA in mice (Klockgether et al., 1996), it can be concluded that an NMDA receptor antagonism by budipine underlies not only its inhibitory effect against seizures evoked by NMDA receptor stimulation, but most probably also against tremor evoked cholinergically. Budipine's interaction with muscarinic receptors in vitro and its effect to attenuate various cholinergically evoked effects in vivo are summarized in Table 3.

Interaction of budipine with the GABAergic system

In slices of rabbit nucleus caudatus labeled with [^{3}H]GABA, budipine (10^{-5}M) significantly inhibited both the spontaneous and electrically evoked overflow of tritium. Inhibitory effects of budipine were also found in this tissue on GABA release induced by potassium chloride (Jackisch et al., 1993). The mechanism underlying this inhibitory effect of budipine on GABA release is not clear at present. In this context it should be noted that budipine exhibited extremely low affinity ($IC_{50} > 10^{-3}$M) at GABA-A receptors labeled with [^{3}H]muscimol and also at benzodiazepine receptors labeled with [^{3}H]diazepam (Przuntek and Stasch, 1985; Stasch et al., 1988); on the other hand, this data may be irrelevant for the presynaptic site, as presynaptic GABA autoreceptors belong to the B subtype (Giralt et al., 1990).

The question whether the inhibitory effect by budipine on striatal GABA release might be useful in the therapy of PD is difficult to answer. However,

Table 3. Interaction of budipine with muscarinic receptors

In vitro studies: Affinity at muscarinic receptor subtypes	References
Native M_1–M_3 receptors: $pA_2 = 6.60 - 7.17$ (10 to 45-fold weaker than biperiden)	1, 2
Human recombinant m1–m5 receptors: $pK_i = 6.30 - 7.45$ (15 to 125-fold weaker than biperiden)	1
In vivo studies: Inhibition of cholinergically evoked tremor	
Inhibition of tremorine-induced tremor in mice (i.v., p.o.) and rats (p.o.): Slightly superior or equipotent to biperiden	2
Inhibition of oxotremorine-induced tremor in mice and rats (p.o.): Equipotent to biperiden	2
Inhibition of physostigmine-induced tremor in mice (p.o.): 5-fold more potent than biperiden	3
In vivo studies: Peripheral antimuscarinic effects	
Mydriasis in mice (s.c.): 135-fold weaker than biperiden	2
Inhibition of carbachol-stimulated salivation in rats (p.o.): 107-fold weaker than biperiden	3
Inhibition of acetylcholine-evoked blood pressure decrease in rats (i.v.): 50-fold weaker than biperiden	3

References: *1* Keim et al. (1998); *2* Menge et al. (1982); *3* Menge and Brand (1985)

it has been reported that neuronal GABAergic activity in striatum is enhanced in PD (Perry et al., 1983). Hence inhibitory effects of drugs like budipine on efferent striatal GABAergic pathways to the external pallidum and to both parts of the substantia nigra, which have been shown to be hyperactive in PD, might be beneficial in this disease (Albin et al., 1989). Furthermore, the disinhibitory GABAergic neuronal circuitry is importantly involved not only in the striatum, but also in other parts of the basal ganglia motor system, especially in the two major projections to the thalamus. Whether the inhibition of GABA release by budipine observed in striatum (Jackisch et al., 1993) also holds true for one (or several) of these other nuclei and thus could positively affect the neuropathological imbalance in PD, including a generally enhanced GABAergic activity, remains to be established.

Interaction of budipine with the noradrenergic system

Budipine evoked a mild central stimulation, detectable by the enhancement of motor activity and vigilance in mice (20–50 mg/kg p.o.). However, in discrimination tests, budipine clearly differed from the classical stimulant, amphetamine, as it did not increase salivation, body temperature and group toxicity (Menge et al., 1982). In rats treated with budipine (50 mg/kg p.o.

single dose; 8 mg/kg p.o. on 5 days), no change in the content and turnover of NA in striatum, thalamus and hypothalamus was detected (Menge et al., 1982). In pithed rats, however, budipine (15 mg/kg i.v.) potentiated the pressor effect to intravenously administered NA and DA (Menge et al., 1982).

In the isolated rat vas deferens, budipine caused contraction of the tissue at relatively high concentrations (10^{-4} M), but at 30 to 100-fold lower concentrations (10^{-6}–3×10^{-6} M) caused potentiation of contractions elicited by NA and tyramine, whereas the NA-uptake inhibitors, desipramine and imipramine, typically potentiated NA- but inhibited tyramine-evoked tissue contraction. The effects of budipine disappeared after chemical denervation by treating the tissue with 6-hydroxydopamine. Following reserpinization of the rats, budipine was still capable of contracting the vas deferens, possibly reflecting the release of NA from a reserpine-resistant storage granule pool rather than relying on inhibition of NA re-uptake (Menge et al., 1982). Similarly, in superfused rabbit cortex slices preincubated with [^{3}H]NA, budipine (10^{-5} M) enhanced basal tritium outflow even in the presence of the NA re-uptake inhibitor cocaine, which mainly consisted of NA metabolites (Jackisch et al., 1993), but did not inhibit uptake of [^{3}H]NA into rat striatal synaptosomes (Offermeier and van Rooyen, 1985). A direct interaction of budipine with α_1- and α_2-adrenoceptors can also be excluded (Menge et al., 1982; Eltze, unpublished results). A possible influence of budipine on NA synthesis has not yet been investigated. Thus, from the results known so far it can be concluded that budipine might act as a weak releaser of NA (Table 4).

Table 4. Interaction of budipine with the noradrenergic system

In vitro studies	References
No direct interaction with α_1- and α_2-adrenoceptors ($>3 \times 10^{-5}$ M)	1, 4
Potentiation of contraction of rat vas deferens to NA and tyramine (at 10^{-6} M)	3
Facilitation of [^{3}H]NA release in rabbit cortex slices (at 10^{-5} M)	2
No inhibition of [^{3}H]NA uptake in rat striatal synaptosomes (at 10^{-4} M)	5
In vivo studies	
Weak central stimulation (motor activity, vigilance) in mice (50 mg/kg p.o.)	3
No amphetamine-like effects (increase in salivation, temperature and group toxicity) in rats (50 mg/kg p.o.)	3
No change in level and turnover of NA in rat striatum and thalamus (50 mg/kg p.o.)	3
Potentiation of blood pressure response to NA and DA in pithed rats (15 mg/kg i.v.)	3

References: *1* Eltze (1982) unpublished; *2* Jackisch et al. (1993); *3* Menge et al. (1982); *4* Menge and Brand (1985); *5* Offermeier and van Rooyen (1985)

Interaction of budipine with the serotonergic system

Although there is a decrease in the 5-HT content at the level of the nucleus caudatus, putamen, pallidum, thalamus, hypothalamus and substantia nigra in brains of patients with PD (Hornykiewicz, 1966), this change is obviously of minor importance and does not add a further imbalance to the well known disturbance of the cholinergic-dopaminergic equilibrium, especially in the substantia nigra. Nevertheless, antiparkinsonian drugs additionally capable of enhancing serotonergic activity could be beneficial for the improvement of mood in PD patients.

In rats treated with budipine (50 mg/kg p.o. as single dose; 8 mg/kg p.o. for 5 days), no change in the content and turnover of 5-HT in striatum, thalamus and hypothalamus was detectable (Menge et al., 1982). In mice, however, repeated administration of budipine (daily dose 35.5 mg/kg i.p., for six days) caused a significant increase in the levels of 5-HT and its metabolite 5-HIAA in the caudate nucleus 5 weeks after drug administration (Mihatsch et al., 1988).

Previous investigations on the ability to interact with 5-HT receptors or to modulate the effects of 5-HT have shown that budipine (5–12.5 mg/kg p.o.) was capable of enhancing the head twitches in rats treated with the 5-HT receptor agonist, quipazine, and preventing those evoked by the 5-HT releaser, p-chloramphetamine (Menge et al., 1982). Similarly, budipine (10 mg/kg i.p.) partially prevented fenfluramine-induced depletion of 5-HT in whole rat brain, but failed to antagonize the hindlimb flexor reflex in spinal cats facilitated by LSD or fenfluramine (Przegalinski et al., 1982). Budipine (20 mg/kg i.p.) did not affect body temperature of rats kept at high ambient temperature or that of rabbits in normothermic surroundings. It slightly reduced (at 10 and 20 mg/kg i.p.), however, fenfluramine hyperthermia in heat-adapted rats, suggesting that budipine has the properties of a weak 5-HT uptake inhibitor rather than stimulating 5-HT receptors directly (Przegalinski et al., 1982).

In isolated human blood platelets, used as a simple model for serotonergic nerve endings, budipine inhibited uptake and caused release of [^{14}C]5-HT at threshold concentrations of 2×10^{-5} M (Eltze, 1980). Budipine was shown to contract the isolated rat stomach strip at a relatively high threshold concentration of 10^{-5} M, the effect of which was attenuable by the 5-HT antagonists, cyproheptadine and methysergide (Menge et al., 1982).

The inhibition of 5-HT uptake mechanisms (Eltze, 1980; Menge et al., 1982; Przegalinski et al., 1982), together with its weak interaction with enkephaline receptors (Neuser et al., 1983; Russ et al., 1983), have been accounted for by the prophylactic effect of budipine (15–60 mg/day) in treating cluster headache (Krüger et al., 1988). Possibly, the additional weak, but nevertheless detectable blockade of voltage-dependent Ca^{2+} channels (IC_{50} = 3×10^{-5} M in rabbit aorta; Eltze, unpublished) also may account for this effect. Possibly, a serotonergic mechanism also explains the mood elevating effect observed in depressed PD patients in early clinical studies (Oppel, 1985).

Table 5. Interaction of budipine with the serotonergic system

In vitro studies	References
Contraction of rat isolated stomach strip by budipine (at 10^{-5} M), attenuable by methysergide or cyproheptadine	2
Inhibition of uptake and facilitation of release of [^{14}C]5-HT in human blood platelets (at 2×10^{-5} M)	1
No inhibition of [^{3}H]5-HT binding in rat brain: $IC_{50} > 10^{-4}$ M	5
In vivo studies	
No change in level and turnover of 5-HT in rat striatum and thalamus (50 mg/kg p.o.)	2
Increase in content of 5-HT and 5-HIAA in mouse nucleus caudatus after 6 weeks (35 mg/kg i.p., for 8 days)	3
Partial prevention of fenfluramine-induced hyperthermia and of 5-HT depletion in rat brain (10 mg/kg i.p.)	4
Enhancement of head twitches to quipazine, but inhibition of those to p-chloramphetamine in rats (5–12.5 mg/kg p.o.)	2

References: *1* Eltze (1980); *2* Menge et al. (1982); *3* Mihatsch et al. (1988); *4* Przegalinski et al. (1982); *5* Stasch et al. (1988)

Altogether, these in-vitro and in-vivo results obtained suggest that budipine might act as a weak 5-HT uptake inhibitor and/or releaser rather than to stimulate 5-HT receptors directly. A summary of these effects is given in Table 5.

Additional interactions of budipine

Interaction with sigma binding sites

Sigma binding sites are pharmacologically distinct from the NMDA receptor/ionophor complex and have been demonstrated to exist in limbic and sensorimotoric structures of the brain that can interact with many psychotropic drugs (Walker et al., 1990). Presently, there is good evidence for the existence of at least two subtypes of sigma binding sites, termed sigma$_1$ and sigma$_2$ (Quirion et al., 1992), the physiological functions of which are not yet well understood, but, in addition to their role in the pathophysiology of schizophrenia, they may be involved in the control of posture and movement (Debonnel, 1995; Walker et al., 1990). However, the hypothesis of the clinical importance of sigma ligands, acting e.g. in "spontaneously" active systems such as substantia nigra, is still in its infancy. Nevertheless, the presence of sigma receptors in the substantia nigra raises the possiblity of the use of sigma ligands in treating PD (Walker et al., 1990).

The affinity of budipine in comparison with amantadine and memantine has been determined at sigma$_1$ binding sites in postmortem human brain tissue. By using the specific ligand [^{3}H](+)pentazocine, a K_i value of

2×10^{-6}M was obtained for budipine, whereas the affinities of both amantadine and memantine ($K_i = 2 \times 10^{-5}$M) were 10-fold lower (Kornhuber et al., 1995).

The different affinities of budipine, amantadine and memantine at NMDA receptors in comparison to $sigma_1$ binding sites have recently been discussed (Kornhuber et al., 1995). The antiparkinsonian effects of amantadine and memantine are thought to be mainly mediated by their uncompetitive antagonism at the NMDA receptor (Kornhuber et al., 1994). Memantine at therapeutic concentrations appears to specifically interact with the NMDA receptor channel, which means its affinity at the PCP binding site exceeds that at the $sigma_1$ binding site by a factor of 40 (affinity rank order: NMDA > $sigma_1$). Amantadine, on the other hand, has an equally potent action at NMDA and $sigma_1$ binding sites (affinity rank order: NMDA = $sigma_1$). When compared to amantadine, budipine has a comparable low affinity at the PCP binding site of the NMDA receptor ($\sim 10^{-5}$M), but a 5-fold higher affinity (2×10^{-6}M) at the $sigma_1$ binding site (affinity rank order: $sigma_1$ > NMDA)(Kornhuber et al., 1995).

Supported by the observed effects of sigma ligands on animal behavior (Walker et al., 1990) and the involvement of $sigma_1$ ligands to modulate responses following NMDA receptor stimulation (Monnet et al., 1992) it has been hypothetized that, in addition to the PCP binding site of the NMDA receptor, the $sigma_1$ binding site may also be involved in the antiparkinsonian effect of amantadine (Kornhuber et al., 1995). This would also explain the better therapeutic responses to amantadine and budipine compared to memantine in this disease and, together with their relatively low affinity at the PCP binding site of the NMDA receptor, suggests a low probability for inducing psychotomimetic side effects by the former two agents (Kornhuber and Weller, 1997). This NMDA-$sigma_1$ binding site interaction could also possibly explain the observations that inhibition of NMDA-evoked acetylcholine release in rabbit striatal slices by budipine occurs at lower concentrations ($IC_{50} = 5 \times 10^{-6}$M) than that necessary for inhibition of [^{3}H]MK-801 binding (4×10^{-5}M) at the NMDA receptor (Jackisch et al., 1994), and could also account for the weaker voltage-dependency of budipine compared with amantadine to block NMDA receptor stimulation-evoked Ca^{2+} inward current in cultured rat neurones (Parsons et al., 1998). Similarly, the preferred interaction of budipine with $sigma_1$ sites may also explain the more pronounced stimulation of aromatic L-amino acid decarboxylase (AADC) by a lower dose of budipine (12.5 mg/kg i.p.) as compared to a 2-fold higher dose (20 mg/kg i.p.) of the preferential NMDA receptor antagonist memantine (Starr et al., 1997).

Interaction with histamine H_1 receptors

A further property of budipine is its competitive interaction with histamine H_1 receptors. Functional studies in guinea-pig ileum yielded an affinity constant (pA_2) of 6.40 (Menge et al., 1982). Although histamine is a neuromodulator in mammalian central nervous system where it regulates functions such as awareness, awakefulness and motor activity, its possible involvement in central disorders such as PD has not been established. Recently, an interaction of histamine with both NMDA receptor-mediated and GABAergic responses has been described. In rat cortical neurones histamine facilitates NMDA-evoked depolarization by two different mechanisms, one involving stimulation of antagonist-sensitive cell surface histamine H_1 receptors, the other suggesting a direct interaction of histamine at the NMDA receptor (Payne et al., 1997). Similar actions of histamine have previously been reported to occur in rat hippocampal neurones, the effect of which was either insensitive or weakly sensitive to histamine H_1 receptor blockade (Brown et al., 1995). In a human astrocytoma cell line, histamine H_1 receptor stimulation releases $[^3H]$GABA by increasing intracellular Ca^{2+} which is attenuable by mepyramine (Soria-Jasso and Arias-Montano, 1996).

If histamine H_1 receptors are involved in NMDA receptor responses and/or GABA release at structures other than cortical and hippocampal neurones, e.g. in basal ganglia resonsible for motor disorders in PD, agents like budipine bearing a moderate histamine H_1 receptor blocking property could possibly have an additional beneficial effect to counteract enhanced NMDA-receptor responses and GABAergic activity within the basal ganglia.

Conclusions

The major components of the mode of action of budipine identified up to now are 1) an indirect dopaminergic activity (via facilitation of DA release, inhibition of synaptosomal DA uptake, inhibition of MAO-B and stimulation of AADC activity) to restore decreased striatal DA levels, 2) blockade of glutamate (NMDA) receptors to counteract increased excitatory glutamatergic activity of corticostriatal neurones and that within the indirect striatal output path, 3) an antimuscarinic activity to block the hyperactivity of striatal cholinergic interneurones, and 4) inhibition of GABA release to suppress the increased striatal GABAergic output activity. By means of these multiple mechanisms, budipine might correct the imbalance of striatal output paths by restoring DA levels in the striatum, and positively influence the secondary functional modifications of other transmitter systems (glutamate, acetylcholine, GABA) observed in basal ganglia circuitry resulting from the nigral cell loss. Thus, budipine combines multiple actions which are otherwise provided only by the combination of agents belonging to different drug classes and in this respect differs from conventional antiparkinsonian drugs. Based on the combination of multiple mechanisms, which may explain its balanced therapeutic activity and good tolerability in PD patients, budipine might be the prototype of a new class of antiparkinsonian agents.

References

Albin RL, Young AB, Penney JB (1989) The functional anatomy of basal ganglia disorders. Trends Neurol Sci 12: 366–375

Ballard P, Tetrud JW, Langston W (1985) Permanent human parkinsonism due to 1-methyl-4-phenyl-1,2,3,6-tetrahydropyridine (MPTP). Seven cases. Neurology 35: 949–959

Bergman H, Wichmann T, DeLong MR (1990) Reversal of experimental parkinsonism by lesions of the subthalamic nucleus. Science 249: 1436–1438

Bien S, Wörz R (1985) Budipine-treatment of neuroleptic-induced parkinsonism. In: Gerstenbrand F, Poewe W, Stern G (eds) Clinical experiences with budipine in Parkinson therapy. Springer, Berlin Heidelberg New York Tokyo, pp 169–175

Brand U, Menge HG (1980) Antagonismus gegen den Tremor nach N-Carbamoyl-2-(2,6-dichlorphenyl)-acetamidin HCl (LON-954) und Oxotremorin. Eine vergleichende Studie zum Wirkungsmechanismus. Arzneim Forsch/Drug Res 30: 1242–1243

Brotchie JM, Mitchell IJ, Sambrook MA, Crossman AR (1991) Alleviation of parkinsonism by antagonism of excitatory amino acid transmission in the medial segment of the globus pallidus in rat and primate. Mov Disord 6: 133–138

Brown RE, Fedorov NB, Haas HL, Reymann KG (1995) Histaminergic modulation of synaptic plasticity in area CA1 of rat hippocampal slices. Neuropharmacology 34: 181–190

Burt DR, Creese J, Snyder SH (1977) Antischizophrenic drugs: chronic treatment elevates dopamine receptor binding in brain. Science 196: 326–328

Carlsson M, Carlsson A (1990) Interactions between glutamatergic and monoaminergic systems within the basal ganglia — implications for schizophrenia and Parkinson's disease. Trends Neurosci 13: 272–276

Coward DM, Dogget NS, Thomas JE (1977) Central transmitter involvement in LON-954-induced tremorgenesis. Neuropharmacology 16: 479–484

Debonnel G (1995) Current hypotheses on sigma receptors and their physiological role: possible implications in psychiatry. J Psychiatr Neurosci 18: 157–172

Di Paola R, Utti RJ (1996) Early detection of Parkinson's disease. Implications for treatment. Drugs Aging 9: 159–168

Ebadi M, Srinivasan SK, Baxi MD (1996) Oxidative stress and antioxidant therapy in Parkinson's disease. Progr Neurobiol 48: 1–19

Eltze M (1980) The effects of prodipine and budipine on [^{14}C]5-HT uptake and release by human blood platelets. Arzneim Forsch/Drug Res 30: 1129–1134

Fisher A, Biggs CS, Starr MS (1997) Evidence that glutamate regulates dopamine synthesis via aromatic L-amino acid decarboxylase. Br J Pharmacol 120 [Suppl]: 239P

Gerlach M, Jutzi P, Stasch JP, Przuntek H (1983) Synthese und pharmakologische Eigenschaften von silierten Dopaminen und 4,4-Diphenylpiperidinen. Z Naturforsch 38b: 237–242

Giralt MT, Bonnano G, Raiteri M (1990) GABA terminal autoreceptors in the pars compacta and in the pars reticulata of substantia nigra are GABA B. Eur J Pharmacol 175: 137–144

Greenamyre JT, O'Brian CF (1991) N-methyl-D-aspartate antagonists in the treatment of Parkinson's disease. Arch Neurol 48: 977–981

Hackmack G, Menge HG, Eistetter K, Krüger U, Schaefer H, Klosa J (1979) 4,4-Diphenylpiperidine, ein neuer Substanztyp für die Therapie des Morbus Parkinson. Österr Chem Ztschr 80: 61–62

Hadjiconstantinou M, Rosetti ZL, Wemlinger TA, Neff NH (1995) Dizozilpine enhances striatal tyrosine hydroxylase and aromatic L-amino acid decarboxylase activity. Eur J Pharmacol 289: 97–101

Hertting G, Zumstein A, Jackisch R, Hoffmann I, Starke K (1980) Modulation by endogenous dopamine of the release of acetylcholine in the caudate nucleus of the rabbit. Naunyn-Schmiedeb Arch Pharmacol 315: 111–117

Hornykiewicz O (1966) Dopamine (3-hydroxytyramine) and brain function. Pharmacol Rev 18: 925–964

Iizuka J, Fischer R (1986) Beeinflussung des Parkinson-Tremors durch Budipin: Eine Vergleichsstudie mit Amantadin. Nervenarzt 57: 184–186

Jackisch R, Huang HY, Reimann W, Limberger N (1993) Effects of the antiparkinsonian drug budipine on neurotransmitter release in central nervous system tissues in vitro. J Pharmacol Exp Ther 264: 889–898

Jackisch R, Kruchen A, Sauermann W, Hertting G, Feuerstein TJ (1994) The antiparkinsonian drugs budipine and biperiden are use-dependent (uncompetitive) NMDA receptor antagonists. Eur J Pharmacol 264: 207–211

Jellinger K, Bliesath H (1987) Adjuvant treatment of Parkinson's disease with budipine: A double-blind trial versus placebo. J Neurol 234: 280–282

Keim C, Mutschler E, Lambrecht G, Eltze M (1998) Affinity of the antiparkinsonian drug, budipine, on native muscarinic M_1-M_3 and human recombinant m1-m5 receptors. Naunyn-Schmiedeb Arch Pharmacol 357 [Suppl]: R23

Kornhuber J, Weller M (1995) Predicting psychotomimetic properties of PCP-like NMDA receptor antagonists. In: Fog R, Gerlach J, Hemmingsen R, Krogsgaard-Larsen P, Thaysen JH (eds) Schizophrenia — an integrated view. Alfred Benzon Symposium 38. Munksgaard, Copenhagen, pp 314–325

Kornhuber J, Weller M (1997) Psychotogenicity and N-methyl-D-aspartate receptor antagonism: Implications for neuroprotective pharmacotherapy. Biol Psychiatry 41: 135–144

Kornhuber J, Weller M, Schoppmeyer K, Riederer P (1994) Amantadine and memantine are NMDA receptor antagonists with neuroprotective properties. J Neural Transm [Suppl] 43: 91–104

Kornhuber J, Herr B, Thome J, Riederer P (1995) The antiparkinsonian drug budipine binds to NMDA and sigma receptors in postmortem human brain tissue. J Neural Transm [Suppl] 46: 131–137

Klockgether T, Turski L (1989) Excitatory amino acids and the basal ganglia: implications for the therapy of Parkinson's disease. Trends Neurosci 12: 285–286

Klockgether T, Turski L (1993) Towards the understanding of the role of glutamate in experimental parkinsonism: agonist-sensive sites in the basal ganglia. Ann Neurol 34: 585–593

Klockgether T, Jacobsen P, Löschmann PA, Turski L (1993) The antiparkinsonian agent budipine is an N-methyl-D-aspartate antagonist. J Neural Transm [P D Sect] 5: 101–106

Klockgether T, Wüllner U, Steinbach JP, Petersen V, Turski L, Löschmann PA (1996) Effect of the antiparkinsonian drug budipine on central neurotransmitter systems. Eur J Pharmacol 301: 67–73

Krüger H, Kohlhepp W, Reimann G, Przuntek H (1988) Prophylactic treatment of cluster headache with budipine. Headache 28: 344–346

Kuhn W, Russ H, Dettner O, Gerlach M, Przuntek H (1987) Interaction of deprenyl and budipine with MAO-B in vivo and in vitro. Neuroscience 22 [Suppl]: S527

Löschmann PA, Lange KW, Kunow M, Rettig KJ, Jähnig P, Honore T, Turski L, Wachtel H, Jenner P, Marsden CD (1991) Synergism of the AMPA-antagonist NBQX and the NMDA-antagonist CPP with L-DOPA in models of Parkinson's disease. J Neural Transm [P D Sect] 3: 203–207

Lupp A, Lücking CH, Koch R, Jackisch R, Feuerstein TJ (1992) Inhibitory effects of the antiparkinsonian drugs memantine and amantadine on N-methyl-D-aspartate-evoked acetylcholine release in the rabbit caudate nucleus. J Pharmacol Exp Ther 263: 717–724

Lupp A, Karge E, Klinger W (1996) Antioxidant properties of NMDA-receptor agonists and antagonists on the microsomal cytochrome P-450 system of rat livers in vitro. Naunyn-Schmiedb Arch Pharmacol 353 [Suppl]: R108

McBain CJ, Mayer ML (1994) N-methyl-D-aspartatic acid receptor structure and function. Physiol Rev 74: 723–760

Menge HG, Brand U (1985) Pharmacologic bases of antiparkinsonian therapy. In: Gerstenbrand F, Poewe W, Stern G (eds) Clinical experiences with budipine in Parkinson therapy. Springer, Berlin Heidelberg New York Tokyo, pp 82–92

Menge HG, Brand U, Dittmann EC, Eltze M, Gernandt W, Gönne S, Müller H, Steinijans V (1982) Zusammenfassende Darstellung der Pharmakologie von Budipin, einem neuen 4,4-Diphenylpiperidin-Derivat für die Parkinson-Therapie. Arzneim Forsch/Drug Res 32: 85–98

Mihatsch W, Russ H, Przuntek H (1988) Intracerebroventricular administration of 1-methyl-4-phenylpyridinium ion in mice: effects of simultaneously administered nomifensine, deprenyl and 1-t-butyl-4,4-diphenylpiperidine. J Neural Transm 71: 177–188

Monnet FP, Debonnel G, DeMontigny C (1992) In vivo electrophysiological evidence for a selective modulation of N-methyl-D-aspartate-induced neuronal activation in rat CA3 dorsal hippocampus by sigma ligands. J Pharmacol Exp Ther 261: 123–130

Neuser D, Stasch JP, Witteler M, Kuhn W, Gerlach M, Jutzi P, Przuntek H (1983) The interaction of 1-alkyl-4,4-diphenylpiperidines with opiate receptors. Eur J Pharmacol 87: 315–318

Niznik HB, Tyndale RF, Sallee FR, Gonzales FJ, Hardwick JP, Inaba T, Kalow W (1990) The dopamine transporter and cytochrome P-450 IID1 (debrisoquine-4-hydroxylase) in brain: resolution and identification of two distinct [^{3}H]GBR-12935 binding proteins. Arch Biochem Biophys 276: 424–432

Offermeier J, van Rooyen JM (1985) The pharmacodynamics of budipine on central neurotransmitter systems. In: Gerstenbrand F, Poewe W, Stern G (eds) Clinical experiences with budipine in Parkinson therapy. Springer, Berlin Heidelberg New York Tokyo, pp 93–105

Oppel F (1985) Long-term treatment with budipine. In: Gerstenbrand F, Poewe W, Stern G (eds) Clinical experiences with budipine in Parkinson therapy. Springer, Berlin Heidelberg New York Tokyo, pp 163–168

Parsons CG, Quack G, Bresink I, Baran L, Przegalinski E, Kostowski W, Krzascik P, Hartmann S, Danysz W (1995) Comparison of the potency, kinetics and voltage-dependency of a series of uncompetitive NMDA receptor antagonists in vitro with anticonvulsive and motor impairment activity in vivo. Neuropharmacology 34: 1239–1258

Parsons CG, Hartmann S, Spielmanns P (1998) Budipine is a low affinity, N-methyl-D-aspartate receptor antagonist: patch clamp studies in cultured striatal, hippocampal, cortical and superior colliculus neurones. Neuropharmacology 37: 719–727

Payne GW, Neuman RS (1997) Effect of hypomagnesia on histamine H_1 receptor mediated facilitation of NMDA responses. Br J Pharmacol 121: 199–204

Perry TL, Javoy-Agid F, Fibiger HC (1983) Striatal GABAergic neuronal activity is not reduced in Parkinson's disease. J Neurochem 40: 1120–1123

Poewe W, Gerstenbrand F, Ransmayr G (1985) Clinical experience with budipine in parkinsonian therapy. In: Gerstenbrand F, Poewe W, Stern G (eds) Clinical experiences with budipine in Parkinson therapy. Springer, Berlin Heidelberg New York Tokyo, pp 122–128

Porter RHP, Greenamyre JT (1995) Regional variation in the pharmacology of NMDA receptor channel blockers: implications for therapeutic potential. J Neurochem 64: 614–623

Przegalinski E, Bigajska K, Lewandowska A (1982) The effect of budipine on the central serotonergic system. Pol J Pharmacol Pharm 34: 309–315

Przuntek H, Stasch JP (1985) Biochemical and pharmacologic aspects of the mechanism of action of budipine. In: Gerstenbrand F, Poewe W, Stern G (eds) Clinical experiences with budipine in Parkinson therapy. Springer, Berlin Heidelberg New York Tokyo, pp 107–112

Przuntek H, Russ H, Henning K, Pindur U (1985) The protective effect of 1-t-butyl-4,4-diphenylpiperidine against the nigrostriatal neurodegeneration caused by 1-methyl-4-phenyl-1,2,3,6-tetrahydropyridine. Life Sci 37: 1195–1200

Quirion R, Bowen WD, Itzhak Y, Junien JL, Musacchio JM, Rothmann RB, Su TP, Tam SW, Taylor DP (1992) A proposal for the classification of sigma binding sites. Trends Pharmacol Sci 13: 85–86

Reimann W, Zumstein A, Jackisch R, Starke K, Hertting G (1979) Effect of extracellular dopamine on the release of dopamine in the rabbit caudate nucleus: evidence for a dopaminergic feedback inhibition. Naunyn-Schmideb Arch Pharmacol 306: 53–60

Ricaurte GA, Langston JW, DeLanney LE, Irwin I, Brooks JP (1985) Dopamine uptake blockers protect against the dopamine depleting effect of 1-methyl-4-phenyl-1,2,3,6-tetrahydropyridine (MPTP) in the mouse striatum. Neurosci Lett 59: 265–270

Russ H, Stasch JP, Witteler M, Neuser D, Przuntek (1983) Der Effekt von Budipin auf das Parkinson Syndrom. Versuch einer neuropharmakologischen Erklärung. In: Seitz D (ed) Verhandlungen der Deutschen Gesellschaft für Neurologie. Springer, Berlin Heidelberg New York Tokyo, pp 757–760

Russ H, Pindur U, Przuntek H (1986) The interaction of 1-alkyl-4,4-diphenylpiperidines with the 1-methyl-1,2,3,6-tetrahydropyridine receptor binding site. J Neural Transm 65: 157–165

Schaefer H, Hackmack K, Eistetter K, Krüger U, Menge HG, Klosa J (1984) Synthese, physikalisch-chemische Eigenschaften und orientierende pharmakologische Untersuchungen von Budipin und verwandten 4,4-Diphenylpiperidinen. Arzneim Forsch/Drug Res 34: 233–240

Schmidt WJ, Bubser M (1989) Anticataleptic effects of the N-methyl-D-aspartate antagonist MK-801 in rats. Pharmacol Biochem Behav 32: 621–629

Siegfried J, Fischer R (1983) La place pour une nouvelle substance chimique dans le traitment de la maladie de Parkinson et la role de la budipine. Med et Hyg 41: 1977–1981

Siegfried J, Fischer R (1985) Budipine: A new chemical substance in the treatment of Parkinson's disease. In: Gerstenbrand F, Poewe W, Stern G (eds) Clinical experiences with budipine in Parkinson therapy. Springer, Berlin Heidelberg New York Tokyo, pp 152–157

Soria-Jasso LE, Arias-Montano JA (1996) Histamine H_1 receptor activation stimulates [^{3}H]GABA release from human astrocytoma U373 MG cells. Eur J Pharmacol 318: 185–192

Spieker S, Löschmann P, Jentgens C, Boose A, Klockgether T, Dichgans (1995) Tremorlytic activity of budipine. a quantitative study with long-term tremor recordings. Clin Neuropharmacol 18: 266–272

Starr MS (1995) Antiparkinsonian actions of glutamate antagonists — alone and with L-DOPA: a review of evidence and suggestions for possible mechanisms. J Neural Transm [P D Sect] 10: 141–185

Starr MS, Fisher A, Biggs CS (1997) Effect of glutamate antagonists on dopamine synthesis. Amino Acids 15: 55

Stasch JP, Ruß H, Schacht M, Witteler M, Neuser D, Gerlach M, Leven M, Kuhn W, Jutzi P, Przuntek H (1988) 4,4-Diphenylpiperidine derivatives and their sila analogues. A comparative study of their interaction with neuronal receptor binding sites and synaptosomal monoamine uptake. Arzneim Forsch/Drug Res 38: 1075–1078

Turski L, Bressler K, Klockgether T, Stephens DN (1990) Differential effects of the excitatory amino acid antagonists, 6-cyano-7-nitroquinoxaline-2,3-dione (CNQX) and 3-(±)-2-carboxypiperazin-4-yl)propyl-1-phosphonic acid (CPP), on spinal reflex activity in mice. Neurosci Lett 113: 66–71

Walker JM, Bowen WD, Walker FO, Matsumoto RR, De Costa B, Rice KC (1990) Sigma receptors: biology and function. Pharmacol Rev 42: 355–402

Wang Y (1985) Observations on the therapeutic effect of budipine on Parkinson's disease. In: Gerstenbrand F, Poewe W, Stern G (eds) Clinical experiences with budipine in Parkinson therapy. Springer, Berlin Heidelberg New York Tokyo, pp 158–162

Zech K, Sturm E, Ludwig G (1985) Pharmacokinetics and metabolism of budipine in animals and humans. In: Gerstenbrand F, Poewe W, Stern G (eds) Clinical experiences with budipine in Parkinson therapy. Springer, Berlin Heidelberg New York Tokyo, pp 113–121

Zhu MY, Juorio AV, Paterson IA, Boulton AA (1992) Regulation of aromatic L-amino acid decarboxylase by dopamine receptors in rat brain. J Neurochem 58: 636–641

Author's address: Dr. M. Eltze, Byk Gulden, P.O. Box 100310, D-78403 Konstanz, Federal Republic of Germany

Parkinson's disease: one disease entity or many?

A. D. Korczyn

Sieratzki Chair of Neurology, Sackler Faculty of Medicine,
Tel-Aviv University, Israel

Summary. Although most neurologists accept that Parkinson's disease (PD) is a unique disorder, we still lack a valid diagnostic biological tool and therefore depend primarily on clinical examination in diagnosis. The distinction between PD and other parkinsonian syndromes is discussed.

Parkinson's disease (PD) is among the most common neurodegenerative brain disorders, and one of the first brain diseases for which a specific rational therapy has been discovered, namely levodopa. Disappointingly, the initial dramatic therapeutic success was followed by only slow and limited further progression, in spite of extensive research into the biology of the disorder. In particular, while levodopa and dopamine agonists have a significant and prolonged symptomatic effect, they seemingly fail to influence the subsequent deterioration. One of the most important hallmarks of PD is the progressive loss of dopaminergic neurons in the substantia nigra pars compacta, and this degeneration continues in patients who are being treated. Moreover, while treatment with levodopa is useful in most parkinsonian patients, there is a group who fail to improve either initially or secondarily.

A central question is whether PD is really a disease or whether it should be regarded as a syndrome (Calne, 1989). In order to define a nosologic entity as a disease, the etiology, pathogenesis as well as pathological and clinical features should be clear and be shared by all the patients (Calne, 1989). Needless to say, these criteria are not fulfilled in PD. Patients are being lumped together based on certain clinical features like hypokinesia, rigidity, tremor and postural instability, making this clinical entity, by definition, a syndrome. This is done in spite of the well recognized fact that similar clinical features may well occur in several other conditions (Table 1). It is reasonable and useful to assume, however, that at least one common disorder exists which fulfills certain diagnostic criteria and is therefore a disease.

The basic clinical definition of PD, which will be agreed upon by most or all neurologists, is given in Table 2. This definition is advantageous for research purposes. Nevertheless, it presents several problems. Firstly, although requiring that several features co-exist, retrospective analysis of patients fulfilling these criteria will easily reveal that at onset usually only one has been

Table 1. Parkinsonian syndromes

Parkinson's disease
Parkinsonism in Alzheimer's disease
Juvenile Huntington's disease
Dopa-responsive dystonia-parkinsonism
Bradykinesia and stooped posture of old age
Multi-system atrophy
Progressive supranuclear palsy
Normal pressure hydrocephalus
Lower body parkinsonism
Vascular parkinsonism
Pharmacologic/toxic parkinsonism
Parkinsonism following encephalitis
Cortico-basal ganglionic degeneration
Wilson's disease
Halervorden-Spatz disease
Guam PD-ALS-dementia complex

Table 2. Clinical definition of PD

An adult-onset, slowly progressive, motor disorder
Combining 2 or more of the following:
Rest tremor
Bradykinesia
Limb rigidity
Gait instability (late)
Dramatic and sustained response to levodopa
Accepted associated phenomena:
Depression (early or late)
Cognitive decline (late)
Limited autonomic involvement (e.g. constipation)

Table 3. Parkinson plus syndromes — Red Flags

Early dementia/apraxia
Early instability and falls
Prominent autonomic impairment
Oculomotor disturbances
Retro/anterocollis
Cerebellar signs
Orofacial dyskinesias on levodopa

satisfied. Thus, an unknown number of PD patients do not fulfill these diagnostic criteria, a problem which is particularly relevant for epidemiological studies.

It is important to note that patients with suspected PD should not only have the clinical manifestations in Table 2 (inclusionary criteria), but should also not have any exclusionary criteria. Some "red flags" are listed in Table 3,

Table 4. Etiology of Parkinsonian syndromes

Infective (Encephalitis lethargica)
Toxic (MPTP, Mn^{++})
Degenerative (PD, MSA, PSP)
Traumatic/lesional (Boxer's parkinsonism, vascular)
Genetic

any of which suggests that the patient actually suffers not from PD but rather from one of the other nosological entities which are listed in Table 1.

One commonly employed criterion for PD regards the response to levodopa (Table 2). The obvious importance of this criterion is that it relates to the basic presumed pathophysiological disorder in PD, the dopamine deficiency. However, the specificity of the levodopa response is low, particularly in early stages of the disease at which time patients with other disorders may also respond rather well. This is the case with multiple-system atrophy and progressive supranuclear palsy. Even at advanced parkinsonism, patients with these disorders may benefit from levodopa. An important caveat could be that peak dose dyskinesias may result in orolingual movements, rather than choreatic movements of the limbs, characteristics of PD.

Several extrinsic factors have been identified which can lead to parkinsonism, including infective, toxic, degenerative, traumatic and vascular (Table 4). In the present paper, we shall abstain from employing the commonly-used term, idiopathic PD, which is a contradiction in terms: If the etiology is unknown, one cannot speak of a disease. Rather, the term PD will be used, assuming, as stated above, that such a disease exists and that our research will eventually discover its cause and pathogenesis. In order to be able to do so, however, an attempt must be made to define — at least tentatively — those patients who are most likely to belong to this nosologic entity. In practical terms, this should be based mainly on clinical features of the disease although biological markers such as Lewy bodies found at autopsy (Rajput et al., 1991; Hughes et al., 1993), PET fluorodopa results (Martin et al., 1989; Leenders et al., 1990) or electrophysiologic data, if available, should also be consistent.

Other disorders in which a marked response to levodopa exists are postencephalitic parkinsonism (Nisipeanu et al., 1997), the pallido-pyramidal syndrome (Nisipeanu et al., 1994), and GM_2-gangliosidosis (Inzelberg and Korczyn, 1994).

The issue of therapeutic response is complicated because some classical parkinsonian symptoms fail to respond to levodopa, including posture, speech, freezing and constipation. Nevertheless, in a patient fulfilling the symptomatic criteria enumerated in Table 2, failure to respond satisfactorily to levodopa is characteristic of multisystem atrophy and progressive supranuclear palsy, probably reflecting the striatal involvement of these disorders, which does not exist in patients with Lewy body pathology. In this context, the response to levodopa should take into account primarily improvement in rigidity and hypokinesia.

Additional indicators suggestive of different etiologies described in PD include familiality and onset age. It is well known that about 10% of PD patients report the occurrence of the disease among family members and in about 15% the first manifestation occurs prior to age 50. These subgroups partially overlap. At present, the significance of these factors are not completely known but once genetic factors are not only identified (Gasser et al., 1998; Polymeropoulos et al., 1997) but also better understood, these could shed light on the etiology and pathogenesis of PD. The recent identification of mutations in the alpha-synuclein gene as being responsible for some familial cases of early onset parkinsonism is relevant because this protein is detected in Lewy bodies. Since Lewy bodies contain alpha-synuclein not only in familial but also in sporadic PD, it is reasonable to assume — in analogy to APP mutations in Alzheimer's disease — that alpha-synuclein is not only a bystander but is actually involved in disease pathogenesis, whether its gene contains a mutation or not.

At present, therefore, the term PD should be used only tentatively since at a later stage it can be found to include several specific disorders, sharing clinical (and possibly even pathological) features but with distinct etiology and pathogenesis (Table 4).

Other clinical features may also be important. The existence of cognitive decline was denied by James Parkinson himself (Parkinson, 1817) and subsequently by many prominent neurologists. However, this situation has lately changed with the identification of cortical Lewy bodies in diffuse Lewy body disease (DLBD). The exact delineation between DLBD and PD still has to be defined, and it is possible that a full spectrum exists with some patients having purely extrapyramidal motor disorder while the other extreme includes patients with cognitive decline but without extrapyramidal features. On the other hand, the existence of tremor has not been demonstrated as a helpful indicator of a separate nosologic entity.

Pathological criteria of PD are widely recognized (Table 5). Nevertheless, it should be realized that in some patients the absence of these markers is not reflected by clinical differences (Rajput et al., 1991; Hughes et al., 1993). Similarly, electrophysiological indicators of PD were sought. Although group differences were reported (Inzelberg et al., 1995), the separation is incomplete and thus cannot yet reliably distinguish PD from similar movement disorders (Orlov et al., 1998). Kinematic studies, while separating PD from normal controls, have not been used in an attempt to separate PD from other movement disorders (Inzelberg et al., 1995).

The place of PD among other movement disorders is constantly revised, based on the elucidation of clinical, pathological and genetic markers and

Table 5. Pathological criteria of PD

Depigmentation of substantia nigra (mainly)
Intracytoplasmic eosinophilic depositions (Lewy bodies), predominantly in substantia nigra and other pigmented brain-stem nuclei

pathogenetic mechanisms. As more data accumulate, PD may justify being upgraded from its present status of a syndrome into a disease, with a unique cause and pathogenesis and, therefore, a treatment that would either prevent the disease or stop its progression.

References

Calne DB (1989) Is "Parkinson's disease" one disease? J Neurol Neurosurg Psychiatry [Suppl]: 18–21

Gasser T, Müller-Myhsok B, Wszolek ZK, Ochlmann R, Calne DB, Bonifati V, Bereznai B, Fabrizio E, Vieregge P, Horstmann RD (1998) A susceptibility locus for Parkinson's disease maps to chromosome 2p13. Letter — Nat Genet 18: 262–265

Hughes AJ, Daniel SE, Blankson S, Lees AJ (1993) A clinicopathologic study of 100 cases of Parkinson's disease. Arch Neurol 50: 140–148

Inzelberg R, Korczyn AD (1994) Parkinsonism in adult-onset GM_2 gangliosidosis. Mov Disord 9: 375–380

Inzelberg R, Flash T, Korczyn AD (1995) Kinematic analysis of complex movements in Parkinson's disease. In: Przuntek H, Kraus PH, Klotz P, Korczyn AD (eds) Instrumental methods and scoring in extrapyramidal disorders. Springer, Berlin Heidelberg New York Tokyo, pp 181–187

Leenders KL, Salmon EP, Tyrrell P, Perani D, Brooks DJ, Sager H, Jones T, Marsden CD, Frackowiak RSJ (1990) The nigrostriatal dopaminergic system assessed in vivo by positron emission tomography in healthy volunteer subjects and patients with Parkinson's disease. Arch Neurol 47: 1290–1298

Martin WRW, Palmer MR, Patlak CS, Calne DB (1989) Nigrostriatal function in humans studied with positron emission tomography. Ann Neurol 26: 535–542

Nisipeanu P, Kuritzki A, Korczyn AD (1994) Familial levodopa responsive parkinsonian-pyramidal syndrome. Mov Disord 9: 673–675

Nisipeanu P, Paleacu D, Korczyn AD (1997) Infectious and postinfectious parkinsonism. In: Watts RI, Koller WC (eds) Movement disorders: neurologic principles and practice. McGraw-Hill, New York, pp 307–312

Orlov E, Grozman GB, Drory VE, Simon E, Shabtai H, Korczyn AD, Giladi N (1998) Electromyography of the anal sphincter in parkinsonism. Neurology 50 [Suppl 4]: A132

Parkinson J (1817) An essay in the shaking palsy. Whittingham and Rowland, London

Polymeropoulos MH, Lavedant C, Leroy E, Ide SE, Dehejia A, Dutra A, Pike B, Root H, Rubenstein J, Boyer R, Stenroos ES, Chandrasekharappa S, Athanassiadou A, Papapetropoulos T, Johnson WG, Lazzarini AM, Duvoisin RC, Di Iorio G, Golbe LI, Nussbaum L (1997) Mutation in the α-synuclein gene identified in families with Parkinson's disease. Science 276: 2045–2047

Rajput AH, Rozdilsky B, Rajput A (1991) Accuracy of clinical diagnosis in parkinsonism — a prospective study. Can J Neurol Sci 18: 275–278

Author's address: Prof. Amos D. Korczyn, Sieratzki Chair of Neurology, Sackler Faculty of Medicine, Tel-Aviv University, Ramat-Aviv 69978, Israel

Parkinson's disease — a multifactorial neurodegenerative disorder

O. Riess[1] and **R. Krüger**[2]

[1]Molecular Human Genetics, Ruhr-University, Bochum, Federal Republic of Germany
[2]Department of Neurology, Ruhr-University, St. Josef-Hospital, Bochum, Federal Republic of Germany

Summary. The pathogenesis of idiopathic Parkinson's disease (PD) is not known, but is thought to be multifactorial, deriving from environmental factors acting on genetically predisposed individuals with aging. Association studies of DNA polymorphisms are able to detect a genetic background predisposing to PD. Mechanisms as oxidative stress, xenobiotica toxicity and altered dopamine metabolism might lead to a selective cell death of most vulnerable nerve cells and represent the primary subject to be studied by DNA analysis. Furthermore, protein aggregation is likely to be a major cause for the disease. Recently it has been shown that α-synuclein is accumulated in Lewy bodies of sporadic PD and mutated in some rare families with an autosomal dominant trait of the disease (ADPD). The identification of further genes responsible for PD will subsequently lead to first insights into the pathogenesis of one of the most common neurodegenerative disorders in humans.

Parkinson's disease (PD) is one of the most common neurological disorders in an aged population and was first described by James Parkinson in 1817. The clinical manifestations of this neurodegenerative disorder include resting tremor, muscular rigidity, bradykinesia, and postural instability. A relatively specific pathological feature accompanying the neuronal degeneration is an intracytoplasmic inclusion body, known as the Lewy body, which is found in many brain regions, including the substantia nigra, locus ceruleus, nucleus basalis, hypothalamus, cerebral cortex, cranial nerve motor nuclei, and the central and peripheral divisions of the autonomic nervous system. Recent findings indicate that the clinical diagnosis of PD does concur with that of pathologically proven PD in about 75% of cases (Hughes et al., 1992, 1993).

The clinical diagnosis of PD is complicated by numerous diseases presenting with akinetic-rigid features (Table 1). In multiple system atrophy, Steele-Richardson-Olszewski syndrome, diffuse Lewy body disease, and intracranial neoplasms, parkinsonism might be pure at the time of presentation. Wilson's

Table 1. Hereditary diseases with Parkinsonism

Disease	Inheritance	Chromosomal localization	Protein function
Parkinson's disease	AD	4q	α-Synuclein
Juvenile parkinsonism	AR	6q	Gene unknown
Frontotemporal dementia and parkinsonism (FTDP)	AD	17q	Gene unknown
Dystonia-parkinsonism of filipinos (Lubag)	X	Xq	Gene unknown (DYT3)
Dopamin-sensitive dystonia	AD	14q	GTP Cyclohydrolase I
Wilson's disease	AR	13q	Cu-binding ATPase
Spinocerebellar ataxia type 3 / MJD	AD	14q	Protein function unknown
Huntington's disease (Westphal's variant)	AD	4p	Protein function unknown

AD autosomal dominant
AR autosomal rezessive
SCA3/MJD Spinocerebellar ataxia Typ3 / Machado-Joseph disease
DRPLA Dentato-rubro-pallido-luysian atrophy

disease should be excluded in patients presenting with an akinetic-rigid syndrome under age of 50. Furthermore, parkinsonism is known to be caused by drugs, toxins as manganese or MPTP, or develops postencephalitic and may mimic idiopathic PD. Parkinsonism is also seen after repeated head injury, particularly in boxers. In all these conditions failure to respond to levodopa is an important diagnostic clue.

The pathogenesis of idiopathic PD is not known, but is thought to be multifactorial, deriving from environmental factors acting on genetically predisposed individuals with aging. This is supported by the common prevalence of the disease ranging between 1.4% among 55 years old and 3.4% among 75 years old. PD either occurs in a common sporadic form or it can be inherited in a much rarer, autosomal dominant form (ADPD).

For a long period a genetic contribution to PD was discussed controversial as twin studies failed to demonstrate increased concordance for PD in monozygotic (MZ) versus dizygotic twins (DZ). However, based on limitations of their designs these studies are not reliable. New studies on concordance rate in twins are based on the highly sensitive estimation of ^{18}fluorodopa uptake into basal ganglia to confirm preclinical PD. The concordance rates were 45% in MZ and 28% in DZ twins (Burn et al., 1992). Repeated clinical investigation during several years would most likely further increase the conclusions drawn from twin studies. However, one should keep in mind that "classical" PD is most likely complex and with the exception of rare families not caused by a defect of a single gene. Polygenic inheritance or multifactorial models are proposed. These models assume that several gene loci as well as various biological and environmental factors (age, sex, immunology, infections, diet, etc.) jointly contribute to PD.

Allelic associations

The overwhelming majority of the PD patients is sporadic as the family history does not indicate further affected individuals in their first degree relatives. However, there is evidence for autosomal dominant inheritance with reduced penetrance in an increasing amount of families. If it holds true that genetic factors play a major role in the pathogenesis of PD one could assume that the disease has been originated in a single or a few founders in a certain population. In this case, the genetic background (DNA sequence) should be similar in all offsprings of the founder(s) which can be investigated by the analysis of DNA polymorphisms. In these studies one compares the frequency of the alleles in a group of PD patients with these of an unaffected control population. One has to ascertain that the control group has the same population characteristics as the patient group, both ethnically and geographically. Major limitations of association studies are (i) the lack of a matched control population, (ii) the small number of analyzed persons and (iii) the analysis of a single polymorphism instead of several loci in the neighborhood of a candidate gene.

Mitochondrial defects in PD

The oxidative phosphorylation consists of 5 enzyme complexes. Four enzymes (complexes I-IV) transport electrons from NADH or succinate to oxygen and pump protons out of the mitochondria to form an electrochemical gradient. The fifth enzyme (complex V) uses the electrochemical gradient to synthesize adenosine triphosphate (ATP) from adenosine diphosphate (ADP). These 5 enzymes are assembled from 13 polypeptides coded by the mitochondrial DNA (mtDNA) and approximately 50 polypeptides coded by the nuclear DNA (nDNA). The mtDNA codes for 7 complex I (NADH:ubiquinone oxidoreductase) subunits, 1 complex III (ubiquino:cytochrome c oxidoreductase) subunit, 3 complex IV (cytochrome c oxidase) subunits, and 2 complex V (ATP synthetase) subunits. In addition, the mtDNA codes 2 ribosomal RNAs (12S and 16S) and 22 transfer RNAs necessary for mtDNA gene expression.

It has been shown that toxins that inhibit oxidative phosphorylation, such as the complex I inhibitor MPTP, produce disorders in which the motor manifestations are similar to those of PD. Recent investigations have shown defects of the mitochondrial oxidative phosphorylation in platelets, muscle, and substantia nigra from PD patients. These observations suggest that defects of the oxidative phosphorylation may have an important role in the pathogenesis of PD. As only a few of the subunits that constitute complex I are encoded by the mitochondrial genome, absence of maternal transmission of PD does not exclude a role for mitochondrial dysfunction in the aetiology of this neurodegenerative disease. However, it is still unclear whether the observed reduction in complex I is a primary or secondary effect. Mitochondrial polymorphisms (mutations) have been reported including a $tRNA^{gln}$ (bp4336), a missense at bp3397, a 9bp insertion in the 12S rRNA and a point mutation at bp 5460. All of these DNA changes have been found in a slightly greater number in samples from patients compared with controls. However, the analysis of 100 brain samples for these polymorphisms showed no significant difference between controls and PD. Therefore there is no convincing evidence for a pathological role of these polymorphisms in the aetiology of PD.

Xenobiotic metabolism

The interest in possible exogenous causes of PD was stimulated by the discovery that exposure to the pyridine derivate MPTP can result in parkinsonism with dopaminergic cell loss in the substantia nigra. However, at least 400 people are knwon to have self-administered MPTP, but only a few have developed symptoms of parkinsonism. Therefore, a genetic predisposition may exist that increases susceptibility to the neurotoxic effects of MPTP and possibly to other toxins, such as organochlorines and alkylated phosphates (Seidler et al., 1996). The presynaptic dopamine transporter in nigral dopaminergic neurons confers susceptibility to the cytotoxic effects of the neurotoxic metabolite of MPTP. Polymorphisms in this or other genes as debrisoquine hydroxylase and monoamine oxidase B might make patients with PD more vulnerable to these toxins by the inability to sufficiently inactivate endogenous or exogenous toxins, leading to increased free radical stress or inhibition of mitochondrial function (Fig. 1). These enzymes influence the toxicity of MPTP. MPP+, the neurotoxic metabolite of MPTP, targets dopaminergic neurons because it is taken up selectively by the presynaptic dopamine transporter. Expression of the dopamine-transporter gene in nonneuronal cell cultures confers MPP+ uptake and toxicity. Mutations in the mouse dopamine-transporter gene, generated by site-directed mutagenesis, increased both the uptake velocity and affinity of MPP+ with little effect on the transport of dopamine (Kitayama et al., 1993). Similar mutations in the human dopamine-transporter gene have not been reported yet. However, an association between a polymorphic repeat motif in the 3'-untranslated region of the dopamine-transporter gene and PD has been found recently (LeCouteur et al., 1997). Indications for the involvement of other detoxification enzymes have been identified. Bandmann et al. (1997) found an association of slow acetylator genotype for N-acetyltransferase 2 with familial PD. N-acetyltransferase 2 catalyses the transfer of an acetyl group from the cofactor acetyl coenzyme A to the amine nitrogen atom of aromatic amines and hydrazines. The degree of activity of this enzyme determines the rate of detoxification of aromatic amines. No association has been found for CYP2D6, CYP2E1, NAD(P)H-menadione reductase (NQ1), and glutathione transferases M1 and T1 (Bandmann et al., 1997).

Oxidative stress

As summarized in the hypothesis of oxidative stress the formation of free radicals during the enzymatic- or autooxidation of dopamine in nigrostriatal

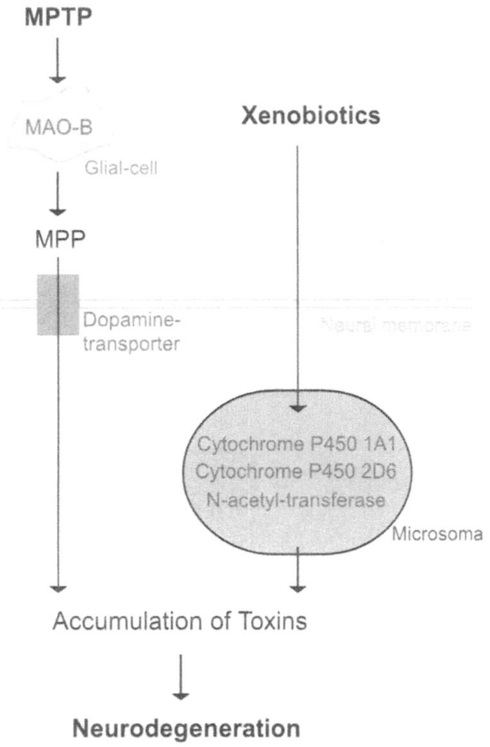

Fig. 1. A schematic representation of the MPTP and xenobiotica metabolism. MPTP is metablized into toxic MPP+ which in turn is transported into the cell via the dopamine transporter. Accumulation of toxins in the cell is also possible by alteration of the detoxication process

neurons is the point of depart for mitochondrial damage and neuronal degeneration. Under normal conditions these cytotoxic radicals are rapidly inactivated by radical-scavengers like Glutathione-peroxidase (GPX), Catalase (CAT), Superoxide-dismutase (SOD) and heme-oxygenase-1 (HO-1) (Jellinger, 1993; Kimpara et al., 1997). An imbalance caused by increased dopamine metabolism or deficient protective mechanisms can lead to lipid-peroxidation, membrane damaging, inhibition of complex I of the respiratory chain and finally degeneration of dopaminergic neurons (Jellinger, 1993) (Fig. 2). The formation of free radicals is catalyzed by increased tissue-levels of iron, found in the SN of PD brains (Jenner et al., 1994). A reduced activity of GPX can be observed in the substantia nigra (SN) of PD patients (Jenner et al., 1994). Linkage studies in three families with ADPD with polymorphic markers in the genes for GPX, CAT and SOD failed to delineate an involvement of these genes in the etiology of PD (Gasser et al., 1994).

A recent study using a bi-allelic polymorphism in the promoter of the *nitric oxide synthase* (*NOS*) gene showed a significant difference in the allelic distribution between PD and control group (Kurth et al., 1997). Because nitric oxide (NO) is a short-lived free radical, these data support the hypothesis of an involvement of oxygen free-radical metabolism in PD.

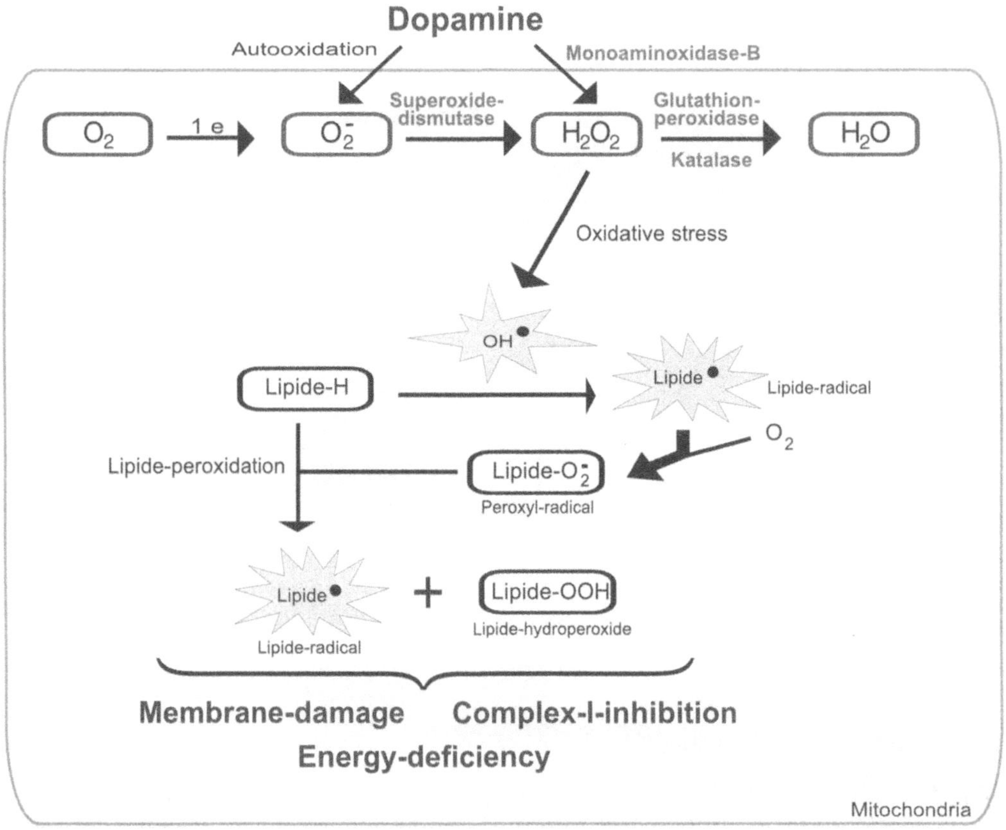

Fig. 2. Dopamine producing cells are most vulnerable to oxidative stress due to the autooxidation and enzymatic metabolism into free radicals which might lead to lipide-peroxidation and subsequently to membrane-damage, complex-I-inhibition and energy-depletion

The dopaminergic neurotransmission pathway in PD

Neuropathologically, PD is primarily due to the decrease of dopamine (DA) in the nigrostriatal pathway. Therefore dopamine transmission, dopamine transporters and dopamine receptors are considered as possible factors in the pathophysiology of PD. Since the discovery that levodopa treatment could alleviate the symptoms of PD it has been assumed that the striatum is the site of action of dopamine formed from levodopa. However, it is now apparent that levodopa can be converted to dopamine and activate dopamine receptors in other parts of the brain as well. Due to the high homology of the dopamine receptors it is technically difficult to determine their region specific expression in the brain. To understand the pathogenesis of PD and the action of levodopa in PD, the effects of dopamine must be understood in the anatomical context of receptor localization. In this regard it is most interesting to know that mice lacking dopamine D2 receptors develop Parkinsonian-like locomotor impairment (Balk et al., 1995). These mice have therapeutic potential as a model for investigating and correcting dysfunctions of the dopaminergic system.

DA receptors are classified into the D1 family (D1 and D5) and the D2 family (D2, D3 and D4) according to their pharmacological characteristics (Seeman et al., 1994). For dopamine receptors D2, D3, D4 and D5, polymorphic markers have been described, making these genes accessible for association studies (Table 2). A recent study was able to show a significant difference in the allelic distribution between PD and a control group using an intronic marker in the D2 gene (Planté-Bordeneuve et al., 1997). This result suggests an involvement of the D2 receptor in the pathogenesis of PD. In contrast, no association was found for the D3, D4 and D5 receptors, respectively (Nanko et al., 1994).

The dopamine transporter (DAT) is responsible for the dopamine re-uptake into presynaptic nerve terminals. The level of expression of DAT is markedly reduced in the substantia nigra of PD patients (Harrington et al., 1996). In addition DAT is involved in the selective uptake of MPP+, the active neurotoxic metabolite of MPTP, into dopaminergic neurons, conferring to a Parkinsonian syndrome similar to PD (Tipton et al., 1993). Using a polymorphic marker in the 3′ untranslated region of the *DAT* gene a recent study identified an allele, which was associated with a tenfold increase in the risk of PD (LeCouteur et al., 1997). Tyrosine hydroxylase (TH) catalyzes the conversion of tyrosine to levodopa, the rate-limiting step for the synthesis of dopamine. Decreased levels of expression could be revealed in surviving dopaminergic neurons in PD brains (Javoy-Agid et al., 1990). Neither association studies in sporadic PD nor linkage studies in familial PD were able to demonstrate a role of TH in the pathogenesis of PD (Planté-Bordeneuve et al., 1994; Gasser at al., 1994).

The dopamine metabolizing enzymes Catechol-o-methyltransferase (COMT) and Monoaminoxidase A and B (MAO A and B) are also discussed as candidates in the etiology of PD. MAO is involved in catabolic reactions. The B-isoenzyme produces free radicals and hydrogen peroxide (H_2O_2) through the oxidative metabolism of dopamine resulting in lipid-peroxidation and mitochondrial damage (Riederer et al., 1989). Furthermore MAO B is able to convert the neurotoxin MPTP into its active metabolite MPP+ (Langston et al., 1984). Some analogues of MPTP are selectively metabolized into the toxic form by MAO A (Heikkila et al., 1988). To date two studies were able to show an allelic association between PD and a two-allele-polymorphism in intron 13 of the MAO B gene (Kurth et al., 1993; Costa et al., 1997). The fact, that several studies are completely contradictory in finding an association with different alleles, underline the limitations of association studies.

Role of other candidate genes in the pathogenesis of PD

Brain-derived neurotrophic factor sustains dopaminergic neurons in vitro and in vivo following cytotoxic insults in a rat model of PD (Hyman et al., 1991; Frim et al., 1994). Neurotensin (NT)-receptors are synthezised in neurons in the SN and expressed in their perikarya and the terminal regions

Table 2. Association and linkage studies in sporadic and familiar PD

Candidate gene	Localisation	Sporadic PD association	ADPD linkage
Dopaminergic neurotransmission:			
Dopamine-D2-receptor	11q22–23	Planté-Bordeneuve et al., 1997; Nanko et al., 1994; Comings et al., 1991	
Dopamine-D3-receptor	3q13.3	Nanko et al., 1994	
Dopamine-D4-receptor	11p15.5	Nanko et al., 1994	
Dopamine-D5-receptor	4p15.1–3		
Dopamine transporter	5p15.3	Le Couteur et al., 1997; Planté-Bordeneuve et al., 1997	
Tyrosine hydroxylase	11p15.5	Planté-Bordeneuve et al., 1997	Gasser et al., 1994
Dopamine-β-hydroxylase	9q34.3		
Catechol-o-methyltransferase	22q12.1	Xie et al., 1997; Hoda et al., 1996	
Monoaminoxidase A	X	Hotamisligil et al., 1994; Kurth et al., 1993	
Monoaminoxidase B	Xp11.3	Costa et al., 1997; Kurth et al., 1993; Ho et al., 1995; Morimoto et al., 1995	
Oxidative stress:			
Glutathion-peroxidase	3q11		Gasser et al., 1994
Katalase	11p13		Gasser et al., 1994
Superoxiddismutase	21q21		Bandmann et al., 1995; Gasser et al., 1994
Heme oxigenase-1	22q12	Kimpara et al., 1997	
Nitric-oxide-synthase	17cen-17q11	Kurth et al., 1997	
Xenobiotika-metabolism:			
Debrisoquine-4-hydroxylase	22q13	Wilhelmsen et al., 1997; Armstrong et al., 1992; Smith et al., 1992; Diederich et al., 1996; Gasser et al., 1996	Gasser et al., 1996; Gasser et al., 1994; Planté-Bordeneuve et al., 1994; Mazzetti et al., 1994
Cytochrome P450IA1	5q22–24	Takakubo et al., 1996	
N-acetyl-transferase		Bandmann et al., 1997	
Further candidates:			
GABA-β1-receptor	4p12–13		
Neurotensin-receptor	20q13		
Brain-derived neurotrophic factor	11p14		Gasser et al., 1994
Amyloid-precursor	21q21		Gasser et al., 1994
alpha-synuclein	4q21.3–22		Polymeropoulos et al., 1997

(Yamada et al., 1995). Significant NT-receptor decreases were found in the SN of PD patients in post mortem brain tissues (Chinaglia et al., 1990). Gamma-amino-butyric acid (GABA) mediates most inhibition in the central nervous system. Changes in the striatal GABA function are likely to contribute to the basal ganglia dysfunction and thus may be related with certain aspects of parkinsonian symptomatology (Kish et al., 1986). A decreased density of GABA receptors has been observed in cortical and subcortical regions of PD brains (Nishino et al., 1988).

Mutations in the amyloid precursor protein (APP) have been identified as resposible for some cases of familial early-onset Alzheimer disease (AD) (Schellenberg, 1995). Because of some links in the neuropathology between AD and PD like e.g. the coexistence of pathological characteristics for PD, the Lewy-bodies, in AD, and of neurofibrillary tangles and amyloid plaques in brains of PD patients, it seems to be worthwhile to consider APP as a candidate gene also in PD (Schmidt et al., 1996).

Juvenile parkinsonism

An autosomal recessive form of juvenile parkinsonism is a levodopa-responsive parkinsonism whose pathological finding is a highly selective degeneration of dopaminergic neurons in the zona compacta of the substantia nigra. The disease has been reported mainly in the Japanese population under different names such as autosomal recessive form of juvenile Parkinsonism, autosomal recessive early-onset parkinsonism with diurnal fluctuation, or a familial form of juvenile parkinsonism. Characteristic clinical features include juvenile-onset (<40 years of age in most patients) of typical parkinsonism with rigidity, resting and/or postural tremor, bradykinesia, postural imbalance, a slow and protracted course, superb response to levodopa, frequent and early occurence of dopa-induced dyskinesia and wearing-off phenomenon, amelioration of symptoms after sleep (sleep benefit), and mild foot dystonia. Extensive reduction of tyrosine hydroxylase activity in the substantia nigra of the brain of affected individuals was reported and is quite similar to the finding in PD. Neuropathological features resemble that of PD but Lewy body formation is missing. The gene of ARJP has been localized to human chromosome 6q25.2–27 close to the SOD2 gene which has finally been excluded as the disease candidate by sequence analysis in affected individuals.

α-Synuclein and its involvement in ADPD

Based on common clinical and pathological findings in PD and Alzheimer's disease (AD), another multifactorial genetic disorder in older persons, these neurodegenerative conditions are likely to result in part from similar pathogenic processes (Uitti and Calne, 1993). For instance, PD patients often develop dementia, and AD patients frequently develop parkinsonian signs. In addition, the age at onset of PD as well as of AD is modulated by apolipoprotein E genotypes (Zareparsi et al., 1997). Also, neuropathologically there are some features in common in both diseases. The protein alpha-synuclein was originally identified as the precursor protein for the non-β amyloid component of Alzheimer's disease amyloid plaques (Uéda et al., 1993) but has now been identified as a major component of the Lewy bodies in PD (Spillantini et al., 1997).

Recently, a gene for one form of ADPD has been mapped to human chromosome 4q21–22 and subsequently an alanine-to-threonine amino-acid substitution at position 53 (A53T) has been identified in a gene encoding the aforementioned α-synuclein. This mutation is segregating with the disease in all but one of the patients in an Italian-American kindred (known as the Contursi family) and in another three apparently unrelated Greek families. Surprisingly, in mouse, rat, and canary threonine is present at position 53 of α-synuclein. This raises the possibility that the A53T substitution represents a rare polymorphism and in its entire consequence that α-synuclein is not involved in the pathogenesis of PD. Screening more than 200 German PD patients, we identified an alanine-to-proline missense mutation at position 30 of the α-synuclein gene in a family with ADPD (Krüger et al., 1998). This substitution occurs in a repeat motif which is highly conserved in animals and in the synuclein gene family. This base pair substitution has not been found in 400 chromosomes of age matched healthy German controls and in another 75 DNA samples of human brains without Lewy body histopathology. As the A30P mutation co-segregates with disease in the family we conclude that it is causative for PD in this family. Together, these results suggest that α-synuclein participates in the pathogenesis of some forms of ADPD. Linkage to chromosome 4q21–23 has been excluded for a total of 107 Caucasian families with ADPD (Scott et al., 1997, Gasser et al., 1997) giving evidence that only a small proportion of those PD families with autosomal dominant inheritance of the disease will be caused by mutations in the α-synuclein gene. Also, the A53T mutation was not detected in our sample of 240 patients with the sporadic form of PD indicating a high degree of complexity and heterogeneity causing Parkinson's disease.

Proteins or fragments of proteins that are mutated in familial neurodegenerative disorders frequently accumulate as deposits that characterize the neuropathology of these disorders. In this respect, it has been shown recently that α-synuclein, but not its related protein β-synuclein, is a main component of the Lewy body in PD (Spillantini et al., 1997). Lewy bodies are a characteristic pathological hallmark of PD and dementia with Lewy bodies (DLB). Besides neurofilaments, ubiquitin is another component of the Lewy bodies and thought to degrade missfolded proteins of naturally short-lived cytoplasmic proteins. A major effect of the known mutations may be to promote the aggregation of α-synuclein into filaments, resulting in the formation of Lewy bodies. Alpha-synuclein aggregation and Lewy-body formation may be important in the aetiology and pathogenesis of all cases of PD.

Conclusion

In the last few years, the genetic contribution to PD has gained major attention and resulted in the identification of the first gene, called α-synuclein, involved in the pathogenesis of the disease. It has been demonstrated that α-synuclein is a major component of Lewy bodies, the neuropathological feature of PD. The elucidation of polygenic changes in the dopamine pathway, mitochondrial dysfunction, and of xenobiotics metabolism is technically now possible by means of association and genotype studies. The increasing knowledge on the pathogenesis of PD at a molecular level will subsequently have important implications for the development of individual therapeutic strategies to intervene in disease progression.

References

Balk J-H, Picetti R, Salardi A, Thiriet G, Dierich A, Depauis A, Le Meur M, Borrelli E (1995) Parkinsonian-like locomotor impairment in mice lacking dopamine D2 receptors. Nature 377: 424–428

Bandmann O, Vaughan J, Holmans P, Marsden CD, Wood NW (1997) Association of slow acetylator genotype for N-acetyltransferase 2 with familial Parkinson's disease. Lancet 350: 1136–1139

Burn DJ, Mark MH, Playford ED, et al (1992) Parkinson's disease in twins studied with ^{18}F-Dopa and positron emission tomography. Neurology 42: 1894–1900

Chinaglia G, Probst A, Palacios JM (1990) Neurotensin receptors in Parkinson's disease and progressive supranuclear palsy: an autoradiografic study in basal ganglia. Neuroscience 39(2): 351–360

Costa P, Chekoway H, Levy D, Smith-Weller T, Franklin GM, Swanson PD, Costa LG (1997) Association of a polymorphism in intron 13 of the monoamine oxidase B gene with Parkinson disease. Am J Med Gen 74: 154–156

Frim DM, Uhler TA, Galpern WR, Beal MF, Breakefield XO, Isacson O (1994) Implanted fibroblasts genetically engineered to produce brain-derived neurotrophic factor prevent 1-methyl-4-phenylpyridinium toxicity to dopaminergic neurons in the rat. Proc Natl Acad Sci USA 91: 5104–5108

Gasser T, Wszolek ZK, Trofatter J, Ozelius L, Uitti RJ, Lee CS, Gusella J, Pfeiffer RF, Calne DB, Breakfield XO (1994) Genetic linkage studies in autosomal dominant Parkinsonism: evaluation of seven candidate genes. Ann Neurol 36: 387–396

Gasser T, Müller-Myhsok B, Wszolek ZK, et al (1997) Genetic complexity and Parkinson's disease. Science 277: 388–389

Golbe LI, Di Iorio G, Bonavita V, Miller DC, Duvoisin RC (1990) A large kindred with adPD. Ann Neurol 27: 276–282

Golbe LI (1993) A large kindred with PD: onset age, segregation ratios, and anticipation. Mov Disord 8: 406

Harrington KA, Augood SJ, Kingsbury AE, Foster OJ, Emson PC (1996) Dopamine transporter (DAT) and synaptic vesicle amine transporter (VMAT2) gene expression in the substantia nigra of control and Parkinson's disease. Mol Brain Res 36(1): 157–162

Heikkila RE, Kindt MV, Sonsalla PK (1988) Importance of monoamine oxidase A in the bioactivation of neurotoxic analogs of 1-methyl-4-phenyl-11,2,5,6-tetrahydropyridine. Proc Natl Acad Sci USA 85: 6172–6176

Hughes AJ, Daniel SE, Kilford L, Lees AJ (1992) Accuracy of clinical diagnosis of idiopathic Parkinson's disease: a clinicopathologic study of 100 cases. J Neurol Neurosurg Psychiatry 55: 181–184

Hughes AJ, Daniel SE, Blankson S, Lees AJ (1993) A clinicopathologic study of 100 cases of Parkinson's disease. Arch Neurol 50: 140–148

Hyman C, Hofer M, Barde YA (1991) BDNF is a neurotrophic factor for dopaminergic neurons of the substantia nigra. Nature 350: 230–232

Javoy-Agid G, Hirsch EC, Dumas S (1990) Decreased tyrosine hydroxylase messenger RNA in the surviving dopamine neurons of the substantioa nigra in Parkinson's disease: an in situ hybridization study. Neuroscience 38: 245–253

Jellinger KA (1993) Pathogenese und Pathophysiologie der Parkinson-Krankheit. Neuropsychiatrie 7(1): 29–37

Jenner P, Schapira AHV, Marsden CD (1992) New insights into the cause of Parkinson's disease. Neurology 42: 2241–2250

Kimpara T, Takeda A, Watanabe K, Itoyama Y, Ikawa S, Watanabe M, Arai H, Sasaki H, Higuchi S, Okita N, Takase S, Saito H, Takahashi K, Shibahara S (1997) Microsatellite polymorphism in the human heme oxygenase-1 promoter and its application in association studies with Alzheimer and Parkinson disease. Hum Genet 100: 145–147

Kish SJ, Rajput A, Gilbert J, Rozdilsky B, Chang LJ, Shannak K, Hornykiewicz O (1986) Elevated gamma-aminobutyric acid level in striatal but not extrastriatal brain regions in Parkinson's disease: correlation with striatal dopamine loss. Ann Neurol 20(1): 26–31

Kitayama S, Wang J-B, Uhl GR (1993) Deopamine transporter mutants selectively increase MPP+ transport. Synapse 15: 58–62

Krüger R, Kuhn W, Müller Th, Woitalla D, Graeber M, Kösel S, Przuntek H, Epplen JT, Schöls L, Riess O (1998) Ala30Pro mutation in the gene encoding α-synuclein in Parkinson's disease. Nat Genet 18: 106–108

Kurth JH, Kurth MC, Poduloso SE, Schwankhaus JD (1993) Association of a monoamine oxidase B allele with Parkinson's disease. Ann Neurol 33: 368–372

Kurth JH, Eggers-Sedlet B, Liebermann A, Kurth MC (1997) Mov Disord 12(5): 835

Langston JW, Irwin I, Langston EB, Forno LS (1984) Pargyline prevents MPTP induced parkinsonism in primates. Science 225: 1480–1482

LeCouteur DG, Leighton PW, McCann SJ, Pond SM (1997) Association of a polymorphism in the dopamine-transporter gene with Parkinson's disease. Mov Disord 12: 760–763

Nishino N, Fujiwara H, Nogushi-Kuno SA, Tanaka C (1988) GABAA receptor but not muscarinic receptor density was decreased in the brain of patients with Parkinson's disease. Jpn J Pharmacol 48(3): 331–339

Planté-Bordeneuve V, Davis MB, Maraganore DM, Marsden CD, Harding AE (1994) Tyrosine hydroxylase polymorphism in familial an sporadic Parkinson's disease. Mov Disord 9: 337–339

Planté-Bordeneuve V, Taussig B, Thomas F, Said G, Wood NW, Marsden CD, Harding AE (1997) Evaluation of four candidate genes encoding proteins of the dopamine pathway in familial and sporadic Parkinson's disease. Neurology 48: 1589–1593

Polymeropoulos M, Lavedan C, Leroy E, et al (1997) Mutation in the α-synuclein gene identified in families with Parkinson's disease. Science 276: 2045–2047

Riederer P, Konradi C, Hebenstreit G, Youdim MB (1989) Neurochemical perspectives to the function of monoamine oxidase. Acta Neurol Scand [Suppl] 126: 41–45

Schellenberg GD (1995) Genetic dissection of Alzheimer disease, a heterogeneous disorder. Proc Natl Acad Sci USA 92: 8552–8559

Schmidt ML, Martin JA, Lee VMY, Trojanowski JQ (1996) Convergence of Lewy bodies and neurofibrillary tangles in amygdala neurons of Alzheimer's disease and Lewy body disorders. Acta Neuropathol 91: 475–481

Scott WK, Stajich JM, Yamaoka LH, Speer MC, Vance JM, Roses AD, Pericak-Vance MA, The Deane Laboratory Parkinson Disease Research Group (1997) Genetic complexity and Parkinson's disease. Science 277: 387–388

Seeman P, Van Tol HMM (1994) Dopamine receptor pharmacology. TIPS 15: 264–270
Seidler A, Hellenbrand W, Robra B-P, Vieregge P, Nischan P, Joerg J, Oertel WH, Ulm G, Schneider E (1996) Possible environmental, occupational and other ecological factors for Parkinson's disease: a case control study in Germany. Neurology 46: 1275–1284
Spillantini MG, Schmidt ML, Lee VM-Y, Jakes R, Goedert M (1997) α-Synucelin in Lewy bodies. Nature 388: 839–840
Tipton KG, Singer TP (1993) Advances in our understanding of the mechanisms of neurotoxicity of MPTP and related compounds. J Neurochem 61: 1191–1206
Uèda K, et al (1993) Proc Natl Acad Sci USA 90: 11282–11286
Uitti RJ, Calne DB (1993) Pathogenesis of idiopathic parkinsonism. Eur Neurol 33: 6–23
Yamada M, Yamada M, Richelson E (1995) Heterogeneity of melanized neurons expressing neurotensin receptor messenger RNA in the substantia nigra and the nucleus paranigralis of control and Parkinson's disease brain. Neuroscience 64(2): 405–417
Zareparsi S, Kaye J, Camicioli R, Grimslid H, Oken B, Litt M, Nutt J, Bird T, Schellenberg G, Payami H (1997) Modulation of the age at onset of Parkinson's disease by apolipoprotein E genotypes. Ann Neurol 42: 655–658

Authors' address: Olaf Riess, M.D., Molecular Human Genetics, Ruhr-University, Universitätsstrasse 150, D-44780 Bochum, Federal Republic of Germany

Mechanism and consequences of nerve cell death in Parkinson's disease

E. C. Hirsch

INSERM U 289, Hôpital de la Salpêtrière, Paris, France

Summary. The etiology of Parkinson's disease remains unknown, making it difficult to develop therapeutical approaches to stop the progression of the disease. The best known treatment to date is based on the use of L-DOPA or dopaminergic agonists. These are merely substitutive therapies and have limitations because of their side effects. Thus, the development of new therapeutical strategies will require a far better knowledge of the mechanism and the consequences of nerve cell death in Parkinson's disease.

Parkinson's disease is characterized by a selective vulnerability of subpopulations of dopaminergic neurons in the mesencephalon. The fact that the neurons which degenerate in Parkinson's disease are already sensitive to oxidative stress in control subjects and the reported increased production of oxygen free radicals in Parkinson's disease suggest that oxidative stress may be involved in the mechanism of nerve cell death. Furthermore, oxygen free radicals are also involved in an oxygen-dependent pro-apoptotic pathway stimulated by the inflammatory reaction observed in Parkinson's disease. These data suggest that anti-oxidant or anti-inflammatory treatments may slow down the progression of the disease.

On the other hand, new substitutive therapies may be developed by trying to restore the activity of the neurons located downstream from the nigrostriatal pathway. Indeed, the nigrostriatal denervation induces a hyperactivity of the output structures of the basal ganglia (internal segment of the globus pallidus and substantia nigra pars reticulata), as demonstrated in various animal models of the disease. These changes in the activity of the output structures of the basal ganglia seem to be directly induced by the hyperactivity of the glutamatergic afferent fibers from the subthalamic nucleus. The fact that L-DOPA treatment or a reduction in the activity of the subthalamic nucleus alleviate the symptoms of the disease and restore the activity of the output structures of the basal ganglia in parkinsonism suggests that these structures play a key role in the pathophysiology of the disease and could represent a potential therapeutic target.

Although Parkinson's disease was first described in the nineteenth century its cause remains a mystery. Recent evidence suggests the involvement of a mutation in the gene coding for a protein called α synuclein (Polymeropoulos et al., 1997) suggesting the contribution of a genetic factor in the etiology of the disease. Yet, familial cases represent only a small proportion of all cases and environmental toxic compounds are also strongly suspected of participating in the etiology of the disease (Barbeau et al., 1985). This absence of a definitive explanation of the etiology of the disease probably explains why no curative treatment for Parkinson's disease has yet been devised.

The clinical manifestation of the disease is mainly due to the loss of the dopaminergic neurons of the nigrostriatal pathway. The most commonly used treatment for Parkinson's disease is aimed at restoring the concentrations of dopamine in the striatum using L-DOPA or dopaminergic agonists. Yet, this therapy is only substitutive and presents side effects such as ON/OFF phenomena or dyskinesia which are painful for the patients (Marconi et al., 1994). Thus, there is a clear need for other therapeutical approaches designed to slow the progression of the disease or correct the symptoms without inducing side effects. In this context, the discovery of new therapeutical targets probably depends on a better understanding of 1) the mechanism by which dopaminergic neurons degenerate and 2) the consequences of the loss of the dopaminergic neurons for the neuronal circuits located downstream to the lesions. Indeed, a better understanding of the mechanism of nerve cell death should allow the development of therapeutical tools aimed at stopping the progression of the disease, whereas an understanding of the consequences of nerve cell death should allow the development of drugs aimed at relieving the symptoms. These two aspects will be successively analysed in this review.

Mechanism of nerve cell death in Parkinson's disease

The cardinal neurochemical characteristic of Parkinson's disease is a loss of dopaminergic neurons in the substantia nigra. Yet, in the mesencephalon of patients with this disease not all dopaminergic neurons degenerate to the same extent (Hirsch et al., 1988; Damier et al., 1997). This suggests that some neurons are more susceptible to the disease than others and implies that the identification of phenotypical differences in the control mesencephalon between the neurons which are susceptible to the disease and those which are not may provide clues to the mechanism of nerve cell death. Interestingly, the most susceptible neurons have been shown to be particularly sensitive to oxidative stress (for review, see Hirsch, 1993). Such neurons produce more oxygen free radicals than those which are less susceptible to the disease; judging from the fact that they contain neuromelanin (Hirsch et al., 1988), the synthesis of which is known to be associated with the production of oxygen free radicals (Graham, 1979). Furthermore, they are less well protected against oxygen free radicals, as they are located in an environment containing few glial cells expressing glutathione peroxidase (Damier et al., 1993), an enzyme that catabolizes hydrogen peroxide, thus preventing the formation of highly toxic hydroxyl radicals. Therefore, because the neurons most sensitive to the disease are poorly protected against an increase in the level of oxygen free radicals, it is tempting to speculate that oxidative stress participate in the mechanism of nerve cell death in Parkinson's disease.

Several lines of evidence suggest an involvement of oxidative stress in Parkinson's disease. For exemple, alterations of lipids, proteins and nucleic acids due to oxygen free radicals have been reported in the substantia nigra of patients with Parkinson's disease (Alam et al., 1997a; Alam et al., 1997b). Yet, the origin of the rise in oxygen free radical production during the evolution of the disease is still unknown. It may involve several mechanisms that are not mutually exclusive and which may potentiate each other.

1) Iron, a metal which catalyzes the formation of hydroxyl radicals through the Fenton reaction, accumulates directly within the dopaminergic neurons in Parkinson's disease (Hirsch et al., 1991; Jellinger et al., 1992; Good et al., 1992).
2) Alterations in mitochondrial functions (deficiency in complex I activity) leading to the accumulation of oxygen free radicals have been evidenced in the substantia nigra of patients with Parkinson's disease (Mizuno et al., 1989; Schapira et al., 1990).
3) Increased manganese-dependent superoxide dismutase activity has been reported in the substantia nigra of patients with Parkinson's disease (Saggu et al., 1996), probably leading to an accumulation of hydrogen peroxide.
4) Glutathione levels are decreased in the substantia nigra of patients with Parkinson's disease (Perry et al., 1982; Owen et al., 1996) and even in presymptomatic cases of Parkinson's disease (Owen et al., 1996), thus reducing the efficacy of defense mechanisms against oxygen free radicals.
5) Nitric oxide levels are probably also increased in the substantia nigra of patients with Parkinson's disease (Hunot et al., 1996), leading by combination with superoxide to the formation of $ONOO^-$ which is highly toxic. This idea should, however, be treated with caution, since both neuroprotective and toxic roles have been described for nitric oxide in the central nervous system (Chao et al., 1993; Dawson et al., 1991; Dawson et al., 1994; Demerlé-Pallardy et al., 1993; Wink et al., 1993).

Taken together, these data suggest a major alteration in the metabolism of oxygen free radicals that may participate in the mechanism of nerve cell death in Parkinson's disease. Yet, oxygen free radicals may also be involved in signal transduction pathways that mediate apoptosis one of which uses the sphingomyelin- or ceramide-dependent signaling pathway. This pathway, first described in the immune system (Kolesnick et al., 1994; Hannun et al., 1995), has recently also been shown to induce apoptosis in neurons (Brugg et al., 1996). This signaling system may be of interest with respect to the pathophysiology of Parkinson's disease, because it can be activated by tumor necrosis factor α (TNFα) (Hannun et al., 1995; Kolesnick et al., 1996), which has been detected in microglial cells in the substantia nigra of patients with Parkinson's disease in the vicinity of dopaminergic neurons expressing receptor for this cytokine (Boka et al., 1994). The involvement of this oxidant-mediated apoptogenic pathway in degeneration of dopaminergic neurons was recently analyzed in vitro in primary cultures of rat mesencephalon (Hunot et al., 1997). Activation of the pathway was shown to produce successively 1) an increased cytoplasmic content of oxygen free radicals, 2) translocation of transcription factor NFκB from the cytoplasm to the nucleus and 3) neuronal degeneration by an apoptotic process. Furthermore, the production of oxygen free radicals was necessary for the activation of this transduction pathway, since the addition of N-acetylcysteine, which prevents an increase in the level of free radicals, abolished the translocation of NFκB and blocked the development of apoptosis. Whether such a pathway is also activated in Parkinson's disease cannot be directly verified post-mortem. However, the 70-fold increase in NFκB translocation in the dopaminergic neurons of patients with Parkinson's disease and the presence of cells displaying the characteristic morphological features of apoptosis argue strongly in favor of TNFα having such an effect (Hunot et al., 1997). More generally, as the number of glial cells expressing other cytokines, such as interleukin-1β or interferon γ, is also increased in the substantia nigra (Hunot et al., 1997), an inflammatory reaction could be involved in the degeneration of dopaminergic neurons (for review, see Hirsch et al., 1997). Furthermore, since these cytokines act synergistically to induce the inflammatory reaction and the production of nitric oxide, which regulates iron homeostasis, they may represent a key target for the development of new therapeutical strategies aimed at slowing the neuronal loss in Parkinson's disease.

While oxidative stress and inflammatory reactions have also been described in other brain disorders (for review see Beal et al., 1996 or Hopkins and Rothwell, 1995), they would appear to be crucially important in Parkinson's disease, where the neurons most susceptible to the disease are particularly sentitive to these factors. Therapy based on preventing the formation of oxygen free radicals or an development of the inflammatory reaction could reduce the progression of the disease. However, whatever the drug used, this would imply specifically targeting the most vulnerable neurons in the substantia nigra. Other potential therapeutic strategies include the delivery of neurotrophic factors or anti-apoptotic compounds within the most vulnerable regions of the brain.

Consequences of nerve cell death in Parkinson's disease

Another potential therapeutic strategy for Parkinson's disease, albeit substitutive, is to try to normalize the activity of the neuronal circuits located downstream from the nigrostriatal pathway which are likely to be altered by the reduction in striatal dopamine content. However, this would require a better knowledge of the consequences of nigrostriatal degeneration.

The loss of dopaminergic neurons in the substantia nigra pars compacta results in decreased dopamine concentrations in the striatum (Hornykiewicz, 1960). These reduced dopamine concentrations provoke a cascade of functional changes in the circuitry of the basal ganglia, which are thought to participate in the development of the motor parkinsonian syndromes (Albin and Young, 1989; Crossman, 1989; Delong, 1990; Wichmann and Delong, 1993). Nigrostriatal denervation, experimentally produced by intoxication with 1-methyl-4-phenyl-1,2,3,6-tetrahydropyridine (MPTP) in non-human primates or 6-hydroxydopamine-induced lesioning of midbrain dopaminergic neurons in the rat, leads to an overactivity of the output structures of the basal ganglia, namely the internal segment of the pallidum and the substantia nigra pars reticulata (GPi/SNpr). This hyperactivity of the GPi/SNpr may, in turn, modify the activity of the thalamocortical pathway and participate in the clinical picture of parkinsonism. The hyperactivity of the output stations of the basal ganglia is now well documented in studies using various techniques, such as electrophysiological recordings (Wichmann and Delong, 1993), metabolic studies using the 2-deoxyglucose method (Mitchell et al., 1989) and, more recently, biochemical analysis. Most of the biochemical studies were based on estimates of the cellular expression of the messenger RNA coding for the 67 kDA isoform of glutamic acid decarboxylase (GAD_{67} mRNA), the synthetic enzyme of GABA and an indirect marker of GABAergic activity since the output structures of the basal ganglia use GABA as neurotransmitter (Soghomonian and Chesselet, 1992; Chesselet et al., 1993; Herrero et al., 1996; Vila et al., 1996a). Another biochemical marker has been introduced to study the changes in the overall metabolic activity in these structures after nigrostriatal denervation in 6-hydroxydopamine-lesioned rats (Porter et al., 1994) and MPTP-intoxicated monkeys (Vila et al., 1996b). This method is based on a regional analysis of the histochemical activity of cytochrome oxidase (CO), the terminal enzyme in the mitochondrial electron transport chain, several studies having indicated that CO activity is directly linked to neuronal activity (for review, see Wong-Riley, 1989). Such a methodology allows an analysis of global changes in the functional activity of neurons, even when markers of the neurotransmission are not available. Using this technique we were able to confirm the hyperactivity of the output structures of the basal ganglia in monkeys rendered parkinsonian by MPTP intoxication (Vila et al., 1996). However, the relative contributions to regional cytochrome oxidase activity of cell bodies and dendrites of intrinsic neurons or terminals of afferent neurons could not be determined. More recently, we used cytochrome oxidase mRNA measurements to perform similar studies (Vila et al., 1997). Unlike cytochrome oxidase enzymatic activity, which is expressed in neuronal perikarya dendrites and terminals, the mRNA coding for both nuclear and mitochondrial subunits of cytochrome oxidase is concentrated mainly in neuronal cell bodies (Hevner and Wong-Riley, 1991) and is a good marker of overall neuronal activity. Using quantitative in situ hybridization of the mRNA coding for cytochrome oxidase mitochondrial-derived subunit I we were able to confirm the increased activity of the internal segment of the globus pallidus and the substantia nigra pars reticulata (Vila et al., 1997). Taken together, all these studies performed using various techniques confirm the hyperactivity of the output station of the basal ganglia in parkinsonian syndromes. Furthermore, these structures are very likely involved in the clinical manifestation of the disease. Thus, restoration of dopamine concentrations using L-DOPA in MPTP-intoxicated monkeys alleviated the clinical manifestation and in the same time reversed the biochemical changes observed in the output structures of the basal ganglia (Vila et al., 1996a; Herrero et al., 1996; Vila et al., 1996b; Vila et al., 1997). However, since the monkeys developed severe dyskinesia, it is likely that even if the overall activity of basal ganglia output structures was globally restored, qualitative alterations in GABAergic neurotransmission may have persisted in these structures.

According to the current model of basal ganglia circuitry, it is likely that hyperactivity of basal ganglia output structures is due to the combined effects of a reduced inhibitory input from the striato-pallido/nigral GABAergic pathway and a hyperactivity of the subthalamo-pallido/nigral glutamatergic neurons (Grossman et al., 1988; Mitchel et al., 1989). Hyperactivity of the subthalamic neurons after nigrostriatal denervation has been confirmed using several techniques such as electrophysiology (Mitchel et al., 1989; Bergman et al., 1994) or cytochrome oxidase activity or mRNA expression (Vila et al., 1996b; Vila et al., 1997). These data suggest that the hyperactivity of the subthalamic nucleus plays a major role in the pathophysiology of parkinsonism and that it represents an effective therapeutic target. Such a concept is supported by several animal studies. A reduction in the activity of subthalamic neurons by injection of excitatory amino acid kynurenic acid-antagonist in reserpinized rats or MPTP-intoxicated monkeys alleviates in a dose-dependent manner the symptomatology induced by the dopaminergic hypoactivity (Graham et al., 1990; Brotchie et al., 1991). Furthermore, the destruction of this nucleus was shown to produce similar effects in monkeys (Bergman et al., 1990; Aziz et al., 1991; Aziz et al., 1992), this effect being associated with restoration of the activity of basal ganglia output structures, measured biochemically (Guridi et al., 1996). More recently, high frequency stimulation applied to the subthalamic nucleus was found to induce an alleviation of rigidity and bradykinesia both in MPTP-intoxicated monkeys (Bennazouz et al., 1993) and patients with Parkinson's disease (Limousin et al., 1995).

Taken together these data indicate that reducing the subthalamic nucleus hyperactivity could represent an effective therapeutic strategy in Parkinson's disease. This suggests that glutamate receptor antagonists acting in the internal segment of the globus pallidus and the substantia nigra pars reticulata might be used to alleviate motor and perhaps other symptoms of the disease. It has recently been shown that one subunit of the AMPA receptor was predominantly expressed in the neuropil of the pallidum, suggesting that antagonists of this receptor may represent the best pharmacological agent (Bernard et al., 1996). In favor of this hypothesis, AMPA receptors antagonists have been shown to alleviate motor symptomatology in animal models of Parkinson's disease produced by a unilateral 6-hydroxydopamine lesion of the substantia nigra in rats or MPTP intoxication in monkeys (Brotchie et al., 1991; Klockgether et al., 1991; Stan, 1995). However, the possibility that other glutamate receptor antagonists acting on NMDA-receptors could be used with similar effects cannot be ruled out.

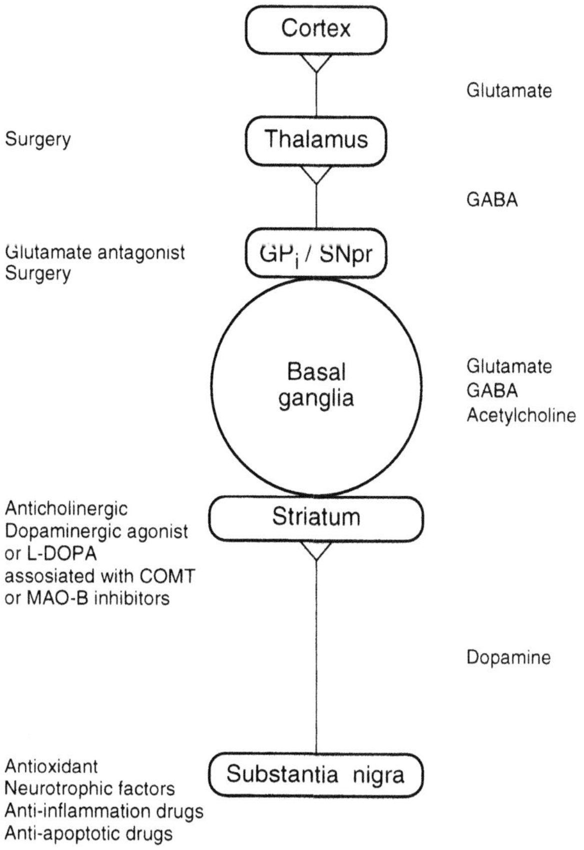

Fig. 1. Diagram illustrating the major targets for treatment of Parkinson's disease. Neurotransmitters involved in neuronal circuitry are shown on the right, some potential targets on the left

Conclusion

The data reviewed here, show that a better understanding of the basic knowledge concerning the mechanism and the consequences of nerve cell death in Parkinson's disease can lead to new therapeutical strategies (Fig. 1). In this context, besides L-DOPA or dopaminergic agonists, anti-oxidants, anti inflammatory and antiglutamate drugs may be of interest. Yet, specific drugs have still to be developed, and targeting problems be solved, before such compounds could be used in the clinic.

References

Alam ZI, Daniel SE, Lees AJ, Marsden CD, Jenner P, Halliwell B (1997a) A generalised increase in protein carbonyls in the brain in Parkinson's but not incidental Lewy body disease. J Neurochem 69: 1326–1329

Alam ZI, Jenner A, Daniel SE, Lees AJ, Cairns N, Marsden CD, Jenner P, Halliwell B (1997b) Oxidative DNA damage in the parkinsonian brain: an apparent selec-

tive increase in 8-hydroxyguanine levels in substantia nigra. J Neurochem 69: 1196–1203
Albin RL, Young AB, Penney JB (1989) The functional anatomy of basal ganglia disorders. Trends Neurosci 12: 366–375
Aziz TZ, Peggs D, Sambrook MA, Crossman AR (1991) Lesions of the subthalamic nucleus for the alleviation of 1-methyl-4-phenyl-1,2,3,6-tetrahydropyridine (MPTP)-induced parkinsonism in the primate. Mov Disord 4: 288–292
Aziz TZ, Peggs D, Agarwal E, Sambrook MA, Crossman AR (1992) Subthalamic nucleotomy alleviates parkinsonism in the 1-methyl-4-phenyl-1,2,3,6-tetrahydropyridine (MPTP)-exposed primate. Br J Neurosurg 6: 575–582
Barbeau A, Roy M (1985) Identification of environemental related foci of Parkinson's disease in the Province of Quebec. Ann Neurol 18: 138
Beal MF (1996) Mitochondria, free radicals and neurodegeneration. Curr Opin Neurobiol 6: 661–666
Benazzouz A, Gross C, Feger J, Boraud T, Bioulac B (1993) Reversal of rigidity and improvement in motor performance by subthalamic high-frequency stimulation in MPTP-treated monkeys. Eur J Neurosci 5: 382–389
Bergman H, Wichmann T, DeLong MR (1990) Reversal of experimental parkinsonism by lesions of the subthalamic nucleus. Science 249: 1436–1438
Bergman H, Wichmann T, Karmon B, DeLong MR (1994) The primate subthalamic nucleus II. Neuronal activity in the MPTP model of parkinsonism. J Neurophysiol 72: 507–520
Bernard V, Gardiol A, Faucheux B, Bloch B, Agid Y, Hirsch EC (1996) Expression of glutamate receptors in the human and rat basal ganglia: effect of the dopaminergic denervation on AMPA receptor gene expression in the striatopallidal complex in Parkinson's disease and rat with 6-OHDA lesion. J Comp Neurol 368: 553–568
Boka G, Anglade P, Wallach D, Javoy-Agid F, Agid Y, Hirsch EC (1994) Immunocytochemical analysis of tumor necrosis factor and its receptors in Parkinson's disease. Neurosci Lett 172: 151–154
Brotchie JM, Mitchell IJ, Sambrook MA, Crossman AR (1991) Alleviation of parkinsonism by antagonism of excitatory aminoacid transmission in the medial segment of the globus pallidus in rat and primate. Mov Disord 6: 133–138
Brugg B, Michel PP, Agid Y, Ruberg M (1996) Ceramide induces apoptosis in cultured mesencephalic neurons. J Neurochem 66: 733–739
Chao CC, Hu S, Molitor TW, Shaskan EG, Peterson PK (1992) Activated microglia mediate neuronal cell injury via a nitric oxide mechanism. J Immuno 149: 2736–2741
Chesselet MF, Mercugliano M, Soghomonian JJ, Salin P, Qin Y, Gonzales C (1993) Regulation of glutamic acid decarboxylase gene expression in efferent neurons of the basal ganglia. Prog Brain Res 99: 143–154
Crossman AR (1989) Neural mechanisms in disorders of movement. Comp Biochem Physiol 93: 141–149
Damier P, Hirsch EC, Zhang P, Agid Y, Javoy-Agid F (1993) Glutathione peroxidase, glial cells and Parkinson's disease. Neuroscience 52: 1–6
Damier P, Hirsch EC, Agid Y, Graybiel AM (1997) Temporospatial progression of the loss in dopaminergic neurons in the substantia nigra in Parkinson's disease. Mov Disord 12 [Suppl 1]: 274
Dawson VL, Dawson TM, London ED, Bredt DS, Snyder SH (1991) Nitric oxide mediates glutamate neurotoxicity in primary cortical cultures. Proc Natl Acad Sci USA 8: 6368–6371
Dawson VL, Brahmbhatt HP, Mong JA, Dawson TM (1994) Expression of inducible nitric oxide synthase causes delayed neurotoxicity in primary mixed neuronal-glial cortical cultures. Neuropharmacology 33: 1425–1430

DeLong MR (1990) Primate models of movement disorders of basal ganglia origin. Trends Neurosci 13: 281–285

Demerlé-Palladry C, Lonchampt MO, Chabrier PE, Braquet P (1993) Nitric oxide synthase induction in glial cells: effect on neuronal survival. Life Sci 52: 1883–1890

Good PF, Olanow CW, Perl DP (1992) Neuromelanin-containing neurons of the substantia nigra accumulate iron and aluminum in Parkinson's disease: a LAMMA study. Brain Res 593: 343–346

Graham DG (1979) On the origin and significance of neuromelanin. Arch Pathol Lab Med 103: 359–362

Graham WC, Robertson RG, Sambrook MA, Crossman AR (1990) Injection of excitatory amino acid antagonists into the medial pallidal segment of a (MPTP) 1-methyl-4-phenyl-1,2,3,6-tetrahydropyridine treated primate reverses motor symptoms of parkinsonism. Life Sci 47: PL-91–7

Guridi J, Herrero MT, Luquin MR, Guillen J, Ruberg M, Laguna J, Vila M, Javoy-Agid F, Agid Y, Hirsch EC, Obeso JA (1996) Subthalamotomy in parkinsonian monkeys. Behavioural and biochemical analysis. Brain 119: 1717–1727

Hannun YA, Obeid LM (1995) Ceramide: an intracellular signal for apoptosis. Trends Biochem Sci 20: 73–77

Herrero MT, Levy R, Ruberg M, Luquin MR, Villares J, Guillen J, Faucheux B, Javoy-Agid F, Guridi J, Agid Y, Obeso JA, Hirsch EC (1996) Consequence of nigrostriatal denervation and L-DOPA therapy on the expression of glutamic acid decarboxylase (GAD) messenger RNA in the pallidum. Neurology 47: 219–224

Hevner RF, Wong-Riley MTT (1991) Neuronal expression of nuclear and mitochondrial genes for cytochrome oxidase (CO) subunits analyzed by in situ hybridization: comparison with CO activity and protein. J Neurosci 11: 1942–1958

Hirsch EC (1993) Does oxidative stress participate in nerve cell death in Parkinson's disease? Eur Neurol 33: 52–59

Hirsch EC, Graybiel AM, Agid Y (1988) Melanized dopaminergic neurons are differentially susceptible to degeneration in Parkinson's discase. Nature 334: 345–348

Hirsch EC, Brandel JP, Galle P, Javoy-Agid F, Agid Y (1991) Iron and aluminium increase in the substantia nigra of patients with Parkinson's disease. An X-ray microanalysis. J Neurochem 56: 446–451

Hirsch EC, Hunot S, Damier P, Faucheux B (1998) Glial cells and the inflammatory reaction in Parkinson's disease: a role in neurodegeneration? Ann Neurol 44 [Suppl]: S115–S120

Hopkins SJ, Rothwell NJ (1995) Cytokines and the nervous system I: expression and recognition. Trends Neurosci 18: 83–88

Hornykiewicz O (1963) Die topische Lokalisation und das Verhalten von Noradrenalin und Dopamin (3-Hydroxytyramin) in der Substantia Nigra des normalen und parkinsonkranken Menschen. Wien Klin Wochenschr 75: 309–321

Hunot S, Boissière F, Faucheux B, Brugg B, Mouatt-Prigent A, Agid Y, Hirsch EC (1996) Nitric oxide synthase and neuronal vulnerability in Parkinson's disease. Neuroscience 72: 355–363

Hunot S, Brugg B, Ricard D, Michel PP, Muriel MP, Ruberg M, Faucheux BA, Agid Y, Hirsch EC (1997a) Nuclear translocation of NF-kB is increased in dopaminergic neurons of patients with Parkinson's disease. Proc Natl Acad Sci USA 94: 7531–7536

Hunot S, Betard C, Faucheux B, Agid Y, Hirsch EC (1997b) Immunohistochemical analysis of interferon-γ and interleukin-lβ in the substantia nigra of Parkinsonian patients. Mov Disord 12 [Suppl 1]: 20

Jellinger K, Kienzl E, Rumpelmair G, Riederer P, Stachelberger H, Ben-Schachar D, Youdim MBH (1992) Iron-melanin complex in substantia nigra of parkinsonian brain: an X-ray microanalysis. J Neurochem 59: 1168–1171

Klockgether T, Turski L (1990) NMDA antagonists potentiate antiparkinsonian actions of L-DOPA in monoamine-depleted rats. Ann Neurol 28: 539–546

Kolesnick R, Golde DW (1994) The sphingomyelin pathway in tumor necrosis factor and interleukin-1 signaling. Cell 77: 325–328

Limousin P, Pollak P, Benazzouz A, Hoffman D, Le Bras J-F, Broussolle E, et al (1995) Effect of parkinsonian signs and symptoms of bilateral subthalamic nucleus stimulation. Lancet 345: 91–95

Marconi R, Lefebvre-Caparros D, Bonnet AM, Vidailhet M, Dubois B, Agid Y (1994) Levodopa-induced dyskinesias in Parkinson's disease phenomenology and pathophysiology. Mov Disord 9: 2–12

Mitchell IJ, Clarke CE, Boyce S, Robertson RG, Peggs DE, Sambrook MA, Crossman AR (1989) Neural mechanisms underlying parkinsonian symptoms based upon regional uptake of 2-deoxyglucose in monkeys exposed to 1-methyl-4-phenyl-1,2,3,6-tetrahydropyridine. Neuroscience 32: 213–226

Mizuno Y, Ohta S, Tanaka M, Takamiya S, Susuki K, Sato T, Oya H, Ozawa Y, Kagawa Y (1993) Deficiencies in complex I subunits of the respiratory chain in Parkinson's disease. Biochem Biophys Res Commun 163: 1450–1455

Owen AD, Schapira AH, Jenner P, Marsden CD (1996) Oxidative stress and Parkinson's disease. Ann N Y Acad Sci 786: 217–223

Perry TL, Godin DV, Hansen S (1982) Parkinson's disease: a disorder due to nigral glutathione deficiency? Neurosci Lett 33: 305–310

Polymeropoulos MH, Lavedant C, Leroy E, Ide SE, Dehejia A, Dutra A, Pike B, Root H, Rubenstein J, Boyer R, Stenroos ES, Chandrasekharappa S, Athanassiadou A, Papapetropoulos T, Johnson WG, Lazzarini AM, Duvoisin RC, Di Iorio G, Golbe LI, Nussbaum RL (1997) Mutation in the α-synuclein gene identified in families with Parkinson's disease. Science 276: 2045–2047

Porter RHP, Greene JG, Higgins Jr DS, Greenamyre JT (1994) Polysynaptic regulation of glutamate receptors and mitochondrial enzyme activities in the basal ganglia of rats with unilateral dopamine depletion. J Neurosci 14: 7192–7199

Saggu H, Cooksey J, Dexter D, Welles FR, Lees AJ, Jenner P, Marsden CD (1989) A selective increase in particulate superoxide dismutase activity in parkinsonian substantia nigra. J Neurochem 53: 692–697

Schapira AHV, Cooper JM, Dexter D, Clark JB, Jenner P, Marsden CD (1990) Mitochondrial complex I deficiency in Parkinson's disease. J Neurochem 54: 823–827

Soghomonian JJ, Chesselet MF (1992) Effects of nigrostriatal lesions on the levels of messenger RNAs encoding two isoforms of glutamate decarboxylase in the globus pallidus and entopeduncular nucleus of the rat. Synapse 11: 124–133

Starr MS (1995) Glutamate/dopamine D1/D2 balance in the basal ganglia and its relevance to Parkinson's disease. Synapse 19: 264–293

Vila M, Herrero MT, Levy R, Faucheux B, Ruberg M, Guillen J, Luquin MR, Guridi J, Javoy-Agid F, Agid Y, Obeso JA, Hirsch EC (1996a) Consequence of nigrostiatal denervation on the GABAergic neurons of substantia nigra pars reticulata and superior colliculus in parkinsonian syndromes. Neurology 46: 802–809

Vila M, Levy R, Herrero MT, Faucheux B, Obeso JA, Agid Y, Hirsch EC (1996b) Metabolic activity of the basal ganglia in parkinsonian syndromes in human and nonhuman primates: a cytochrome oxidase histochemistry study. Neuroscience 71: 903–912

Vila M, Levy R, Herrero MT, Ruberg M, Faucheux B, Obeso JA, Agid Y, Hirsch EC (1997) Consequences of nigrostriatal denervation on the functioning of the basal ganglia in human and nonhuman primates: an in situ hybridization study of cytochrome oxidase subunit I mRNA. J Neurosci 17: 765–773

Wichmann T, Bergman H, DeLong MR (1994) The primate subthalamic nucleus. I. Functional properties in intact animals. J Neurophysiol 72: 494–506

Wink DA, Hanbauer I, Krishna MC, DeGraff W, Gamson J, Mitchell JB (1993) Nitric oxide protects against cellular damage and cytotoxicity from reactive oxygen species. Proc Natl Acad Sci USA 90: 9813–9817

Wong-Riley MTT (1989) Cytochrome oxidase: an endogenous metabolic marker for neuronal activity. Trends Neurosci 12: 94–101

Author's address: Prof. Dr. Etienne C. Hirsch, Recherche CNRS, Hôpital de la Sâlpêtrière, 47 boulevard de l'Hôpital, F-75013 Paris Cedex 13, France

Functional imaging of Parkinson's disease: is it possible to detect brain areas for specific symptoms?

D. J. Brooks

MRC Cyclotron Unit, Imperial College School of Medicine, Hammersmith Hospital, and Institute of Neurology, London, United Kingdom

Summary. $H_2{}^{15}O$ PET activation studies enable the brain systems involved in controlling different aspects of motor function to be defined. In Parkinson's disease (PD) freely chosen limb movements are performed slowly. This bradykinesia is associated with selective underactivity of the supplementary motor area and dorsal prefrontal cortex, frontal association areas that receive subcortical input principally from the basal ganglia. At the same time there is compensatory overactivity of the lateral premotor and parietal cortex, areas that have a primary role in facilitating motor responses to visual and auditory cues. This finding explains why PD patients find it easier to perform cued as opposed to freely chosen actions. Levels of activation of the supplementary motor area and dorsal prefrontal cortex in PD can be restored with dopaminergic medication, implants of fetal mesencephalic tissue, internal pallidotomy or high frequency electrical subthalamic stimulation.

Activation studies suggest that Parkinsonian rest tremor arises from a combination of inappropriate overactivity of cerebellar connections and loss of dopaminergic function. When tremor is relieved by ventral thalamotomy or thalamic stimulation this cerebellar overactivity is corrected but at the expense of reducing levels of primary motor cortex activation.

It has been hypothesised that dyskinesias in PD arise due to altered dopamine receptor binding following chronic exposure to levodopa stimulation. Functional imaging findings, however, are against this hypothesis and rather suggest that downstream increases in basal ganglia opioid neurotransmission are more likely to be relevant.

Introduction

Functional imaging provides a sensitive means of detecting and characterising regional changes in brain blood flow, metabolism and receptor binding in Parkinson's disease. There are three main approaches to functional imaging: Positron emission tomography (PET) allows quantitative examination of regional cerebral blood flow (rCBF), glucose and oxygen metabolism (rCMRGlc, $rCMRO_2$), and receptor binding. Single photon emission tomography (SPECT) also provides measurements of rCBF and receptor binding though is less sensitive than PET. Magnetic resonance imaging and spectroscopy can be used to detect cerebral activation and metabolite changes. This review concentrates on PET findings but compares SPECT data where relevant.

There are two basic approaches to determining the changes in cerebral function associated with PD: First, abnormalities in the patterns of resting levels of regional cerebral blood flow, metabolism, and neuroreceptor binding can be examined. Second, by monitoring levels of rCBF, abnormal patterns of brain activation associated with the presence of tremor and dyskinesias or performance of motor tasks can be detected.

Akinesia

Functional anatomy of motor control

When normal subjects make arm movements in freely selected directions $H_2{}^{15}O$ PET can detect associated rCBF increases in the contralateral sensorimotor cortex (SMC) and basal ganglia while there is bilateral activation of anterior cingulate, mesial premotor cortex — also known as the supplementary motor area (SMA), lateral premotor cortex (PMC), and dorsal prefrontal cortex (DPFC) (Playford et al., 1992). Self-paced extensions of the index finger result in a similar pattern of activation (Jahanshahi et al., 1995). The dorsal prefrontal cortex is not activated, however, if finger extension is cued in an unpredictable fashion by an auditory tone. This suggests that the DPFC has a primary responsibility for motor decision making.

When novel sequences of finger movements are being learned (for example playing piano music for the first time) the DPFC is strongly activated but it is not required when prelearned sequences of finger movement are performed automatically (such as playing familiar piano music) (Jenkins et al., 1994). The DPFC is also activated during problem solving tasks such as the Tower of London puzzle (Owen et al., 1996). These findings suggest that the DPFC also has a primary responsibility for motor explicit learning and planning.

The SMA appears to behave as separate rostral and caudal functional regions. Both caudal and rostral SMA are activated when the direction and timing of arm movements are freely chosen or can be prepared in advance but only caudal SMA is activated when movements are cued in an unpredictable fashion (Jenkins et al., 1996). If increasingly complex prelearned sequences of finger movements are performed but finger movements are kept at a constant rate, rostral SMA becomes increasingly activated while caudal SMA blood flow remains at a steady level (Boecker et al., 1996). These findings suggest that the rostral SMA has a particular role in preparing limb movements and in facilitating complex sequential patterns of movement while the caudal SMA facilitates prepared movement patterns.

When PD patients off dopaminergic medication perform joystick movements in freely chosen directions in response to an auditory tone their response time is about 20% longer than normal. If they are scanned with $H_2{}^{15}O$ PET during performance of this task, or while making self-paced index finger extensions, the SMA and DPFC are both underactive while the primary motor and lateral premotor cortex both activate normally (Playford et al., 1992; Jahanshahi et al., 1995). Similarly, if PD patients are scanned during performance of the Tower of London problem solving task underactivity of prefrontal areas is evident (Dagher et al., 1997). Inability to activate the SMA and DPFC in PD explains the increased time required by these patients when deciding or preparing volitional movements (akinesia) or solving problems and their slowness when performing sequences of limb movements (bradykinesia) (Benecke et al., 1987; Dick et al., 1989).

Selective impairment of SMA and DLPFC activation in PD is in line with current models of basal ganglia connectivity. The medial globus pallidus (GPM) is the main output relay of the basal ganglia and sends inhibitory projections to the SMA, arcuate premotor, and dorsal prefrontal areas via the ventral thalamus. Loss of striatal dopamine leads to disinhibition of GPM by both direct (striatum — internal pallidum) and indirect (striatum — external pallidum — subthalamus — internal pallidum) pathways resulting in excessive pallidal GABAergic output (Penney Jr and Young, 1986; Alexander et al., 1990; Crossman, 1990).

While the SMA and DLPFC receive their major subcortical input from the basal ganglia, the lateral parietal and premotor areas are targeted by cerebellar projections. If SMA and dorsal prefrontal activation is impaired in PD, it might be predicted that these akinetic patients would adapt by making greater use of cerebellar-lateral parietal-lateral premotor connections. Overactivity of this circuit has indeed been demonstrated in PD when patients perform sequential finger movements with one or both hands (Samuel et al., 1997a; Rascol et al., 1997) see Fig. 1. The lateral parietal and premotor areas are thought to preferentially facilitate cued rather than freely chosen movements (Passingham, 1987). The ability of PD patients to overactivate these lateral cortical areas helps to explain why they are better able to perform visually cued as opposed to freely chosen actions.

Functional effects of treatment

In order to demonstrate that underfunctioning of the SMA and DLPFC is responsible for akinesia in untreated PD, it is necessary to show that effective treatment selectively restores their levels of activation. Apomorphine, a combined dopamine D_1 and D_2 agonist, improves the response times of PD patients when they perform paced joystick movements in freely chosen directions. This improvement of akinesia is associated with significant increases in both SMA and DLPFC activation while levels of primary motor and lateral premotor cortex blood flow remain unaltered (Jenkins et al., 1992; Brooks et al., 1993). Striatal implantation of fetal dopaminergic cells also selectively

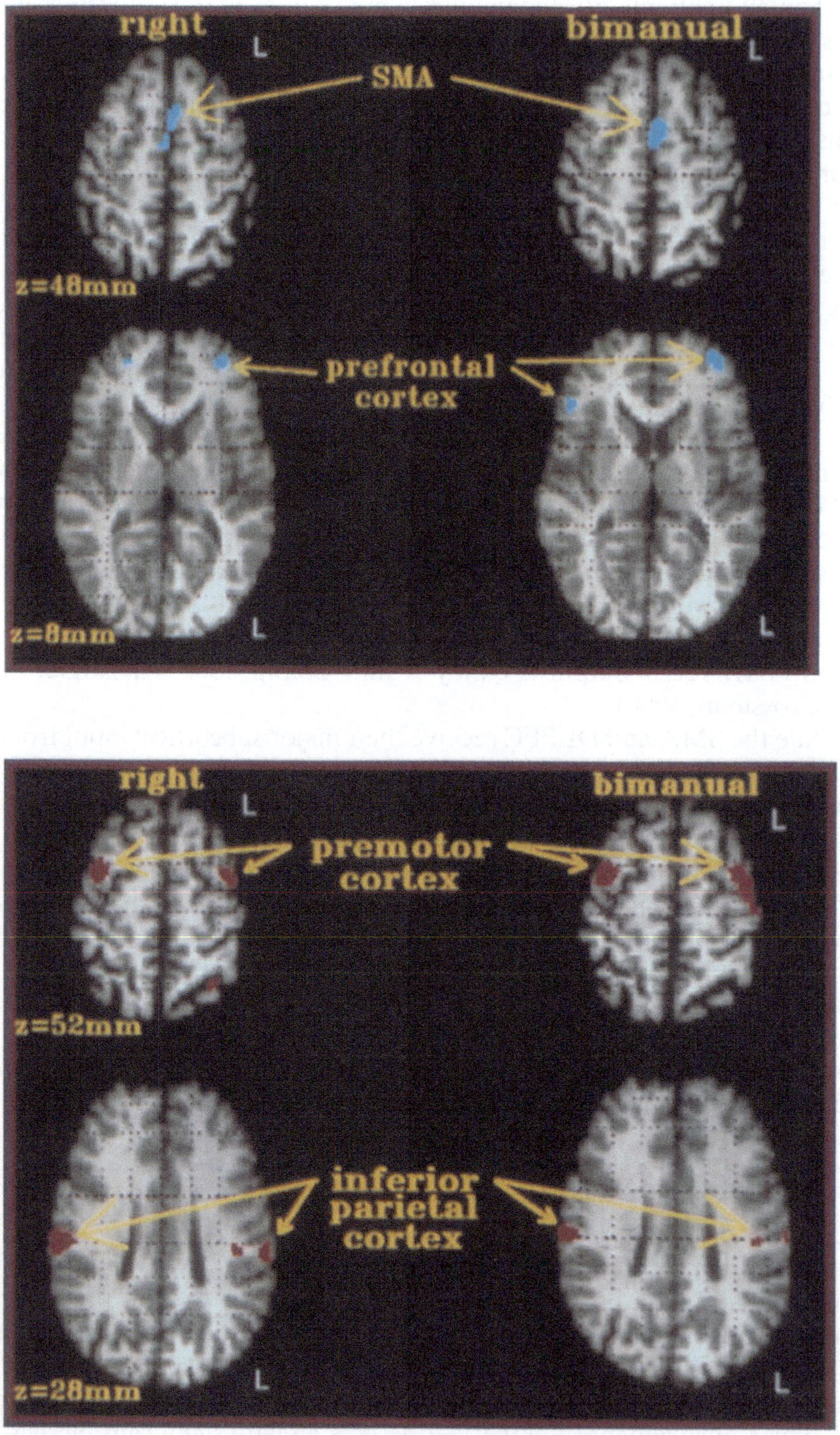

Fig. 1a,b. Images showing areas of statistically reduced (blue) prefrontal and SMA and increased (red) lateral premotor and parietal activation in a group of PD patients performing sequential bimanual finger movements paced at one movement every 3 seconds (photos courtesy of M. Samuel), $p < 0.01$

increases levels of SMA activation during freely chosen limb movements in PD (Ceballos-Baumann et al., 1996). These findings confirm that SMA and DLPFC activation is required for efficient generation of motor programmes and that their impaired function results in akinesia.

A number of clinical series have now established that medial pallidotomy dramatically reduces levodopa-induced abnormal involuntary movements in chronically medicated PD patients (Laitinen et al., 1992; Dogali et al., 1995; Baron et al., 1996; Samuel et al., 1998; Fazzini et al., 1997). Pallidotomy also improves "off" bradykinesia, tremor, and rigidity scores by 30–40% when rated with the Unified Parkinson's Disease Rating Scale (UPDRS). High frequency electrical pallidal stimulation has also been reported to be efficacious (Arcusa et al., 1996). The mechanisms underlying these therapeutic responses, however, are still unclear.

The rationale underlying pallidotomy is that the loss of striatal dopamine in PD results in reduced inhibition of the GPM and intra-operative single-cell recordings in patients have confirmed the presence of increased burst firing (Hutchison et al., 1994; Sterio et al., 1994). It is argued that lesioning the motor GPM with either stereotactic thermocoagulation or high frequency electrical stimulation reduces the excessive inhibition of the ventral thalamus and frontal projection areas and so facilitates movement in PD.

There have been two published PET activation series concerning the functional effects of unilateral medial pallidotomy. Significant post-operative increases in activation of SMA, lateral premotor cortex, and dorsal prefrontal cortex were reported for a group of six PD patients studied off medication while performing paced joystick movements in freely selected directions (Samuel et al., 1997b) — Fig. 2. These patients showed concommitant significant post-operative improvements in dyskinesia score (75%), "off" contralateral wrist rigidity (83%) and contralateral bradykinesia (56%) when assessed with the UPDRS. A second PET activation pallidotomy study involved six PD patients who were scanned off medication at rest and then while reaching out to grasp different lighted targets arranged in a row at 3 second intervals (Grafton et al., 1995). Pallidotomy resulted in no overall improvement in disability but performance of the motor task resulted in increased levels of caudal SMA and ventral lateral premotor activation after surgery. Despite the lack of clinical improvement, levels of SMA activation before and after pallidotomy correlated with reaction times in this patient group.

One PET report, to date, has demonstrated a beneficial effect of high freqency electrical pallidal stimulation on regional brain function in PD. This study (Davis et al., 1997) measured levels of resting rCBF with the stimulator switched off, switched on at a sub-effective intensity, and switched on at an effective intensity. Clinically effective GPM stimulation improved contralateral bradykinesia and rigidity when patients were rated off medication and this was associated with improved levels of resting SMA and lentiform nucleus rCBF. Stimulation of the GPM at a low intensity did not result in either clinical improvement or increased SMA rCBF.

Current models for basal ganglia connectivity suggest that the subthalamic nucleus (STN) acts to excite the neurones of the GPM via glutamatergic

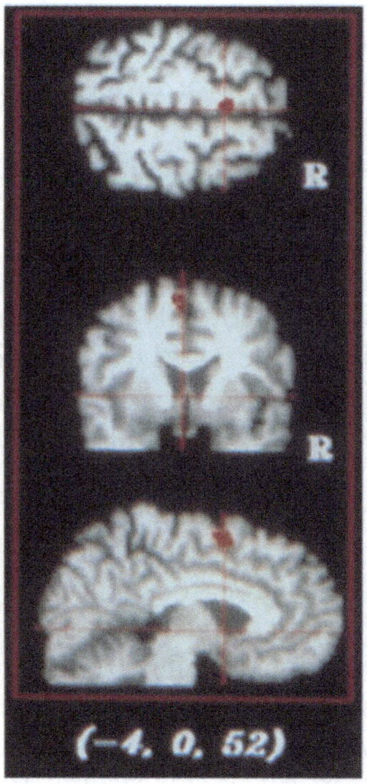

Fig. 2. MRI surface renderings of the cerebral hemispheres in standard stereotactic space for a group of six PD patients before and after right medial pallidotomy with the PET activation rCBF increases superimposed. The red areas show voxels in the rostral SMA that are significantly more activated ($p < 0.01$) after pallidotomy while patients perform paced joystick movements in freely chosen directions. (photos courtesy of M. Samuel)

projections. As a consequence, while the excessive inhibitory output of the GPM can be directly reduced by medial pallidotomy, an alternative approach is to reduce the excitatory input to GPM from the subthalamic nucleus via high-frequency electrical stimulation. This approach has been shown to result in dramatic relief of bradykinesia and rigidity though appears to be less effective than pallidotomy in relieving dyskinesias (Limousin et al., 1996).

To date, there have been two reports of $H_2{}^{15}O$ PET activation fingings in PD patients before and after STN stimulation. In both studies levels of rCBF were measured in patients with PD when off medication during performance of paced joystick movements in freely selected directions. Both studies (Ceballos-Baumann et al., 1997; Limousin et al., 1997) reported increased activation of rostral SMA, lateral premotor, and dorsolateral prefrontal cortex along with relief of akinesia when the STN stimulator was switched on. In contrast to pallidotomy, however, STN stimulation was also associated with reduced levels of motor cortex activation. The mechanism causing the decreased motor cortex activation remains unclear though antidromic stimulation of direct projections to STN is a possible explanation.

In summary, pallidotomy and high frequency electrical pallidal and subthalamic stimulation all appear to increase levels of SMA and dorsal prefrontal activation in PD while improving motor response times. These findings add further support to the suggestion that dysfunction of SMA and DPFC is directly responsible for the akinesia in PD.

Tremor

Activation findings

Parkinson's disease is characterised by a 3–5 Hz rest tremor but is just as frequently associated with a 4–8 Hz postural tremor. This has led to some debate over the relationship between PD and essential tremor. While the pathology of PD targets dopaminergic cells in the substantia nigra compacta and ventral midbrain tegmentum, pure lesions of the substantia nigra in non-human primates do not result in tremor (Larochelle et al., 1971). Lesioning olivo-rubro-cerebellar connections results in a 4–8 Hz tremor in harmaline exposed animals. This tremor reverts to a typical parkinsonian 3–5 Hz tremor if the nigrostriatal fibres are additionally lesioned (Poirier et al., 1966). This animal model would, therefore, suggest that parkinsonian rest tremor results from a combination of abnormal cerebellar and dopaminergic function. Cells firing at 3–5 Hz have been identified in the contralateral ventral thalamus and sensorimotor cortex of primates with parkinsonian tremor, even when their limbs are deafferented or paralysed with gallamine (Lamarre and Joffroy, 1979). Cooling of the motor cortex abolishes tremor but does not abolish thalamic firing; this has led to the postulate that parkinsonian rest tremor is generated by disinhibited firing of a ventral thalamic generator and conducted via thalamo-cortical projections (Lamarre, 1984).

$H_2{}^{15}O$ PET activation studies have reported that, even at rest when tremor free, essential tremor (ET) patients show abnormally raised levels of cerebellar blood flow (Jenkins et al., 1993; Liao et al., 1992; Wills et al., 1994). Arm extension leading to involuntary postural tremor results in further bilateral rises in cerebellar and red nuclear blood flow along with increased contralateral striatal, thalamic, and sensorimotor cortex activation. Mimicking a tremor, however, results only in significant activation of the ipsilateral cerebellar hemisphere. These activation findings support involuntary postural tremor arising from overactive cerebellar-thalamic projections.

High frequency electrical stimulation of the thalamic ventral intermediate (Vim) nucleus abolishes treatment-resistant parkinsonian tremors (Benabid et al., 1991). There have been three PET studies on the functional effects of Vim stimulation in drug-resistant tremulous PD. Deiber et al. (1993) measured cerebellar rCBF in six patients with unilateral parkinsonian tremor in the presence of clinically effective and ineffective stimulator frequencies. Suppression of unilateral tremor with effective Vim stimulation was associated with a bilateral reduction in cerebellar blood flow. Ineffective lower stimulator frequencies reduced contralateral sensorimotor cortex, but not cerebellar rCBF. Parker et al. (1992) performed a similar study with seven unilateral tremulous PD patients. Again, abolition of unilateral tremor with Vim stimulation led to a bilateral cerebellar vermis and nuclear rCBF reductions and contralateral rCBF reductions in sensorimotor cortex. Davis et al. (1997) reported that abolition of unilateral parkinsonian tremor with effective Vim stimulation reduced ipsilateral cerebellar rCBF. Taken together, these activation findings support parkinsonian as well as essential tremor originating from inappropriate overactivity of cerebellar projections to thalamus. High frequency Vim thalamic stimulation relieves tremor by suppressing firing of this system.

If thalamic output is excitatory, as is believed, thalamotomy would be predicted to impair activation of the motor cortical areas. In support of this suggestion, Boecker et al. (1997) have reported that two parkinsonian patients showed reduced motor, lateral premotor, and lateral parietal cortex activation during arm movements in freely chosen directions after the tremor in that arm was abolished by Vim thalamotomy. Interestingly, while joystick response times to external pacing were improved by thalamotomy, both patients reported that their fine finger manipulations and writing were impaired.

Dopaminergic function

^{18}F-dopa PET allows the integrity of the nigrostriatal dopaminergic system to be examined in vivo. In one study ^{18}F-dopa PET findings for 20 patients with isolated, predominantly postural tremor (eight familial, 12 sporadic) and 11 patients with isolated parkinsonian rest tremor were reported (Brooks et al., 1992). Striatal ^{18}F-dopa uptake fell in the normal range for all eight familial ET patients. Two of the group of 12 sporadic postural tremor patients had subnormal putamen ^{18}F-dopa uptake, one falling in the PD range. This patient subsequently became parkinsonian. More recently, Asenbaum and coworkers have reported that striatal ^{123}I-β-CIT uptake, a marker of dopamine transporter function, was normal in all their ET cases (Asenbaum et al., 1996). These PET and SPECT findings suggest that the majority of essential tremors are not associated with dopaminergic dysfunction.

In contrast to ET, patients with isolated parkinsonian rest tremor invariably show reduced putamen ^{18}F-dopa uptake (Brooks et al., 1992). Playford and co-workers performed ^{18}F-dopa PET on three patients who originally had essential tremor but developed a transient parkinsonian rest tremor on exposure to neuroleptics (Playford et al., 1995). Striatal ^{18}F-dopa uptake was normal in all these subjects. This finding supports Lamarre's hypothesis that a parkinsonian rest tremor requires dysfunction of both olivocerebellar and dopaminergic connections, the latter being interrupted either structurally or via dopamine receptor blockade, while essential tremor only requires the former (Lamarre, 1984).

Although dopaminergic function is intact in most cases of ET, it is well recognised that isolated postural tremor may, on occasion, be a presenting feature of PD particularly when late onset and asymmetrical. Additionally, some relatives and co-twins of PD patients have an isolated postural tremor on examination and a proportion of these show reduced levels of putamen ^{18}F-dopa uptake (Burn et al., 1992; Piccini et al., 1997b). In our unit we have compared the prevalence of dopaminergic dysfunction in co-twins and relatives of PD patients who did, or did not, have isolated postural tremor on examination. Dopaminergic dysfunction was as common in PD relatives without a postural tremor as those with tremor. This argues against isolated postural tremor in PD kindreds arising as a consequence of nigrostriatal dopaminergic cell loss. It is likely that these relatives have abnormal overactivity of cerebellar connections similar to that demonstrated in ET.

In summary, the postural tremor seen in PD probably arises from abnormal overactivity of cerebellar connections explaining its refractoriness to levodopa. Parkinsonian rest tremor results from a combination of cerebellar connection and nigrostriatal dopaminergic dysfunction.

Dyskinesia

Involuntary choreic or dystonic movements appear in up to 80% of PD patients after 5 years of exposure to levodopa therapy. It has been suggested that their onset requires both a loss of dopamine terminal function and changes in the binding properties of striatal dopamine receptors. Dopamine receptors broadly fall into two classes: D_1-type (D_1,D_5), which are adenyl cyclase dependent and D_2-type (D_2,D_3,D_4) which are not. The striatum contains mainly D_1 and D_2 receptors and these play a primary role in modulating locomotor function. In rodent and non-human primate models it is known that lesioning the nigrostriatal system leads to an upregulation of D_2 receptor binding and it has been hypothesised that in PD D_2 receptor supersensitivity might underlie dyskinesia onset. This, however, seems unlikely as the initial upregulation of striatal D_2 binding that follows nigral lesioning in animal models normalises spontaneously over a few months or more rapidly following exposure to levodopa and D_2 agonists (Fuxe et al., 1981; Reches et al., 1984; Murata and Kanazawa, 1993).

Bromocriptine monotherapy is rarely associated with dyskinesias in PD patients (Lees and Stern, 1981). As this D_2 agonist is a weak antagonist at D_1 sites the suggestion has been made that dyskinesias arise as a consequence of D_1 supersensitivity (Mouradian et al., 1989; Boyce et al., 1990b). Both selective D_1 and D_2 agonists can, however, induce dyskinesias in MPTP-lesioned monkeys when used as de novo monotherapy so it seems unlikely dyskinesias arise as a consequence of changes in binding properties of a particular dopamine receptor subtype (Bedard et al., 1992; Blanchet et al., 1993; Luquin et al., 1994).

Using ^{11}C-SCH23390 and ^{11}C-raclopride PET, striatal D_1 and D_2 binding has been reported in levodopa treated PD patients divided into groups with and without dyskinesias (Turjanski et al., 1997). The treated groups were matched for duration of clinical disease (3–10 years) and levodopa exposure (0.5–8 years). Levels of striatal D_1 and D_2 binding were similar whether dyskinesias were present or not. This finding suggests that onset of dyskinesias is not primarily determined by changes in striatal dopamine receptor availability.

If dyskinesias do not arise as a primary consequence of altered dopamine receptor binding, their aetiology may involve changes in downstream basal ganglia neurotransmitter function. Non-human primate studies suggest that chorea arises due to indirect basal ganglia pathway dysfunction (Augood et al., 1989; Boyce et al., 1990a). Opioid peptides are present in high concentration in the basal ganglia. Striatal-external pallidal projections transmit enkephalin while striato-internal pallidal projections transmit dynorphin. PD patients with a dyskinetic response to levodopa could, therefore, have altered levels of basal ganglia enkephalin and dynorphin — and hence altered levels of opioid binding — compared with patients who show a sustained non-dyskinetic response.

In favour of this hypothesis, animal models of Parkinson's disease have implicated derangement of opioid transmission in the development of dyskinesias. Following destruction of nigrostriatal dopamine projections, levels of enkephalin increase in the striatum while dynorphine levels are reduced (Gerfen et al., 1991; Engber et al., 1991; Mocchetti et al., 1987; Henry and Brotchie, 1996). Chronic pulsatile levodopa exposure leading to dyskinesias or hyperkinetic behaviour is associated with a further increase in enkephalin levels while dynorphin levels become normalised or elevated (Mocchetti et al., 1987; Henry and Brotchie, 1996; Young III et al., 1986; Li et al., 1990).

Enkephalin acts on δ opioid receptors which, when stimulated, inhibit GABA release in the external pallidum. Dynorphin acts on κ opioid receptors which, when stimulated, inhibit glutamate release in the internal pallidum. Henry and Brotchie (1996) have suggested that the net effect of a rise in enkephalin and dynorphine levels, following chronic exposure of animals with dopaminergic lesions to levodopa, is to decrease the levels of inhibitory internal pallidal output. This results in disinhibition of the thalamus and frontal cortex and so leads to involuntary movements. In contrast, exposure of dopamine depleted animals to bromocriptine normalises the raised preproenkephalin levels and does not elevate dynorphin expression (Henry and Brotchie, 1996). This differential effect may explain why bromocriptine therapy is associated with a very low incidence of dyskinesias compared with levodopa.

PD patients without dyskinesias show normal basal ganglia uptake of ^{11}C-diprenorphine, a non-selective opioid antagonist (Burn et al., 1995). In contrast, striatal, thalamic, and anterior cingulate ^{11}C-diprenorphine binding is significantly reduced in dyskinetic PD patients (Piccini et al., 1997c) — Fig. 3. Such a reduction in basal ganglia ^{11}C-diprenorphine binding would be compatible with the presence of raised levels of enkephalin and dynorphin reducing opioid receptor availability. These PET findings when combined with animal studies suggest that the role of opioid antagonists in the treatment of dyskinesias needs to be re-examined. There is some evidence that intravenous naloxone can suppress dyskinesias (Trabucchi et al., 1982) but, to date, trials of opioid antagonists administered orally have probably not used high enough doses to achieve therapeutic brain levels.

There has been one PET activation study reported in abstract on the functional abnormalities associated with dyskinesias (Piccini et al., 1997a). In

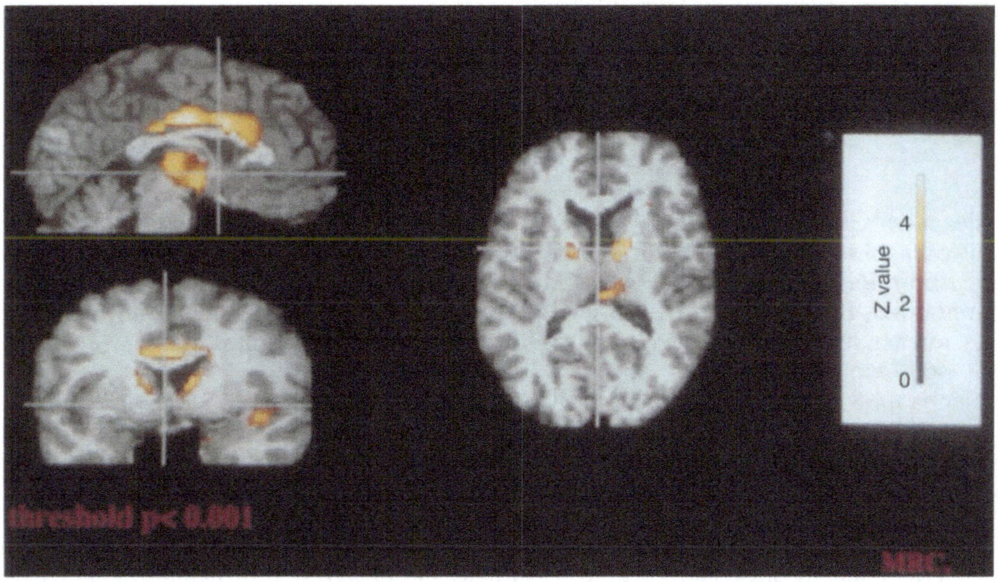

Fig. 3. An SPM overlayed on an MRI template showing areas of significantly decreased ^{11}C-diprenorphine binding in dyskinetic compared with non-dyskinetic PD patients with equivalent disease duration (photos courtesy of P. Piccini)

this study, PD patients who developed focal limb dyskinesias on levodopa exposure had rCBF measurements while moving a joystick in the presence and absence of their involuntary movements. Lentiform nucleus blood flow was raised when dyskinesias were present and levels correlated with their severity. This finding suggests that dyskinesias are associated with inappropriate striatal overactivity.

In summary, there is now strong evidence that dyskinesias in PD result when basal ganglia opioid neurotransmission becomes deranged due to a combination of disease progression and chronic exposure to levodopa. Striatal overactivity results in increased pallidal levels of enkephalin and dynorphin and breakthrough of unwanted movements.

Conclusions

In this review functional imaging evidence has been presented to support the hypothesis that akinesia in PD arises from underfunctioning of the frontal association areas receiving major input from the basal ganglia (the supplementary motor area and dorsal prefrontal cortex). This underfunctioning is reversed when dopaminergic therapy is administered or by performing a pallidotomy or implanting a subthalamic stimulator. The rest tremor that characterises PD results from a combination of abnormal overactivity of cerebellar-thalamic connections and loss of dopaminergic projections while postural tremor simply reflects the former. Dyskinesias arise when striatal overactivity leads to excess opioid neurotransmission allowing breakthrough of unwanted movements.

References

Alexander GE, Crutcher MD, Delong MR (1990) Basal ganglia thalamo-cortical circuits: parallel substrates for motor, oculomotor, "prefrontal" and "limbic" functions. Prog Brain Res 85: 119–146

Arcusa MJ, Barcia-Salorio JL, Burguera A, Pascual-Leone A (1996) Chronic, high frequency stimulation of the globus pallidus internus in Parkinson's disease. Neurology 46 [Suppl 2]: A490

Asenbaum S, Brucke T, Pirker W, Muller C, Wober C, Podreka I (1996) [^{123}I]β-CIT-SPECT in Parkinson's disase and essential tremor. Mov Disord 11 [Suppl 1]: 112

Augood SJ, Emson PC, Mitchell IJ, Boyce S, Clarke CE, Crossman AR (1989) Cellular localisation of enkephalin gene expression in MPTP-treated cynomolgus monkeys. Mol Brain Res 6: 85–92

Baron MS, Vitek JL, Bakay RAE, Green J, Kaneoke Y, Hashimoto T, et al (1996) Treatment of advanced Parkinson's disease by posterior GPi pallidotomy: 1-year results of a pilot study. Ann Neurol 40: 355–366

Bedard PJ, Gomez Mancilla B, Blanchette P, Gagnon C, Di Paolo T (1992) Levodopa-induced dyskinesia: facts and fantasy. What does the MPTP monkey tell us? Can J Neurol Sci 19: 134–137

Benabid AL, Pollak P, Gervason C, Hoffman D, Gao DM, Hommel M, et al (1991) Long-term suppresion of tremor by chronic stimulation of the ventral intermediate thalamic nucleus. Lancet 337: 403–406

Benecke R, Rothwell JC, Dick JP, Day BL, Marsden CD (1987) Disturbance of sequential movements in patients with Parkinson's disease. Brain 110: 361–379

Blanchet P, Bedard PJ, Britton DR, Kebabian JW (1993) Differential effects of selective D-1 and D-2 dopamine agonists on levodopa-induced dyskinesia in 1-methyl-4-phenyl- 1,2,3,6-tetrahydropyridine-exposed monkeys. J Pharmacol Exp Therap 267: 275–279

Boecker H, Ceballos-Baumann A, Dagher A, Samuel M, Passingham RE, Friston KJ, et al (1996) Central processing of increasingly complex learned finger sequences: Correlational analysis of 3D $H_2{}^{15}O$ PET data. Neurology 46 [Suppl]: A382

Boecker H, Wills AJ, Ceballos-Baumann A, Samuel M, Thomas DGT, Marsden CD, et al (1997) Stereotactic thalamotomy in tremor-dominant Pakinson's disease: A $H_2{}^{15}O$ PET motor activation study. Ann Neurol 41: 108–111

Boyce S, Clarke CE, Luquin R, Peggs D, Robertson RG, Mitchell IJ, et al (1990a) Induction of chorea and dystonia in Parkinsonian primates. Mov Disord 5: 3–7

Boyce S, Rupniak NMJ, Steventon MJ, Iversen SD (1990b) Differential effects of D_1 and D_2 agonists in MPTP-treated primates: Functional implications for Parkinson's disease. Neurology 40: 927–933

Brooks DJ, Playford ED, Ibanez V, Sawle GV, Thompson PD, Findley LJ, et al (1992) Isolated tremor and disruption of the nigrostriatal dopaminergic system: An ^{18}F-dopa PET study. Neurology 42: 1554–1560

Brooks DJ, Jenkins IH, Passingham RE (1993) Positron emission tomography studies on regional cerebral control of voluntary movement. In: Mano N, Hamada I, DeLong MR (eds) Role of the cerebellum and basal ganglia in voluntary movement. Excerpta Medica, Amsterdam, pp 267–274

Burn DJ, Mark MH, Playford ED, Maraganore DM, Zimmerman Jr TR, Duvoisin RC, et al (1992) Parkinson's disease in twins studied with ^{18}F-dopa and positron emission tomography. Neurology 42: 1894–1900

Burn DJ, Rinne JO, Quinn NP, Lees AJ, Marsden CD, Brooks DJ (1995) Striatal opioid receptor binding in Parkinson's disease, striatonigral degeneration, and Steele-Richardson-Olszewski syndrome: An ^{11}C-diprenorphine PET study. Brain 118: 951–958

Ceballos-Baumann AO, Jenkins IH, Morrish PK, Brooks DJ, Rehncrona S, Odin P, et al (1996) Increase of movement-related activation of the supplementary motor area after embryonic mesencephalic grafting in Parkinson's disease. Neurology 46 [Suppl]: A458

Ceballos-Baumann AO, Bartenstein P, Von Falkenhayn I, Boecker H, Riescher H, Schwaiger M, et al (1997) Parkinson's disease ON and OFF subthalamic nucleus stimulation: A PET activation study. Neurology 48 [Suppl 2]: A250

Crossman AR (1990) A hypothesis on the pathophysiological mechanisms that underlie levodopa- or dopamine agonist-induced dyskinesia in Parkinson's disease: implications for future strategies in treatment. Mov Disord 5: 100–108

Dagher A, Owen AM, Brooks DJ (1997) Neuronal circuits involved in planning and spatial working memory in Parkinson's disease and normal controls: A PET study. J Cereb Blood Flow Metabol 17 [Suppl 1]: S682

Davis KD, Taub E, Houle S, Lang AE, Dostrovsky JO, Tasker RR, et al (1997) Globus pallidus stimulation activates the cortical motor system during alleviation of parkinsonian symptoms. Nat Med 3: 671–674

Deiber M-P, Pollak P, Passingham R, Landais P, Gervason C, Cinotti L, et al (1993) Thalamic stimulation and suppression of parkinsonisn tremor. Evidence of cerebellar deactivation using positron emission tomography. Brain 116: 267–279

Dick JPR, Rothwell JC, Day BL, Cantello R, Buruma O, Gioux M, et al (1989) The Bereitschaftspotential is abnormal in Parkinson's disease. Brain 112: 233–244

Dogali M, Fazzini E, Kolodny E, Eidelberg D, Sterio D, Devinsky O, et al (1995) Stereotaxic ventral pallidotomy for Parkinson's disease. Neurology 45: 753–761

Engber TM, Susel Z, Kuo S, et al (1991) Levodopa replacement therapy alters enzyme activities in striatum and neuropeptide content in striatal output neurons of 6-hydroxydopamine lesioned rats. Brain Res 552: 113–118

Fazzini E, Dogali M, Sterio D, Eidelberg D, Beric A (1997) Stereotactic pallidotomy for Parkinson's disease: A long-term follow-up of unilateral pallidotomy. Neurology 48: 1273–1277

Fuxe K, Agnati LF, Kohler C, Kuonen D, Ogren SO, Andersson K, et al (1981) Characterisation of normal and supersensitive dopamine recetors: effects of ergot drugs and neuropeptides. J Neural Transm 51: 3–37

Gerfen CR, McGinty JF, Young III WS (1991) Dopamine differentially regulates dynorphin, substance P, and enkephalin expression in striatal neurons: in vivo hybridisation histochemical analysis. J Neurosci 11: 1016–1031

Grafton ST, Waters C, Sutton J, Lew MF, Couldwell W (1995) Pallidotomy increases activity of motor association cortex in Parkinson's disease — a positron emission tomographic study. Ann Neurol 37: 776–783

Henry B, Brotchie JM (1996) Potential of opioid antagonists in the treatment of levodopa-induced dyskinesias in Parkinson's disease. Drugs Aging 9: 149–158

Hutchison WD, Lozano AM, Davis KD, Saint Cyr JA, Lang AE, Dostrovsky JO (1994) Differential neuronal activity in segments of globus pallidus in Parkinson's disease patients. NeuroReport 5: 1533–1537

Jahanshahi M, Jenkins IH, Brown RG, Marsden CD, Passingham RE, Brooks DJ (1995) Self-initiated versus externally-triggered movements: Measurements of regional cerebral blood flow and movement-related potentials in normals and Parkinson's disease. Brain 118: 913–933

Jenkins IH, Fernandez W, Playford ED, Lees AJ, Frackowiak RSJ, Passingham RE, et al (1992) Impaired activation of the supplementary motor area in Parkinson's disease is reversed when akinesia is treated with apomorphine. Ann Neurol 32: 749–757

Jenkins IH, Bains PG, Colebatch JG, Thompson PD, Findley LJ, Frackowiak RSJ, et al (1993) A PET study of essential tremor: evidence for overactivity of cerebellar connections. Ann Neurol 34: 82–90

Jenkins IH, Brooks DJ, Nixon PD, Frackowiak RSJ, Passingham RE (1994) Motor sequence learning: A study with position emission tomography. J Neurosci 14: 3775–3790

Jenkins IH, Jahanshahi M, Brown R, Jueptner M, Marsden CD, Passingham RE, et al (1996) Frontal activation during self-initiated finger movements and externally paced movements triggered by predictable and unpredictable stimuli. Neurology 46 [Suppl]: A382

Laitinen LV, Bergenheim AT, Hariz MI (1992) Leksell's posteroventral pallidotomy in the treatment of Parkinson's disease. J Neurosurg 76: 53–61

Lamarre Y (1984) Animal models of tremor. In: Findley LJ, Capildeo R (eds) Movement disorders: tremor. MacMillan Press, London, pp 183–194

Lamarre Y, Joffroy AJ (1979) Experimental tremor in the monkey: activity of thalamic and precentral cortical neurons in the absence of peripheral feedback. Adv Neurol 24: 109–122

Larochelle L, Bedard P, Poirier LJ, Sourkes TL (1971) Correlative neuro-anatomical and neuropharmacological study of tremor and catatonia in the monkey. Neuropharmacology 10: 273–288

Lees AJ, Stern GM (1981) Sustained bromocriptine therapy in previously untreated patients with Parkinson's disease. J Neurol Neurosurg Psychiatry 44: 1020–1023

Li SJ, Jiang HK, Satchowiak MS, et al (1990) Influence of nigrostriatal dopaminergic tone on the biosynthesis of dynorphin and enkephalin in rat striatum. Brain Res Mol Brain Res 8: 219–225

Liao K-K, Zeffiro T, Kertzman C, Hallett M (1992) Regional cerebral blood flow abnormalities in essential tremor. Mov Disord 7 [Suppl 1]: 46

Limousin P, Pollak P, Benazzouz A, et al (1996) Effect on parkinsonian signs and symptoms of bilateral subthalamic nucleus stimulation. Lancet 345: 91–95

Limousin P, Greene J, Polak P, Rothwell JC, Benabid AL, Frackowiak RSJ (1997) Positron emission tomography (PET) study of modulation of cerebral activity by subthalamic nucleus (STN) and internal globus pallidus (GPi) stimulation in Parkinson's disease. Neurology 48 [Suppl 2]: A249

Luquin MR, Guillen J, Martinez-Vila E, Laguna J, Martinez Lage JM (1994) Functional interaction between dopamine D_1 and D_2 receptors in 'MPTP' monkeys. Eur J Pharmacol 253: 215–224

Mocchetti I, Naranjo J, Costa E (1987) Regulation of striatal enkephalin turnover in rats receiving antagonists of specific dopamine subtypes. J Pharmacol Exp Ther 241: 1120–1124

Mouradian MM, Heuser IJE, Baronti F, Fabbrini G, Juncos JL, Chase TN (1989) Pathogenesis of dyskinesias in Parkinson's disease. Ann Neurol 25: 523–526

Murata M, Kanazawa I (1993) Repeated L-dopa administration reduces the ability of dopamine storage and abolishes the supersensitivity of dopamine receptors in the striatum of intact rat. Neurosci Res 16: 15–23

Owen AM, Doyon J, Petrides M, Evans AC (1996) Planning and spatial working memory — a positron emission tomography study in humans. J Neurosci 8: 353–364

Parker F, Tzourio N, Blond S, Petit H, Mazoyer B (1992) Evidence for a common network of brain structures involved in parkinsonian tremor and voluntary repetitive movement. Brain Res 584: 11–17

Passingham RE (1987) Two cortical systems for directing movement. Motor areas of the cerebral cortex. Ciba Foundation 132: 151–164

Penney Jr JB, Young AB (1986) Striatal inhomogeneities and basal ganglia function. Mov Disord 1: 3–15

Piccini P, Boecker H, Weeks RA, Brooks DJ (1997a) Dyskinesia correlated regional blood flow changes in Parkinson's disease. Neurology 48 [Suppl 2]: A327
Piccini P, Morrish PK, Turjanski N, Sawle GV, Mark MH, Maraganore DM, et al (1997b) Dopaminergic function in familial Parkinson's disease: A clinical and ^{18}F-dopa PET study. Ann Neurol 41: 222–229
Piccini P, Weeks RA, Brooks DJ (1997c) Opioid receptor binding in Parkinson's patients with and without levodopa-induced dyskinesias. Ann Neurol 42: 720–726
Playford ED, Jenkins IH, Passingham RE, Nutt J, Frackowiak RSJ, Brooks DJ (1992) Impaired mesial frontal and putamen activation in Parkinson's disease: a PET study. Ann Neurol 32: 151–161
Playford ED, Britton TC, Thompson PD, Brooks DJ, Findley LJ, Marsden CD (1995) Exacerbation of postural tremor with emergence of parkinsonism following neuroleptic administration. J Neurol Neurosurg Psychiatry 58: 487–489
Poirier LJ, Sourkes TL, Bouvier G, Boucher R, Carabin S (1966) Striatal amines, experimental tremor, and the effect of harmaline in the monkey. Brain 89: 37–52
Rascol O, Sabatini U, Fabre N, Brefel C, Loubinoux I, Celsis P, et al (1997) The ipsilateral cerebellar hemisphere is overactive during hand movements in akinetic parkinsonian patients. Brain 120: 103–110
Reches A, Wagner HR, Jackson-Lewis V, Yablonski-Alter E, Fahn S (1984) Chronic levodopa or pergolide administration induces downregulation of dopamine receptors in denervated striatum. Neurology 34: 1208–1212
Samuel M, Ceballos-Baumann AO, Blin J, Uema T, Boecker H, Brooks DJ (1997a) Evidence for lateral premotor and parietal overactivity in Parkinson's disease during sequential and bimanual movements: A PET study. Brain 120: 963–976
Samuel M, Ceballos-Baumann AO, Turjanski N, Boecker H, Gorospe A, Linazasoro G, et al (1997b) Pallidotomy in Parkinson's disease increases SMA and prefrontal activation during performance of volitional movements: An $H_2{}^{15}O$ PET study. Brain 120: 1301–1313
Samuel M, Caputo E, Brooks DJ, Scaravilli T, Branston N, Rothwell JC, et al (1998) A study of medial pallidotomy for Parkinson's disease: Clinical outcome, MRI location and complications. Brain (In press)
Sterio D, Beric A, Dogali M, Fazzini E, Alfaro G, Devinsky O (1994) Neurophysiological properties of pallidal neurons in Parkinson's disease. Ann Neurol 35: 586–591
Trabucchi M, Bassi S, Frattola L (1982) Effect of naloxone on the "on-off" syndrome in patients receiving long-term levodopa therapy. Arch Neurol 39: 120–121
Turjanski N, Lees AJ, Brooks DJ (1997) PET studies on striatal dopaminergic receptor binding in drug naive and L-dopa treated Parkinson's disease patients with and without dyskinesia. Neurology 49: 717–723
Wills AJ, Jenkins IH, Thompson PD, Frackowiak RSJ, Findley LJ, Brooks DJ (1994) Red nuclear and cerebellar but no olivary activation associated with essential tremor: A positron emission tomography study. Ann Neurol 36: 636–642
Young III WS, Bonner TI, Brann MR (1986) Mesencephalic dopamine neurons regulate the expression of neuropeptide mRNAs in the rat forebrain. Proc Natl Acad Sci USA 83: 9827–9831

Author's address: David J. Brooks, Hartnett Professor of Neurology, MRC Cyclotron Unit, Hammersmith Hospital, Du Cane Road, London W12 0NN, U.K.

Multiple system atrophy

W. H. Oertel and **O. Bandmann**

Department of Neurology, Philipps University Marburg, Federal Republic of Germany

Summary. Multiple system atrophy (MSA) is a degenerative central nervous system disease of unclear origin. Patients affected typically show symptoms attributable to the combined involvement of the extrapyramidal, pyramidal, cerebellar and autonomic nervous systems. At onset patients mostly see a doctor because of extrapyramidal or — more rarely — cerebellar symptoms. Evidence of autonomic nervous system involvement is often not apparent, at least to the neurologist, before the history is taken. In later stages, by contrast, involvement of all of the above systems is clinically detectable.

Historical review and terminology

In his "Essay on the shaking palsy" James Parkinson may well have been the first author to describe a patient with MSA (Parkinson, 1817). His "case no. 1" initially presented with tremor and general fatigability. These were followed by frequent falls, dysphagia and speech problems as well as incontinence. 100 years later von Stauffenberg first associated the cerebellar and autonomic symptoms as well as the akinesia and rigidity he saw clinically in one of his patients with lesions of the cerebellum, basal ganglia, pons and olives (Stauffenberg, 1918). More recently, in 1960, Shy and Drager again pointed to a combination of extrapyramidal and autonomic disorders. The two patients they described at some length complained of symptoms suggestive of an impairment of the cerebellar and pyramidal systems (Shy and Drager, 1960). The term "multiple system atrophy" was coined by Graham and Oppenheimer in 1969 (Graham and Oppenheimer). In the past few years the clinical and diagnostic criteria proposed by Quinn for the condition have increasingly gained ground (Quinn, 1994). Generally speaking, full-fledged clinically overt MSA may present in one of 3 ways: (1) striatonigral degeneration which is mainly associated with akinesia and rigidity; (2) isolated (non-familial) olivopontocerebellar atrophy combining lesions of the cerebellum, brainstem and basal ganglia; and (3) primary postural hypotension.

Clinical signs and symptoms

Manifestations and course

Patients with MSA typically start complaining of symptoms in the 5th to 7th decades of life. The onset of the disease peaks between years 45 and 60. The earliest onset of autopsy-proven MSA was observed at age 31, the latest at age 78 years (Wenning et al., 1994a).

Wenning et al. studied the initial symptoms and the further clinical course of 100 patients with MSA (Wenning et al., 1994a). Of these, 42% initially developed akinesia and rigidity. These were later associated with extrapyramidal motor disturbances in 91%. In the overwhelming majority of patients akinesia and rigidity were asymmetrical (74%); tremor was present in less than one third (29%); pill rolling typical of idiopathic parkinsonism was only seen in 9% of cases. In the early stage of the condition 29% of the patients responded well or excellently to L-dopa. But a sustained response was seen in less than half of them (13%). One quarter of the patients treated with L-dopa developed orofacial dyskinesias. Patients with early-onset MSA (before age 49 years) responded best to L-dopa treatment.

In 41% of cases autonomic dysfunction caused the initial symptoms. These were impotence in males and urinary incontinence as the most typical complaint in females. Although symptomatic postural hypotension was present in 68% of the patients in the further course, only 15% suffered 3 or more syncopes. Urinary urge incontinence was much more common than urinary retention necessitating catheterization (77% of all female and 68% of all male patients). Only 2 patients reported fecal incontinence. About one third of the patients complained of respiratory problems with inspiratory stridor (34%).

Cerebellar disorders were initially only present in 5% of cases. Over the course of their illness almost 50% developed symptoms attributable to a cerebellar lesion. Only in a small proportion were the cerebellar symptoms the only (9%) or predominant (9%) movement disorders.

Pyramidal tract involvement was clinically detectable in the majority of cases (61%).

Dementia is not typically present in MSA, but systematic neuropsychological examinations usually show deficits in the so-called frontal lobe tests (Robbins et al., 1992). While apraxia is commonly seen in PSP (75%), Leiguarda et al. did not find it in their 10 patients with MSA (Leiguarda et al., 1997).

Quinn proposed distinguishing between striatonigral or olivopontocerebellar MSA as a possible, probable or definitive diagnosis. The diagnostic criteria are listed in Table 1. While isolated progressive autonomic failure as the only manifestation of MSA has been described, it is probably extremely rare (Quinn, 1994).

Of the patients studied by Wenning et al. more than 40% were severely disabled or in a wheel-chair 5 years after the onset of the disease. The mean survival time was 9.5 years (Wenning et al., 1994a). Studying various prognostic factors, Ben Shlomo et al. showed that patients with late-onset MSA carried a poorer prognosis than younger patients (Ben Shlomo et al., 1997)

Table 1. Diagnostic criteria of MSA

Diagnosis	Striatonigral type = MSA − SND (mainly parkinsonian)[a]	Olivopontocerebellar type = MSA − sOPCA (mainly cerebellar)[a]
MSA possible	sporadic adult-age onset, akinesia, rigidity no/poor response to L-dopa[b]	sporadic adult-age onset, cerebellar syndrome with associated extrapyramidal symptoms
MSA probable	see above + established autonomic dysfunction[c] or cerebellar or pyramidal signs or abnormal sphincter EMG	cerebellar dysfunction with or without associated pyramidal or extrapyramidal signs + severe autonomic dysfunction or abnormal sphincter EMG
MSA certain	lesions typical of multiple system atrophy post mortem	lesions typical of multiple system atrophy post mortem

[a]No DMS III dementia, generalized loss of reflexes, severe supranuclear palsy on downward gaze or other known aetiologies

[b]Moderate or satisfactory, but often subsiding response to L-dopa. In patients responding to L-dopa several clinical signs not compatible with Parkinson's disease should be present

[c]Syncopes and/or urinary incontinence or retention or incontience of unclear origin

and that patients with cerebellar involvement tended to have a somewhat better prognosis. Neither the gender nor the presence of akinesia and rigidity or pyramidal symptoms had any effect on the prognosis. The type of MSA, i.e. striatonigral or olivopontocerebellar, was also irrelevant.

Ancillary studies

There are no studies which can confirm or rule out the diagnosis beyond any doubt. But the presumptive clinical diagnosis can be corroborated with 3 tests: (1) blood pressure measurements with the patient reclining and standing; (2) MRI of the skull; (3) sphincter electromyography.

Postural blood pressure measurements help to detect postural hypotension. In the past the definitions of what constitutes hypotension have varied widely (Quinn, 1989). The Schellong test, which has become firmly established in the German-speaking countries is comparatively accurate, but rarely used because it takes a considerable time. In English-speaking countries, the general rule is that a drop of more than 20 mm Hg in standing systolic pressure is considered abnormal (Shulman and Weiner, 1997). More detailed autonomic nervous system testing is best done in specialized cents.

MRI of the skull should always be ordered if MSA is suspected. T_2-weighted images obtained at high field strength (1.5 Tesla) often show cerebellar and brainstem atrophy as well as hypo-intense areas in the putamen (Schwarz et al., 1996). However, these have also been seen in patients with typical Parkinson's syndrome. Hyperintensity at the junction between the putamen and claustrum appears to be a more specific sign. In post-mortem studies this MRI abnormality has been found to be due to local gliosis (Schwarz et al., 1996).

Neuropathologically, patients with MSA often shown a lesion of the nucleus of Onuf. This nucleus is made up of specialized anterior horn neurons in segments S2–S4 which innervate the striate muscles of the external bladder and rectal sphincters. Eardley et al. (1989) first reported signs of denervation as well as re-innervation of the sphincter muscles in MSA patients reflected by an increased amplitude, a polyphasic pattern and prolonged activity of some motor units in the EMG. Palace et al. found 82% of 126 patients suspected of having MSA to have an abnormal sphincter EMG (Palace et al., 1997). But Schwarz et al. (1997) reported that prolonged motor unit and increased polyphasic activities may also be present in patients with idiopathic Parkinson's syndrome and suggested that the presence of abnormal spontaneous activity was a more specific marker of MSA. Similar abnormalities have, also been recorded in some patients (42%) with progressive supranuclear palsy (Valldeoriola et al., 1995).

In specialized centres, further evidence of MSA can be obtained by more sophisticated imaging techniques, particularly those used in research projects. One example is 18F-fluorodopa positron emission tomography (PET), which shows a reduced radionuclide uptake in the putamen and caudate nucleus. Another is 18F-fluorodesoxyglucose PET, which helps to detect reduced cerebellar and striatal metabolism. Using different ligands, dopamine D_2 receptor density assays have shown binding to the receptor to be either normal or reduced in patients with MSA (Schwarz et al., 1992; Brooks et al., 1990). Goldstein et al. found that in patients with isolated autonomic failure and in those with idiopathic Parkinsonism (IPS) associated with autonomic deficits, 18F-fluorodopamine PET of the sympathetic cardiac nerves showed a loss of sympathetic nerve endings in the myocardium (Goldstein et al., 1997). In MSA patients with autonomic deficits, by contrast, the sympathetic nerve endings were intact, but there was evidence of reduced axonal plasma flow in sympathetic nerves.

Other investigations, such as EEG, evoked potentials or EMG of the limb muscles have no or at best little importance for the diagnosis (Abbruzese et al., 1997; Pramstaller et al., 1995). CCT may show cerebellar atrophy in some patients, but its specificity and sensitivity are poor.

Differential diagnosis

Clinically, MSA is most often mistaken for idiopathic Parkinson's syndrome (IPS), but there are a number of distinctive clinical features. These include falls early in the course of the disease, rapid progression and complete or partial absence of an L-dopa response. Abnormal eye movements, dysarthria, irregular tremor and myoclonus as well as antecollis are suggestive of MSA (Quinn, 1994). Unlike patients with MSA, the palsy in patients with progressive supranuclear palsy (PSP) typically involves the vertical downward gaze. PSP is also suggested by nuchal and axial hypertonicity combined with hyperextension of the head, lid retraction and, at times, a quizzical facial expression. Severe pallilalia is also suggestive of PSP. Corticobasal degeneration can best be distinguished from MSA by the early presence of a cortical deficit, particularly apraxia or loss of cortical sensitivity, and by marked myoclonus as well as by evidence of asymmetrical perirolandic frontoparietal cortical atrophy on CCT or MRI. Like MSA, multiple vascular infarctions may cause lesions in different brain areas. But a carefully taken history and the detection of typical lesions by suitable imaging techniques should help in the differential diagnosis. In SCA3/Machado-Joseph ataxia, which is inherited as an autosomal dominant trait, the combination of extrapyramidal and cerebellar dysfunction is not uncommon and the age at onset is often the same as that of MSA (Giunti et al., 1995). However, autonomic dysfunction is extremely rare in hereditary ataxias and a carefully taken family history often shows an ealier onset in other members of the family. In a study of 80 patients with clinically and autopsy-proven MSA an examination of the SCA3 gene showed a normal number of base triplets (Bandmann et al., 1997).

Treatment

As already mentioned before, a minority of patients with MSA of the SND type responds to L-dopa: Almost one third of the patients examined by Wenning et al. (1994a) reported improvement of the motor symptoms by at least 50%, but excellent responses with an improvement by more than 70% were only recorded in 11% of cases. At subsequent follow-ups only 13% of the patients reported a sustained improvement by more than 50% (11%) or by more than 70% (2%). Irrespective of the initial response to L-dopa, mobility deteriorated in more than half of the patients (52%) after discontinuing the drug. This deterioration often was not immediate, but occurred after several days.

Consequently, trial treatment with L-dopa at a dose of up to 1,000 mg daily (L-dopa/peripheral decarboxylase inhibitor) appears to be justified in patients with MSA (see Table 2). Half of the patients treated are found to develop orofacial dyskinesias, but atypically, these tend to be asymmetrical, unilateral or dystonic. Patients not responding to L-dopa or who develop severe side effects may be candidates for medication with amantadine, budipine, anticholinergics and antidepressants.

Antecollis is the most common dystonic symptom in MSA. It may be relieved by locally injecting botulinum toxin A. This may also be helpful in patients with severe inspiratory stridor. The dystonic symptoms may, at times, also be relieved by anticholinergics.

Postural hypotension may occur or become worse after L-dopa treatment is instituted. Patients are initially prescribed compression stockings and increased salt intake. In severe cases sympathomimetics like yohimbine or midodrine or fludrocortisone may be needed (see Table 2).

Table 2. Drug therapy for multiple system atrophy

Akinesia — rigidity	—	L-dopa/peripheral decarboxylase inhibitor up to 1,000 mg daily (apomorphin test, L-dopa test, if needed)
	—	amantadine by infusion or per os, budipine
	—	anticholinergics
	—	antidepressants
Dystonia	—	anticholinergics
	—	botulinum toxin A by injection
Postural hypotension	—	midocrine: Gutron® 2.5 mg tbls. or drops (1 ml : 10 mg); 1–2 tbls. or 7–14 drops q.d. mornings or evenings
	—	yohimbine: (A) Pasuma® sugar-coated tbls.; (CH) not approved; (D) Yohimbin "Spiegel", Yocon-Glenwood®, 1 25 mg tbls.t.i.d.
	—	Fludrocortisone: (CH) Florinef®; (A, D): Astonin H®, Fludrocortisone 0.1 mg tbls. b.i.d. to t.i.d. initially to be increased to maximally 0.5 mg daily. Monitor electrolytes!

Oxybutynin is the drug of choice for the treatment of detrusor muscle hyperreflexia. If urinary retention is present, the patient or a relative should learn self-catheterization.

Physiotherapy, ergotherapy and, if necessary, a wheelchair, provide further help. Patients with severe dysphagia may need a gastric stoma which can be placed by percutaneous endoscopic gastroscopy (PEG).

Pathology

In MSA patients post-mortem studies of the brain have shown the following brain regions to be involved: striate body, substantia nigra, locus ceruleus, pontine nuclei, Purkinje cells, inferior olive and intermediolateral column. In olivopontocerebellar MSA the inferior olives and the pontine nuclei are more severely affected than the cerebellar Purkinje cells. Within the cerebellum the vermis is more atrophic than the hemispheres (Wenning et al., 1996). Quinn has postulated that, for the diagnosis to be certain, cell loss and gliosis should be present at least in 2 of the above structures and oligodendroglial inclusion bodies should be demonstrable histologically by Bilschowsky-silver or anti-ubiquitin staining (Quinn, 1994). While typical of MSA (Lantos and Papp, 1994), oligodendroglial inclusion bodies have also been found to be present in a number of other neurodegenerative diseases (Daniel et al., 1995). If possible, glial fibrillary acid protein (GFAP) should be used for detecting gliosis.

Pathogenesis

The causes underlying MSA are poorly understood, however as with the insights gained into the pathogenesis of idiopathic Parkinson's disease oxidative stress, mitochondrial dysfunction, a genetic component and an inflammatory/autoimmune origin have been proposed as potential mechanisms.

As in other neurodegenerative conditions affecting the basal ganglia, the iron content is increased in the substantia nigra and in the striate body (Dexter et al., 1991). The increased iron content may promote the production of oxygen free radicals in the socalled Haber-Weiss reaction (Olanow, 1993). In patients with MSA the ratio between reduced and oxidated glutathion in the substantia nigra has been found to be lower than in normals. This albeit non-significant reduction was thought to reflect oxidative stress (Sian et al., 1994a). But Cohen has suggested that associated L-dopa treatment may well be responsible (Cohen, 1994). Most of the enzymes involved in the metabolism of glutathion show a normal activity with the exception of glutathion peroxidase, which is increased in the lateral globus pallidus and the caudate nucleus (Sian et al., 1994b).

Studies of mitochondrial function in the substantia nigra and in platelets have failed to produce any evidence of respiratory chain abnormalities (Schapira et al., 1990; Benecke et al., 1993). But Blin et al. has reported that complex I was reduced by 30% in the muscles of MSA patients (Blin et al., 1994). As in his studies in patients with idiopathic Parkinson's disease, Swerdlov found the activity of complex I to be reduced in the so-called cybrid cells. These cells are produced by fusing the mitochondria of patients with mitochondria-free cells of normals. However, as yet only preliminary data are available (Swerdlow et al., 1997).

Typically, MSA occurs sporadically, i.e. without other family members being affected. But the only case control study on MSA done so far suggested that a genetic component may be involved in the development of the disease (Nee et al., 1991). Interest in a potential genetic factor was heightened by evidence showing a genetic basis for other diseases affecting the basal ganglia, e.g. IPS and PSP.

Allelic association assays of HLA antigens and debrisoquine CYP2D6L polymorphism have show a significant association in a small number of cases (Wood, 1997; Conrad et al., 1997) which has, however, not been confirmed in larger samples (Bannister et al., 1983; Iwahashi et al., 1995; Bandmann et al., 1995, 1997). Reporting on a family with autosomal dominant ataxia and an abnormally high number of base triples in the SCA1 gene, Gilman et al. found one member of the family had developed a gait disorders at the age of 60 followed by dysarthria and poor coordination of the hands (Gilman et al., 1997). Sphincter dysfunction was absent, but the patient had become impotent at the age of 57. He died at 74. On autopsy massive atrophy of the cerebellum, medulla oblongata and pons was associated with glial cytoplasmic inclusion bodies. This prompted Gilman et al. to postulate an overlap between MSA and SCA1. The family studied, however, showed some features that were quite unusual for MSA: The age at onset was 19 years in one individual affected and 30 years in anothers whereas MSA usually occurs between 35 and 70 years. The patient described only showed cerebellar involvement without any associated pyramidal or extrapyramidal abnormalities. Also, the sexual impotence he developed may well have been diabetogenic or hypertensive in origin. Thus, he does not fulfill the diagnostic criteria of MSA proposed by Quinn (Quinn, 1994). As Bandmann et al. (1997) found the base triplets in the SCA1 gene to be normal in all of the 80 MSA patients they investigated, an overlap between SCA1 and MSA is thought to be unlikely.

Neither the natural history nor the clinical signs and symptoms of MSA are compatible with an autoimmune disease. But interestingly enough, Schwarz et al. found a severe microgliotic reaction both in the putamen and in other brain regions showing cell loss (Schwarz et al., 1996).

Outlook

Despite extensive research efforts the causes underlying MSA are still poorly understood. Although technically demanding, biochemical assays of basophilic inclusion bodies may offer new insights into the pathogenesis of the disease. Future research will show whether — like in PSP or IPS (Wood, 1997) — genetic risk factors can be identified for MSA. Anecdotal reports on the efficacy of drugs like L-threo-DOPS will have to be confirmed in placebo-controlled trials. The development of animal models will hopefully help to test new therapeutic approaches such as transplantation and the use of nerve growth factors (Wenning et al., 1997).

References

Abbruzese G, Marchese R, Trompetto C (1997) Sensory and motor evoked potentials in multiple system atrophy: a comparative study with Parkinson's disease. Mov Disord 12: 315–321

Bandmann O, Wenning GK, Quinn NP, Harding AE (1995) Arg296 to Cys296 polymorphism in exon 6 of cytochrome P-450-2D6 (CYP2D6) is not associated with multiple system atrophy. J Neurol Neurosurg Psychiatry 59: 557

Bandmann O, Vaughan J, Holmans P, Marsden C, Wood N (1997) Multiple system atrophy is genetically distinct from identified hereditary causes of spinocerebellar degeneration. Neurology 9: 1598–1604

Bannister R, Mowbray J, Sidgwick A (1983) Genetic control of progressive autonomic failure: evidence for an association with an HLA antigen. Lancet 1: 1017

Benecke R, Struemper P, Weiss H (1993) Electron transfer complexes I and IV of platelets are abnormal in Parkinson's disease but normal in Parkinson-plus syndromes. Brain 116: 1451–1463

Ben Shlomo Y, Wenning GK, Tison F, Quinn NP (1997) Survival of patients with pathologically proven multiple system atrophy: a meta-analysis. Neurology 48: 384–393

Blin O, Desnuelle C, Rascol O, Borg M, Peyro Saint Paul H, Azulay J, Bille F, Figarella D, Coulom F, Peillissier J (1994) Mitochondrial respiratory failure in skeletal muscle from patients with Parkinson's disease and multiple system atrophy. J Neurol Sci 125: 95–101

Brooks D, Ibanez V, Sawle GC, Quinn N, Lees AJ, Bannister R, Marsden CD, Frackowiak RS (1990) Differing patterns of striatal 18F-dopa uptake in Parkinson's disease, multiple system atrophy, and progressive supranuclear palsy. Ann Neurol 28: 547–555

Cohen G (1994) The brain on fire? Ann Neurol 36: 333–334

Conrad C, Andreadis A, Trojanowski JQ, Dickson DW, Kang D, Chen X, Wiederholt W, Hansen L, Masliah E, Thal LJ, Katzman R, Xia Y, Saitoh T (1997) Genetic evidence for the involvement of tau in progressive supranuclear palsy. Ann Neurol 41: 277–281

Daniel S, Geddes J, Revesz T (1995) Glial cytoplasmic inclusions are not exclusive to multiple system atrophy. J Neurol Neurosurg Psychiatry 58: 262

Dexter D, Carayon A, Javoy-Agid F, Agid Y, Wells F, Daniel S, Lees A, Jenner P, Marsden C (1991) Alterations in the levels of iron, ferritin and other trace metals in Parkinson's disease and other neurodegenerative diseases affecting the basal ganglia. Mov Disord 114: 1953–1975

Eardley I, Quinn NP, Fowler CJ, Kirby RS, Parkhouse HF, Marsden CD, Bannister R (1989): The value of urethral sphincter electromyography in the differential diagnosis of parkinsonism. Brit J Urol 64: 360–362

Gilman S, Sima A, Junck L, Kluin K, Koeppe R, Lohman M, Little R (1996) Spinocerebellar ataxia type 1 with multiple system degeneration and glial cytoplasmic inclusions. Ann Neurol 39: 241–255

Giunti P, Sweeney M, Harding A (1995) Detection of the Machado-Joseph disease/ spinocerebellar ataxia three trinucleotide repeat expansion in families with autosomal dominant motor disorders, including the Drew family of Walworth. Brain 118: 1077–1085

Goldstein DS, Holmes C, Cannon RO 3rd, Eisenhofer G, Kopin IJ (1997) Sympathetic cardioneuropathy in dysautonomias. NEJM 336: 696–702

Graham J, Oppenheimer D (1969): Orthostatic hypotension and nicotine sensitivity in a case of multiple system atrophy. J Neurol Neurosurg Psychiatry 32: 28–34

Iwahashi K, Miyatake R, Tsuneoka Y, Matsuo Y, Ichikawa Y, Hosokawa K, Sato K, Hayabara T (1995) A novel cytochrome P-450IID6 (CYPIID6) mutant gene associated with multiple system atrophy. J Neurol Neurosurg Psychiatry 58: 263–264

Lantos P, Papp M (1994) Cellular pathology of multiple system atrophy: a review. J Neurol Neurosurg Psychiatry 57: 129–133

Leiguarda RC, Pramstaller PP, Merello M, Starkstein S, Lees AJ, Marsden CD (1997) Apraxia in Parkinson's disease, progressive supranuclear palsy, multiple system atrophy and neuroleptic-induced parkinsonism, Brain 120: 75–90

Nee L, Gomey M, Dambrosia J, Bale S, Eldridge R, Polinski R (1991) Enviromental-occupational risk factors and familial associations in multiple system atrophy: a preliminary investigation. Clin Auton Res 1: 9–13

Olanow C (1993) A radical hypothesis for neurodegeneration. Trends Neurosci 16: 439–444

Palace J, Chandiramani VA, Fowler CJ (1997): Value of sphincter electromyography in the diagnosis of multiple system atrophy. Muscle Nerve 20: 1396–1403

Parkinson J (1817): An essay on the shaking palsy. Sherwood, Neely and Jones, London

Pramstaller P, Wenning GK, Smith SJ, Beck RO, Quinn NP, Fowler CJ (1995) Nerve conduction studies, skeletal muscle EMG, and sphincter EMG in multiple system atrophy. JNNP 58: 618–621

Quinn N (1989) Multiple system atrophy — the nature of the beast. J Neurol Neurosurg Psychiatry 52: 78–89

Quinn N (1994): Multiple system atrophy. In: Marsden C, Fahns S (eds) Movement disorders 3. Butterworth-Heinemann, London, S262–281

Robbins TW, James M, Lange KW, Owen AM, Quinn NP, Marsden CD (1992) Cognitive performance in multiple system atrophy. Brain 115: 271–291

Schapira AHV, Mann VM, Cooper JM, Dexter D, Daniel SE, Jenner P, Clark JB, Marsden CD (1990) Anatomic and disease specificity of NADH CoQl reductase (complex I) deficiency in Parkinson's disease. J Neurochem 55: 2142–2145

Schwarz J, Tatsch K, Arnold G, Gasser T, Trenkwalder C, Kirsch CM, Oertel WH (1992) 123I-iodobenzamide-SPECT predicts dopaminergic responsiveness in patients with de novo parkinsonism. Neurology 42: 556–561

Schwarz J, Weis S, Kraft E, Tatsch K, Bandmann O, Mehraein P, Vogl T, Oertel W (1996) Signal changes on MRI and increases in reactive microgliosis, astrogliosis, and iron in the putamen of two patients with multiple system atrophy. J Neurol Neurosurg Psychiatry 60: 98–101

Schwarz J, Kornhuber M, Bischoff C, Straube A (1997) Electromyography of the external anal sphincter in patients with Parkinson's disease and multiple system atrophy: frequency of abnormal spontaneous activity and polyphasic motor unit potentials. Muscle Nerve 20: 1167–1172

Shulman L, Weiner W (1997) Multiple system atrophy. In: Watts R, Kollers W (eds): Movement disorders: neurologic principles and practice. McGraw-Hill, New York, pp 297–304

Shy G, Drager G (1960) A neurologic syndrome associated with orthostatic hypotension. Arch Neurol 2: 511–527

Sian J, Dexter D, Lees A, Daniel S, Agid Y, Javoy-Agid F, Jenner P, Marsden C (1994a) Alterations in glutathione levels in Parkinson's disease and other neurodegenerative disorders affecting the basal ganglia. Ann Neurol 36: 348–355

Sian J, Dexter D, Lees A, Daniel S, Jenner P, Marsden C (1994b) Glutathione-related enzymes in brain in Parkinson's disease. Ann Neurol 36: 356–361

Stauffenberg V (1918): Zur Kenntnis des extrapyramidalen motorischen Systems und Mitteilung eines Falles von sog. "Atrophie olivo-pontocerebelleuse", Z Gesamte Neurol Psychiatrie 39: 1–55

Swerdlow R, Parks JK, Miller SW, Tuttle JB, Trimmer PA, Sheehan JP, Bennett JP, Davis RE, Parker WD (1997) As in Parkinson's disease, a bioenergetic defect transfers with mitochondrial DNA of patients with multisystem atrophy. Mov Disord 12 S1: 3

Valldeoriola F, Valls-Sole J, Tolosa ES, Marti MJ (1995) Striated anal sphincter denervation in patients with progressive supranuclear palsy. Mov Disord 10: 550–555

Wenning G, Ben Shlomo Y, Magalhaes M, Daniel S, Quinn N (1994a) Clinical features and history of multiple system atrophy. Brain 117: 835–845

Wenning G, Jager R, Kendall B, Kingsley D, Daniel SE, Quinn NP (1994b) Is cranial computerized tomography useful in the diagnosis of multiple system atrophy? Mov Disord 9: 333–336

Wenning G, Tison F, Elliott L, Quinn NP, Daniel SE (1996) Olivopontocerebellar pathology in multiple system atrophy. Mov Disord 11: 157–162

Wood N (1997) Genes and parkinsonism. J Neurol Neurosurg Psychiatry 62: 305–309

Authors' address: Prof. Dr. W. Oertel, Department of Neurology, Philipps-University Marburg, Rudolf-Bultmann-Strasse 8, D-35039 Marburg, Federal Republic of Germany

Tremorlytic activity of budipine in Parkinson's disease

S. Spieker, S. Breit, T. Klockgether, and **J. Dichgans**

Department of Neurology, University of Tübingen, Federal Republic of Germany

Summary. In order to objectively quantify the tremorlytic activity of budipine in Parkinson's disease (PD) we performed longterm tremor recordings in a subset of patients enrolled in two clinical trials.

Eleven PD patients with marked tremor participating in an open-label study underwent longterm recording before and during medication. Nine patients completed the study. Tremor occurrence was reduced from 52 ± 18.6% to 34.7 ± 19.3% ($p < 0.05$); tremor intensity decreased from 15.3 ± 4.8 (SNR)to 11.3 ± 4.8 ($p < 0.01$). UPDRS tremor subscores were also significantly improved.

Fourteen patients who enrolled in a multicenter, double-blind, placebo-controlled study underwent longterm tremor analysis in addition to the Columbia University Rating Scale (CURS). Tremor occurrence was improved in the budipine group ($n = 7$) from 24.7 ± 15.5% to 14.8 ± 14.5% ($p < 0.05$). Tremor intensity decreased from 9.1 ± 2.5 (SNR) to 7.2 ± 1.6. However, the latter result was statistically not significant, probably due to the small patient number. In the placebo-group ($n = 7$) there was no reduction of tremor occurrence or of tremor intensity. The CURS sum score was improved from 5.7 to 3.0 in the budipine group, whereas there was only a smaller improvement in the placebo group (from 7.1 to 5.5).

These data suggest that budipine is an effective tremorlytic agent in PD, which may by used as an alternative to anticholinergics.

Introduction

Previous open-label studies have suggested, that budipine may have a beneficial effect on tremor of Parkinson's disease (Ketz, 1985; Ulm, 1985; Iizuka, 1985). These studies have used subjective, clinical rating scales, such as the "Webster Rating Scale" (Webster, 1968), the "Unified Parkinson's Disease Rating Scale" (UPDRS) (Fahn et al., 1988), or the "Columbia University Rating Scale" (CURS) (Duvoisin, 1971), for determination of tremorlytic activity. Objective data concerning the effect of budipine on tremor have not been obtained sofar. We have previously developed the method of long-term EMG for objective quantification of tremor (Bacher et al., 1989). This method seems well suited to measure the effect of new tremorlytic drugs such as budipine.

By means of this method we studied a subset of patients enrolled in two clinical trails with budipine. The first trial (FKF 701/01 A) was an open-labelled, multicenter study, which was initiated to demonstrate the safety of budipine. We recorded 11 patients enrolled in this study before and during medication with budipine. The second trial (FKF 004) was a double-blind, placebo-controlled, multi-center study. Fourteen patients from three centers were recorded in addition to the standard study protocol.

Methods

Tremor recordings

Lonterm EMG was performed as described in Bacher et al., 1989. In short, surface EMGs were obtained from the extensor carpi radialis and the flexor carpi ulnaris muscles of both forearms with a portable tape recorder. We recorded for about 24 hours. The patients were free to move around and maintain their usual activities at home. They came back on the next day when the data were digitalized and transferred to a PC for analysis. Recording frequency was 200 Hz. 10 hours of the waking time were analyzed from the extensor muscle of the most affected arm. This period was devided into 15 s intervals. Each interval was analyzed separately. Data were first high-pass filtered to smooth baseline shifts and were then demodulated. A FFT of the demodulated signal gave the frequency spectrum. Two parameters were obtained from this spectrum: the main frequency and the intensity at this main frequency. Intensity was defined as the signal-to-noise ratio (SNR) of the peak in the spectrum. If the main frequency was between 3.7 and 10 Hz and the intensity greater than four, the entire interval was defined as having tremor. From all 2343 intervals the overall result was calculated: 1) the percentage of intervals containing tremor as a measure of tremor occurrence, 2) the mean of all main frequencies and 3) the mean intensity.

Patients

Open-label study

Eleven tremor dominant PD patients (six women, five men) paticipated. Mean age was 66 $\pm$ 10 yrs. Disease duration was 5.9 $\pm$ 2.2 yrs. Two patients were Hoehn and Yahr (H&Y) I, six were H&Y II, and three were H&Y III. Three patients only suffered from a tremor at rest. Six had a slight postural tremor in addition. Two exhibited action tremor and postural tremor as well as tremor at rest. Three patients were untreated for PD. The others were on different combinations of L-dopa, bromocriptine, selegiline, and amantadine. Anticholinergic treatment, if present, was discontinued one month prior to onset of budipine therapy.

Controlled study

A total of 84 patients participated in the entire study (44 in the placebo group, 40 in the budipine group). Mean age was 66 $\pm$ 6 yrs in the placebo group, 62 $\pm$ 8 yrs in the budipine group. Mean levodopa equivalance dose was 472 $\pm$ 200 mg and 506 $\pm$ 224 mg in the placebo and budipine group respectively.

Table 1. Clinical characteristics of patients enrolled in the controlled study

	All patients		Patients with EMG recordings	
	placebo	budipine	placebo	budipine
n	44	40	7	7
Age (yrs)	66 ± 6	62 ± 8	63 ± 3	60 ± 6
Sex m	61.4%	67.5%	4	5
w	38.6%	32.5%	3	2
Levodopa-equivalence	472 ± 200 mg	506 ± 224 mg	548 ± 250 mg	392 ± 124 mg
CURS	38 ± 8	36 ± 8	35 ± 8	40 ± 8

Of these patients a subgroup of 14 (nine men, five woman), who were recruited from three different centers, underwent long-term EMG recording. Seven patients received placebo (mean age 63 ± 3 yrs), and seven budipin (mean age 60 ± 6 yrs). All patients were on a combination of levodopa and bromocriptine. Mean levodopa equivalance dose was 548 ± 250 mg and 392 ± 124 mg in the placebo and budipine group, respectively. Levodopa equivalance was defined as the levodopa dose plus ten times the bromocriptine dose. Table 1 shows that this subgroup of patients did not differ significantly from the entire patient group.

Study design

Open-label study

The only inclusion criterion was insufficiently controled tremor. Budipine was given as "add-on" therapy in an open design to patients taking various combinations of conventional antiparkinson drugs, except for anticholinergics. The remaining medication was kept stable during the study. The dosage of budipine was adjusted individually according to treatment effect and drug tolerance. Before and under treatment (three to five months after initiation of therapy) tremor quantification and UPDRS scoring were performed. All patients gave informed consent.

Controlled study

The study was conducted as a randomized, double-blind parallel group design. Study medication was given as add-on therapy to patients with PD and a value between 24 and 50 on the CURS. All patients were on a stable dose of levodopa and bromocriptine. Within six weeks medication was increased to 60 mg (budipine or placebo). In case of adverse advents medication was reduced to 40 mg. Clinical rating was obtained with the Columbia University Rating Scale (CURS) twice before medication and in bi- or four-weekly intervals under medication for four months. A tremor subscore was calculated from the tremor-related items of the CURS. The arithmetic mean of the tremor subscore at weeks −4 and 0 (pretreatment score) and of weeks 12 and 16 (treatment score) was calculated. Longterm EMG was obtained in the subgroup of 14 patients in the week before administration of the study medication and after four months of treatment.

Statistics

Statistical analysis of the UPDRS (open study) and CURS (controlled study) data before and under treatment was performed using Wilcoxon's rank sum test. The EMG data were analysed using a two-sided Student's t test.

Results

Open-label study

Adverse effects, drop-outs, and dosages administered

Side effects consisted mainly of a dry mouth (seven patients), which was tolerated well by all of them. Slight memory impairment was complained of by two patients. two patients dropped out of the study because of adverse reactions, one of which was probably drug related (nausea), the other probably not (cardiac arrhythmias). Of the nine patients who concluded the study, the final daily doses were 40mg (10mg qid) in two patients, 60mg (20mg tid) in six, and 80mg (20mg qid) in one patient (mean dose 58 +/− 12mg).

Clinical ratings

Total UPDRS scores (Fig. 1, left) decreased from 40 to 32.5 (reduction of 20% (7.4 +/− 6.1, $p < 0.01$)). Two of the tremor parameters of the UPDRS were reduced significantly: factor 16 (history of tremor) from 2.4 to 1.4 ($p < 0.001$) (Fig. 1, middle); factor 20 (tremor at rest on examination) from 3.6 to 2.2 ($p < 0.05$) (Fig. 1, right). Factor 21 (action or postural tremor on examination) was not affected (1.8 on both rating sessions).

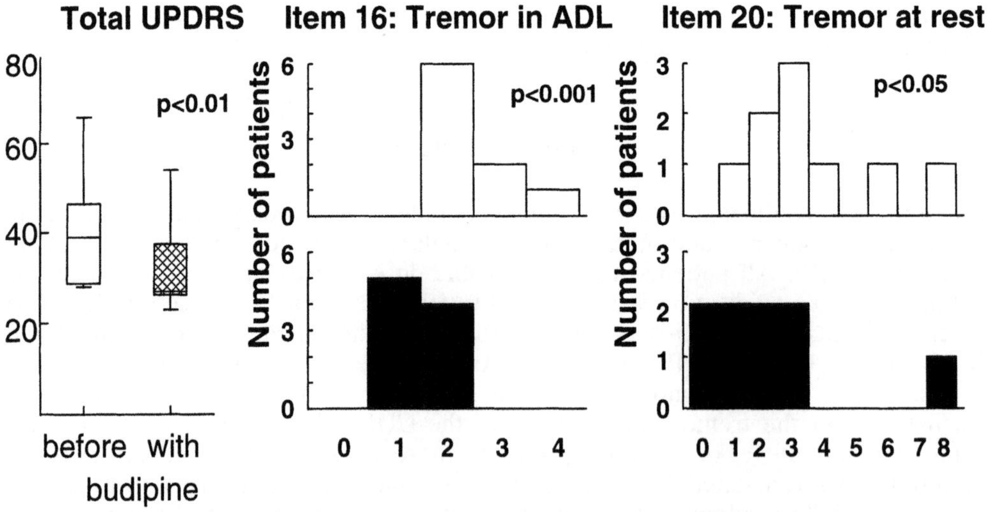

Fig. 1. Improvement of UPDRS ratings in the open label study

Tremor recordings

Mean "tremor occurrence" was reduced from 52.0% to 34.7% (reduction of 33.7 +/− 27.6% ($p < 0.05$)) (Fig. 2, left). Mean "tremor intensity" decreased from 15.3 to 11.3 (reduction of 25.2 +/− 13.1% ($p < 0.01$)) (Fig. 2, right). Except for two cases, both parameters changed proportionally (data not shown). Mean tremor frequencies (5.2 Hz) remained unchanged by medication.

Controlled study

Adverse effects, drop-outs, and dosages administered

In the subgroup of patients with tremor recordings there were no side effects leading to a termination of the study. Two patients in the budipine group prematurely terminated the study because of lack of efficacy. In the budipine group six patients were treated with 60 mg and one with 40 mg. Data concerning adverse effects, drop-outs, and dosages for the entire group are presented by Müller and Przuntek (this issue).

Clinical ratings

Figure 3 shows the change of the tremor subscore from the CURS during each visit for the entire group. The pretreatment and treatment scores resp. decreased from 6.4 ± 3.4 to 5.3 ± 3.9 in the placebo group and from 6.1 ± 2.5 to 3.5 ± 2.6 in the budipine group (Fig. 4, left). The group difference was statistical significant ($p < 0.005$). In the subgroup of patients with EMG recordings the UPDRS tremor subscore decreased from 7.1 ± 3.5 to 5.5 ± 2.9 in the placebo group (n.s.), and from 5.7 ± 2.9 to 3.0 ± 3.2 in the budipine group ($p < 0.05$) (Fig. 4, right). The difference between the treatment arms in this subgroup of patients was not significant.

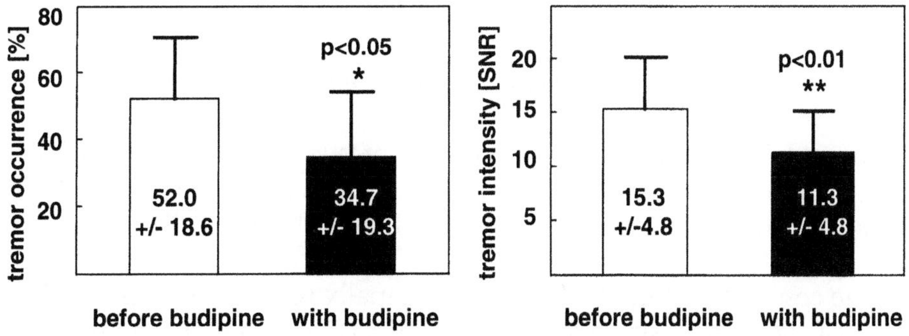

Fig. 2. Improvement of the EMG scores in the open label study

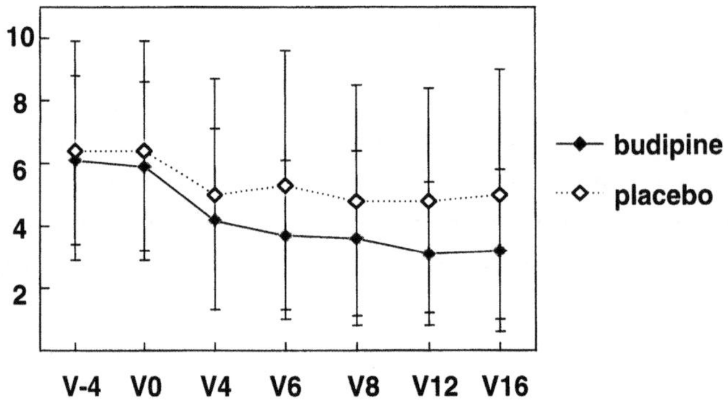

Fig. 3. CURS tremor subscore in the entire patient group

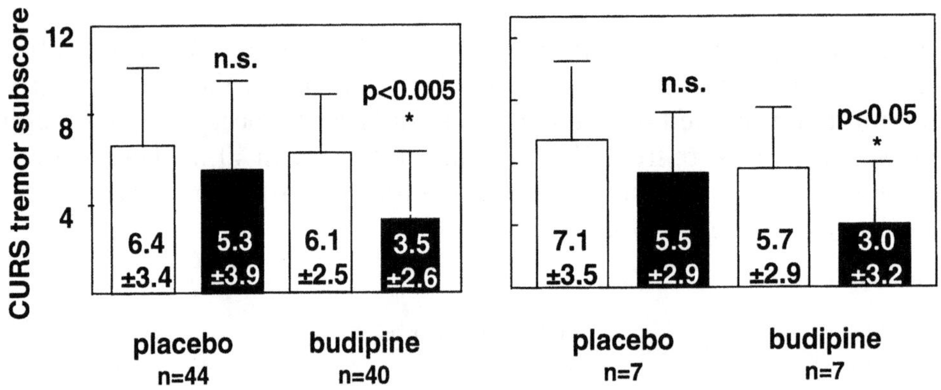

Fig. 4. Pretreatment (white boxes) and treatment (black boxes) tremor subscores in the entire group (left) and in the subgroup of patients with EMG recordings (right)

Tremor recordings

The parameter "tremor occurrence" changed from 26.6 ± 21.1 to 24.7 ± 15.5 in the placebo group (n.s.) and from 24.7 ± 15.5 to 14.8 ± 14.5 in the budipine group ($p < 0.05$) (Fig. 5, left). "Tremor intensity" (Fig. 5, right) did not change in the placebo group (10.1 ± 4.2 to 10.3 ± 3.7; n.s.). The change in the budipine group (9.1 ± 2.5 to 7.2 ± 1.6) just failed to reach significance. The difference between both treatment groups was not significant.

Correlation between clinical ratings and tremor recordings

In the open-label study, with the exception of two patients, both EMG parameters, tremor occurrence and intensity changed in the same direction (data not shown). The same applies to the data of the controlled study (Fig. 6, left).

In the controlled study EMG data principally responded in parallel to the changes of the tremor subscore of the CURS (Fig. 6, right). The only exceptions were two patients of the placebo group. They were clinically judged as

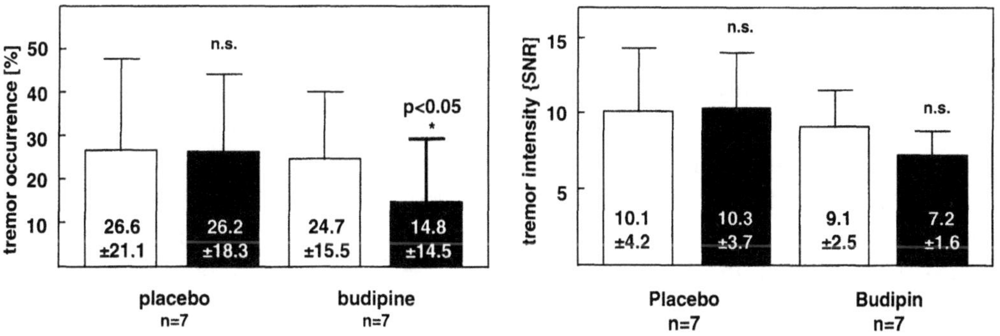

Fig. 5. EMG parameters (left: tremor occurrence; right: tremor intensity) in the subgroup of patients with EMG recordings in the controlled study. White boxes: pretreatment; black boxes: with treatment

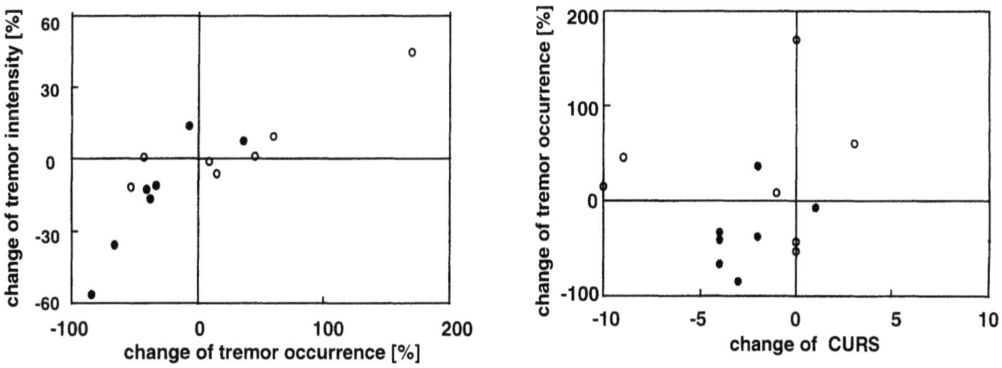

Fig. 6. Left: relation of changes in the two EMG parameter (tremor occurrence and tremor intensity) with treatment (changes given in percentage); right: comparison of changes of EMG data with changes in the CURS tremor subscore. ○ placebo, • budipine

having improved to a great extent, but were measured as being unchanged or as having deteriorated, respectively.

Discussion

The present studies demonstrate that budipine has good tremorlytic activity, both when measured with clinical tremor scores and with an objective method. The open study showed an improvement in the UPDRS tremor score and in the longterm-EMG. In the controlled study clinical scores were significantly better in the budipine group. In those patients with additional EMG recordings both CURS scores and EMG improved significantly in the budipine group and not in the placebo group. However, due to the small patient number the difference between the two groups was not significant. In the light of the good tolerability, especially the low occurrence of cognitive side effects, it seems likely, that budipine will replace anticholinergics for treatment of tremor in PD. On the basis of the present findings a larger controlled multicenter study has been started, which will hopefully confirm these first results.

Both EMG parameters, tremor occurrence and tremor intensity, improved in parallel to each other. Thus, although these two measures conceptually measure different entities, they seem to be redundant and the determination of only one of them should be sufficient in further clinical studies. As tremor occurrence has greater validity than tremor intensity when measuring tremor in PD (Spieker et al., 1997), it should be the preferred parameter. There is a good correlation of the response of the EMG data with the change on the clinical scales under treatment. Interestingly, the only two outliers were in the placebo group. They were clinically evaluated as having improved. According to the EMG, however, they deteriorated, which is much more likely to be correct, as they received placebo. We conclude, that longterm EMG is not only valid and reliable for the quantification of tremor, as we have shown before (Spieker et al., 1995, Spieker et al., 1997); it is also sensitive in measuring drug-induced changes, i.e. it has a good responsitivity (Kirshner and Guyatt, 1985). This issue, however, will also have to be further studied in a larger patient series.

References

Bacher M, Scholz E, Diener HC (1989) 24 hour continuous tremor quantification based on EMG recording. Electroencephalogr Clin Neurophysiol 72: 176–183

Duvoisin RC (1971) The evaluation of extrapyramidal disease. In: Ajuriaguerra J, Gauthier G (eds) Monoamines noyaux gris centraux et syndrome de Parkinson. Georg & Cie, Genf, pp 313–325

Fahn S, Tolosa E, Marin C (1988) Clinical rating scale for tremor. In: Jankovic J, Tolosa E (eds) Parkinson's disease and movement disorders. Urban & Schwarzenberg, Baltimore, pp 225–234

Iizuka J (1985) Controlled single-blind trial of budipine, L-dopa, and amantadine in 102 patients with Parkinson's disease. In: Gerstenbrand WPF, Stern G (eds) Clinical experiences with Budipine in Parkinson therapy. Springer, Berlin Heidelberg New York Tokyo, pp 129–139

Ketz E (1985) Open clinical trial of budipine in 25 tremor dominant parkinsonian patients. In: Gerstenbrand F, Poewe W, Stern G (eds) Clinical experiences with budipine in Parkinson therapy. Springer, Berlin Heidelberg New York Tokyo, pp 140–141

Kirshner B, Guyatt G (1985) A methodological framework for assessing health indices. J Chronic Dis 38: 27–36

Spieker S, Jentgens C, Boose A, Dichgans J (1995) Reliability, specificity and sensitivity of long-term tremor recordings. Electroencephalogr Clin Neurophysiol 97: 326–331

Spieker S, Ströle V, Sailer A, Boose A, Dichgans J (1997) Validity of longterm EMG in the quantification of tremor. Mov Disord 12: 985–991

Ulm G (1985) Effect of Budipine on Parkinsonian tremor resistant to other Parkinsonian mediation. In: Gerstenbrand WPF, Stern G (eds) Clinical experiences with Budipine in Parkinson therapy. Springer, Berlin Heidelberg New York Tokyo, pp 142–151

Webster DD (1968) Critical analysis of the disability in Parkinson's disease. Mod Treat 5: 257–282

Authors' address: Sybille Spieker, MD, Department of Neurology, University of Tübingen, Hoppe-Seyler-Strasse 3, D-72074 Tübingen, Federal Republic of Germany

Current management of motor fluctuations in patients with advanced Parkinson's disease treated chronically with levodopa

E. Melamed, J. Zoldan, R. Galili-Mosberg, I. Ziv, and **R. Djaldetti**

Department of Neurology, Rabin Medical Center, Beilinson Campus, Petah Tikva, and Sackler School of Medicine, Tel-Aviv University, Israel

Summary. Motor fluctuations after long-term administration of levodopa may be due to central pharmacodynamic mechanisms such as reduced striatal synthesis and storage of dopamine from exogenous levodopa and subsensitization of postsynaptic dopaminergic receptors. Peripheral pharmacokinetic mechanisms may be equally important, particularly in motor fluctuations of the "delayed on" (increased time latencies from dose intake to start-up of clinical benefit) and "no-on" (complete failure of a levodopa dose to exert an "on" response) types. Levodopa itself has a very poor solubility. In addition, there is delayed gastric emptying in many advanced patients. Therefore, an oral dose of levodopa may remain in the stomach for long periods of time before it passes into the duodenum where there is immediate absorption. Consequently, in order to overcome response fluctuations caused by impaired pharmacokinetic mechanisms and to improve its absorption, we recommend that levodopa be taken in multiple small doses, on an empty stomach, preferably crushed and mixed with a lot of liquid. Protein intake should be minimized. Prokinetic drugs such as prepulsid (Cisaprid) could be used to facilitate gastric motility and levodopa transit time. Administration of crushed levodopa through nasoduodenal or gastrojejunostomy tubes may be helpful in certain circumstances. Bypassing the stomach with subcutaneous injections of apomorphine may provide dramatic rescue from difficult "off" situations. Oral and s.c. administration of novel, extremely soluble prodrugs of levodopa, e.g., levodopa ethylester, may offer a new approach to overcome difficulties in levodopa absorption. Addition of dopamine agonists, MAO-B inhibitors, COMT inhibitors and controlled release levodopa preparations may be helpful in prolonging the duration of efficacy of each single levodopa dose.

Levodopa, administered orally, usually combined with peripheral dopa decarboxylase inhibitors, continues to be the most widely-used and most effective pharmacological treatment for Parkinson's disease (Melamed, 1987). Undoubtedly, the outstanding therapeutic success of levodopa represents a dramatic and revolutionary breakthrough in medicine, in general, and in neurology, in particular. Although, since the introduction of levodopa, there have been many additional pharmacological and even surgical antiparkinsonian strategies, it still stands out as a mandatory axis of treatment in the majority of patients (Steigler and Quinn, 1992). Indeed, levodopa therapy improves, sometimes markedly, the motor signs and symptoms of the illness, the functional capacity and quality of life and perhaps also life expectancy of the afflicted patients. It is therefore unfortunate that after an initial problem-free period of successful, smooth and stable clinical benefit from levodopa that lasts about two to five years, the responsiveness of many patients worsens with the emergence of a variety of complications (Marsden et al., 1982; Hardie et al., 1984). These adverse reactions include dyskinesias and dystonias, psychotic problems and, particularly, the troublesome motor fluctuations (Marsden and Parkes, 1977; Marsden, 1994). The latter phenomenon may be particularly complex, limiting and disabling. It is believed that most patients on long-term levodopa therapy will, sooner or later, develop response fluctuations of varying types and severity (Riley and Lang, 1993). Because of the serious impact of these phenomena on the quality of life and function of the patients, many efforts are now being undertaken to identify the responsible mechanisms and to devise preventive and therapeutic measures.

Characteristics of motor fluctuations

The natural history of the effectiveness of levodopa in patients with Parkinson's disease may be divided into two major parts, i.e., an initial "levodopa honeymoon" and a later "paradise lost" period.

"Levodopa honeymoon"

This is the good initial period that is induced at the start of levodopa therapy. It is usually temporary, lasts for 2–5 years in most patients, may be shorter or longer in the minority of parkinsonians, and is only rarely eternal. This period is characterized by a successful beneficial effect of levodopa which may range in quality from good to excellent. The patient, his family, and treating physicians are all satisfied. Patients take levodopa in 2–4 daily doses. Overall benefit is smooth and stable throughout the day. Patients do not feel the good effect of each single levodopa dose nor do they experience a beginning or an end of an improvement after intake of individual levodopa doses. They do not feel the need to take the next scheduled dose and can even miss a dose or two without any clinical worsening. The effect of levodopa therapy is global.

"Paradise lost"

This "post-honeymoon" period may be divided into two stages, i.e., an initial simple and a late complex period of response fluctuations. The termination of the good initial period is characterized by a rather drastic change in the pattern of responsiveness to levodopa which may occur abruptly or more gradually. The patients now begin to feel that the beneficial effect is linked to the intake of the individual doses of levodopa. They experience the start of relief after a certain latency period from ingestion of the drug and also the end of the beneficial effect. Patients now feel the necessity to take the next scheduled levodopa dose after the good effect of the previous dose has dissipated. They can no longer tolerate missing single doses and become dependent on regular self-administration for being able to function. Patients now require more daily levodopa doses, e.g., 6–8 for smoother and satisfactory performance. If, during the "honeymoon" period, the nights were usually problem-free, the patients may now begin to require nightly doses to overcome nocturnal bradykinesia and stiffness. When patients report these peculiar change of events, the experienced treating neurologist knows that the satisfactory and optimally-beneficial "levodopa honeymoon" is forever gone and that the patient is entering a period of simple motor fluctuations and that even this phase is temporary and eventually he will lapse into the final and irreversible stage of the natural history of levodopa efficacy, a period with complex and difficult to tame motor fluctuations.

The more advanced period of response fluctuations associated with long-term levodopa therapy is characterized by an increasing severity, frequency and complexity of oscillations in the response to doses of levodopa. Fluctuations at this phase may be predictable and dose-related or completely random and unexpected. Daily motor activity and functional capability is characterized and governed by remarkable shifts or fluctuations between "off" and "on" states. "Off" phases are when the basic parkinsonian symptomatology takes over, sometimes in the form of a real crisis with severe slowness, rigidity and tremor, leaving patients immobile and incapacitated. "On" periods are characterized by a beneficial effect of levodopa taking over with release from the parkinsonian signs when patients can perform. There are several commonly encountered subtypes of response fluctuations.

a. The "wearing off" phenonemon — this signifies shortening of the duration of clinical relief induced by the individual doses of levodopa. If at the initial phase following the termination of the "honeymoon" each dose may work for several (3–5) hours, there is a gradual and progressive shrinkage to 1–2 hours lasting benefit.

b. The "delayed on" phenomenon — this signifies increased latency periods from intake of an oral dose of levodopa to start up of a clinical benefit (Melamed and Bitton, 1986). If, initially when the patient first becomes a fluctuator, such latency may be brief and range from 5–20 minutes, there is a gradual and progressive increase. Patients may sometimes wait for as long as 30 minutes to 1½ hours for an onset of clinical relief.

c. The "no-on" phenomenon — this signifies occasional total failure of an oral dose of levodopa to induce an "on" (Melamed et al., 1986). Such episodic unresponsiveness is often described by the patients as if the unsuccessful dose of levodopa has not been taken at all.

d. The "on-off" phenomenon — although initially this term was used to generally describe the motor fluctuations, it currently refers particularly to abrupt and mostly random and unpredictable loss of benefit which occurs during a successful "on" phase. This may be manifested by a severe parkinsonian crisis associated with sympathetic overactivity e.g., palpitations, hot flushes, sweating and panic. Such an attack may last from minutes to hours and may subside spontaneously or only after intake of an additional dose of levodopa.

The daily performance of the patient is now dominated by swings between various "off" and "on" phases and response pattern may become completely chaotic and out of control. The effect of single doses of levodopa may vary from one dose to the other and from day to day. In most patients, the response to morning doses are somewhat better than to those taken in the second half of the day and particularly in the afternoons. Quality, duration and predictability of efficacy is better after the first doses of the day. Patients need to take levodopa at night to avoid nocturnal "offs" during sleep or wakefulness. "On" periods are almost invariably associated with dyskinesias which may, at this stage, be particularly violent and incapacitating (Lesser et al., 1979; Nutt, 1994). The dyskinesias may be of the "peak dose" or the "beginning and end of dose" subtypes. In many patients, the therapeutic window between pure clinical relief and dyskinesias exerted by single levodopa doses becomes very narrow or non-existent so that when a dose successfully turns them on, they will have obligatory dyskinesias in an "all or none" manner. In many older patients, penalty for success of levodopa in alleviating the motor symptoms is the exacerbation of parkinsonian psychosis manifested mainly by visual hallucinations, paranoid delusions and confusion. Levodopa does not halt the progression of the illness which progresses relentlessly under coverage of this treatment. Therefore, despite the severity and complexity of the motor fluctuations, in most patients, the disease has advanced and they are dependent upon and cannot perform without continued levodopa treatment.

The mechanisms responsible for the development of motor fluctuations

The causes of the change in pattern of responsiveness and emergence of motor fluctuations after chronic treatment with levodopa are only partly understood. Apparently, more than one and possibly several mechanisms are operative in producing this complex syndrome. The success of levodopa in inducing clinical benefit in Parkinson's disease is, in all likelihood, linked to its absorption from the gut, its entry through the blood-brain barrier into the striatum and its conversion to functional dopamine molecules that reach and activate postsynaptic striatal dopaminergic receptors and thus restore the deficient nigrostriatal transmission (Hefti and Melamed, 1980). By deduction, it may be assumed that the "off" parts of response fluctuations are due to lack or termination of stimulation of receptors by the dopamine formed in striatum by an oral dose of exogenous levodopa. It is obvious that, in general, motor fluctuations may be due to many obstacles along the complicated metabolic route of a levodopa dose including failures in its delivery to the brain, in its striatal conversion to dopamine and its ability to interact with the receptors (Nutt, 1987; Wooten, 1988). It presently seems that there are two major groups of responsible mechanisms, i.e., central pharmacodynamic (Fabrini et al., 1988; Mouradian et al., 1988) and peripheral pharmacokinetic causes (Muenter and Tyce, 1971; Fahn, 1977, Djaldetti et al., 1996).

Possible central mechanisms

a. Progression of nigrostriatal degeneration and illness — The remaining dopaminergic nerve terminals constitute a major locus for decarboxylation of exogenous levodopa to dopamine and its vesicular storage for regulated release in the parkinsonian striatum. Continued loss of such terminals decreases the capacity of the striatum to generate and store dopamine from levodopa. Conversion of levodopa to dopamine now shifts to non-dopaminergic compartments that, although they contain dopa decarboxylase, lack vesicular dopamine storage sites. The initial "buffered" effect of levodopa which characterizes the "honeymoon" is lost and dopamine formed from a single dose of levodopa is immediately spilled, activates the receptors and since levodopa is rapidly washed out, it induces only a brief "on" response.

b. Alterations in dopaminergic receptors — Fortunately, for patients with Parkinson's disease, the striatal postsynaptic receptor machinery remains intact and may even be supersensitive due to the denervation of the presynaptic nigrostriatal neurons. It is possible that chronic bombardment of these receptors with large amounts of levodopa and the unregulated formed dopamine causes desensitization and down regulation of these receptors. In time, there could also be an increased aberrant synthesis of certain metabolites of levodopa and dopamine that act as blocking false transmitters, occupy the receptors and prevent their stimulation by the dopamine formed from levodopa. Another possibility is that overstimulation of the receptors by excess unregulated dopamine may induce an "off" state by causing a hyperpolarization block. Since patients fluctuate and have "on" states and also dyskinesias, it should be assumed that the above phenomena are not permanent but are rather oscillatory.

c. Levodopa itself, given chronically, may suppress its own utilization and its conversion to dopamine for instance by a local striatal depletion of co-factors required for the decarboxylation process.

Possible peripheral mechanisms

Pharmaceutically, levodopa is a rather undesirable drug since it is barely soluble, has a very short plasma half-life after oral intake and its clearance from the body is rapid. Levodopa is produced by hydroxylation of tyrosine and is, itself, a neutral amino acid. As such, it competes with the other neutral amino acids such as valine, leucine, and isoleucine on absorption from gut to blood and on penetration from the systemic circulation into the striatum through the blood-brain barrier. A protein-rich meal can adversely affect the success of a levodopa dose due to competition with the amino acids formed in the gut by its breakdown. Levodopa is metabolized peripherally via several mechanisms. Firstly, it can be decarboxylated to dopamine but this route is negligible since most patients receive levodopa in combination with peripheral dopa decarboxylase inhibitors. It can be converted to 3-0-methyldopa by the enzyme catechol-0-methytransferase (COMT). Although this metabolite is inert, its formation can reduce the efficacy of single levodopa doses since it can compete with levodopa at the blood-brain barrier and decrease it entry into striatum. Another, less explored possible mechanism, is enhanced deamination and inactivation by liver enzymes. More importantly, many patients develop atony and reduced motility of the stomach. We now believe that more than all of the above, delayed gastric emptying is a major obstacle in the absorption of oral doses of levodopa. The drug is mainly absorbed in the duodenum. The sluggish movement of the stomach, combined with the poor solubility of levodopa, interferes with its passage from stomach to duodenum and leads to gastric stagnation of the drug (Djaldetti et al., 1996). Protein-containing meals and just food bulk worsen the dissolution of levodopa and further prevents its transfer from stomach into the duodenum. That is probably why levodopa doses taken after food are much less successful in inducing an "on" response. On one hand, the delayed gastric emptying may be due to additional degeneration of the cholinergic plexuses innervating the stomach as part of the pathology of Parkinson's disease. Another possibility is that stomach motility is altered by the chronic bombardment of gastric dopaminergic receptors by levodopa.

It is becoming clear that the various subtypes of response fluctuations may be caused by different specific mechanisms and should therefore be treated accordingly. For instance, the "wearing-off" and "on-off" phenomena may be mainly due to aberrations in central pharmacodynamic mechanisms. By contrast, the "delayed on" and "no-on" phenomena may be mainly or even exclusively due to erratic absorption of orally-ingested levodopa linked to impairment in pheripheral pharmacokinetic mechanisms and particularly delayed gastric emptying.

Practical guidelines for the management of motor response fluctuations

1. General approaches (Waters, 1997; Stocchi et al., 1997)

a. Rescheduling of levodopa daily dosages — in order to counteract the "wearing off" phenomenon and to prolong total duration of daily "on" hours, it is no longer possible to give levodopa in 3–4 doses. The drug has to be administered in multiple, usually 6–8, doses. Many patients have to take levodopa regularly once every 2–2½ hours during the waking hours. Interestingly, the individual doses of levodopa (with dopa-decarboxyase inhibitors such as carbidopa or benserazide) generally do not require increases. In fact, patients may progressively become sensitive and the single doses may even have to be tapered. Each dose of levodopa has to be tailored to the patients' needs. Regularly, the first dose in the morning should be higher to produce ignition. Also, the afternoon doses which usually fail to induce an "on" effect, should be increased. In general, response to the first morning doses of the day is better than that in the second half (afternoons and evenings). The neurologist should strive to achieve as many good daily "on" hours with the least amount of side effects such as dyskinesias, as possible. Often, the "therapeutic window" of each single dose becomes very narrow so that each dose, when successful in eliciting a clinical benefit, will also induce dyskinesias in an "all or none" fashion. A daily schedule and dosage may be optimal for a certain period and then has to be retailored again because there is suddenly either too much or too little of levodopa.

b. Use of controlled-release levodopa preparations (Poewe et al., 1986) — Theoretically, the currently available slow, controlled-release levodopa preparations such as sinemet CR or madopar HRS or LP, may have advantages over the regular short-term forms of levodopa in that they can counteract the "wearing off" phenonemon and prolong duration of beneft of single doses (e.g., from 2 to 4 hours). However, there are certain disadvantages. Firstly, there is an increased delay in time from oral intake to "turning on" and the patient misses the "kick in" phenonemon. Furthermore, in practicality, these formulations also have to be taken in multiple daily doses and the individual doses usually have to be increased by 33–50% to be effective. Therefore, replacement of regular levodopa doses by doses of controlled release preparations is usually unsuccessful. There may be some benefit to adding them on top of the regular levodopa schedule. However, this requires a lot of patience and sophistication by patients and is usually frustrating.

These preparations are best used to counteract nocturnal akinesia. When a controlled-release levodopa is taken at night, before sleep, it may often improve mobility, reduce rigidity, ease pains and promote sleep. If it acts long enough, it may even beneficially affect early morning akinesia and foot dystonia which are due to lack of levodopa in striatum at that particular time.

c. Add deprenyl, a monoamine oxidase (MAO)-B inhibitor (Golbe et al., 1988) — if a fluctuating patient is still not on deprenyl (selegiline, jumex), its addition (5–10mg daily) may be helpful. It blocks degradation of the dopamine formed from exogenous levodopa in the parkinsonian caudate and putamen nuclei. It, therefore, enhances concentrations and duration of dopamine in central synapses and thus prolongs duration of "on" periods and reduces the "wearing off" phenomenon. It may sometimes promote psychotic phenomena and dyskinesias.

d. Add dopamine agonists (Goetz et al., 1996; Montastruc et al., 1993; Jenner, 1995) — these agents, including bromocriptine, lisuride, pergolide, ropinirole, pramipexole and cabergoline, directly stimulate striatal postsynaptic dopa-minergic receptors and enhance nigrostriatal dopaminergic neurotransmission by bypassing the presynaptic nigral projections. When patients have already become fluctuators, these drugs are not sufficiently powerful to replace levodopa. However, their oral administration can serve as a very useful adjunct to levodopa and may act to successfully prolong daily "on" hours. The list of potential side effects includes allergic reactions, acroedema, nausea, orthostatic hypotension, dyskinesias and psychotic phenomena such as visual hallucinations and paranoid delusions. The type of agonist and dosage should be chosen individually for each patient.

Apomorphine is a most powerful dopamine agonist and the only one with a clinical effect similar to that of levodopa. It cannot be given orally but only by subcutaneous injections. It is a very strong emetic drug and therefore could not be used until the advent of the antiemetic domperidone, a peripheral dopamine receptor antagonist. Cotreatment of the patients with domperidone permits utilization of apomorphine. Tolerance against emesis soon develops and domperidone can be discontinued. Apomorphine is best used as a rescue strategy to induce a rapid "on" during difficult "off" situations (Stibe et al., 1988). A s.c. dose of 2–6 mg dramatically relieves the patient within 5–10 minutes. Duration of effect is brief and lasts, on the average, 45–90 minutes.

e. Add COMT inhibitors (Roberts et al., 1997, Kurth et al., 1997; Rajput et al., 1997; Kaakkola et al., 1994) — the two recently available, tolcapone and entecapone, may be beneficial as an add on treatment to levodopa. By blocking formation of the metabolite 3-0-methyldopa, they can improve the duration of benefit induced by single doses of levodopa, reduce "off" time and improve the duration of daily "on" hours.

2. Strategies to improve absorption of levodopa

a. instruct patients to take levodopa before meals, on an empty stomach (Baruzzi et al., 1987);
b. instruct patients to avoid protein-containing meals and also bulk-forming foods (Pincus and Barry, 1987; Carter et al., 1989);
c. instruct patients not to lie down and move around (if they can) after dose ingestion;
d. instruct patients to crush levodopa into fine powder and take dose as a suspension in a sparkling (soda) liquid (Kurth et al., 1993);
e. use dispersible forms of levodopa (e.g., dispersible madopar);
f. use oral soluble prodrugs such as levodopa methylester and levodopa ethylester (Stocchi et al., 1997);
g. use prokinetic drugs to facilitate gastric emptying and transit of levodopa into the duodenum (Djaldetti et al., 1995);
h. in cases of emergency, administer levodopa suspension via nasoduodenal or gastrojejunostomy tubes;
i. bypass gastrointestinal route and administer soluble prodrugs of levodopa e.g. levodopa methyl- and ethylester (Djaldetti and Melamed, 1996) subcutaneously or intravenously or apomorphine subcutaneously.

3. Surgical approaches

a. Thalamotomy and pallidotomy — these can improve contralateral tremor and stiffness and pallidotomy is particularly effective against dyskinesias, but the basic motor fluctuations are not greatly improved;

b. deep brain stimulation — subthalamic stimulation by implanted electrodes can produce dramatic and prolonged benefit with quality of relief similar to that of levodopa but without fluctuations and dyskinesias.

c. transplantation of fetal nigral tissue into the basal ganglia — outgrowth and sprouting of axons of implanted fetal dopaminergic neurons and the establishment of functional synapses can greatly improve the basic motor parkinsonian signs and symptoms. Also, enrichment of the striatum with a new widespread network of dopaminergic nerve terminals that contain both the enzyme dopa decarboxylase and dopamine storage vesicles increases the decarboxylating capacity of exogenous levodopa and of storage of the formed dopamine for more controlled and slower release. Therefore, this surgical approach may improve or even abolish the motor fluctuations and restore the original lost "levodopa honeymoon".

There is no question that treatment of the complex motor fluctuations in patients with advanced Parkinson's disease requires skill and sophistication (Stern, 1997). Although, with the many tools available to us today, we can help our patients better and gain significantly more daily "on" hours, the state of affairs is still far from ideal. There is definitely an urgent need to better understand the responsible mechanisms so that strategies can be designed to prevent emergence of these phenomena and, if they develop, to offer better countermeasures.

Acknowledgement

Supported, in part, by the National Parkinson Foundation, U.S.A.

References

Baruzzi A, Cantin M, Riva R (1987) Influence of meal ingestion time on pharmacokinetics of orally-administered levodopa in parkinsonian patients. Clin Neuropharmacol 10: 527–537

Carter JH, Nutt JH, Woodward WR (1909) amount and distribution of dietary protein affects clinical response to levodopa in Parkinson's disease. Neurology 39: 552–556

Djaldetti R, Melamed E (1996) Levodopa ethylester: A novel rescue therapy for response fluctuations in Parkinson's disease. Ann Neurol 39: 401–404

Djaldetti R, Koren M, Ziv I, Achiron A, Melamed E (1995) Effect of Cisapride on response fluctuations in Parkinson's disease. Mov Disord 10: 81–84

Djaldetti R, Ziv I, Melamed E (1996a) Impaired absorption of oral levodopa: A major cause for response fluctuations in Parkinson's disease. Isr J Med Sci 32: 1224–1227

Djaldetti R, Baron J, Ziv I, Melamed E (1996b) Gastric emptying in Parkinson's disease: Patients with and without response fluctuations. Neurology 46: 1051–1054

Fabrini G, Mouradian MM, Juncos JL, Schlegel J, Mohr E, Chase TN (1988) Motor fluctuations in Parkinson's disease: Central pathophysiological mechanisms, Part I. Ann Neurol 24: 366–371

Fahn S (1977) Episodic failure of absorption of levodopa: a factor in the control of clinical fluctuations in the treatment of parkinsonism. Neurology 27: 390–392

Goetz GG, Diedrich NJ (1992) Dopaminergic agonists for the treatment of Parkinson's disease. Neurol Clin 10: 527–540
Golbe LI, Lieberman AN, Muenter MD, Ahlskog JE, Gopinathan G (1988) Deprenyl in the treatment of symptom fluctuations in advanced Parkinson's disease. Clin Neuropharmacol 11: 45–55
Hardie RJ, Lees AJ, Stern GM (1984) On-off fluctuations in Parkinson's disease: A clinical and neuropharmacological study. Brain 107: 487–506
Hefti F, Melamed E (1980) L-dopa's mechanism of action in Parkinson's disease. Trends Neurosci 3: 229–231
Jenner P (1995) The rational for use of dopamine agonists in Parkinson's disease. Neurology 45 [Suppl 3]: S6–S12
Kaakkola S, Teravainen H, Ahtila S, Rita H, Gordin A (1994) Effect of entecapone, a COMT inhibitor, on clinical disability and levodopa metabolism in parkinsonian patients. Neurology 44: 77–80
Kurth MC, Tetrud LW, Irwin I, Lyness WH, Langston JW (1993) Oral levodopa/carbidopa solution versus tablets in Parkinson's patients with severe fluctuations — a pilot study. Neurology 43: 1036–1039
Kurth MC, Adler CH, Saint-Hillaire M (1997) Tolcapone improves motor function and reduces levodopa requirement in patients with Parkinson's disease experiencing motor fluctuations: a multicenter double-blind randomized placebo-controlled trial. Neurology 48: 81–87
Lesser RP, Fahn S, Snider SR, Cote L (1979) Analysis of the clinical problems in parkinsonism and the complications of long-term levodopa therapy. Neurology 29: 1253–1266
Marsden CD (1994) Parkinson's disease. J Neurol Neurosurg Psychiatry 57: 672–681
Marsden CD, Parkes JD (1977) Success and problems of long-term levodopa therapy in Parkinson's disease. Lancet 1: 345–349
Marsden CD, Parkes JD, Quinn NP (1982) Fluctuations of disability in Parkinson's disease — clinical aspects. In: Marsden CD, Fahn S (eds) Movement disorders. Butterworth, London, pp 96–122
Melamed E (1987) L-dopa. In: Adelman G (ed) Encyclopedia of neuroscience. Birkhauser-Boston, Cambridge, pp 553–554
Melamed E, Bitton V (1986) Delayed onset of responses to individual doses of levodopa in parkinsonian fluctuators on long-term L-dopa therapy. Clin Neuropharmacol 9: 182–186
Melamed E, Bitton V, Zelig O (1986) Episodic unresponsiveness to single doses of L-dopa in parkinsonian fluctuators. Neurology 36: 1001–1003
Montastruc JL, Rascol O, Senard JM (1993) Current status of dopamine agonists in Parkinson's disease management. Drugs 46: 384–393
Mouradian MM, Juncos JL, Fabrini G, Schlegel J, Bartko JJ, Chase TN (1988) Motor fluctuations in Parkinson's disease: Central pathophysiological mechanisms. Part II. Ann Neurol 24: 372–378
Muenter MD, Tyce GM (1971) L-dopa therapy of Parkinson's disease: Plasma L-dopa concentrations therapeutic response and side effects. Mayo Clin Proc 46: 231–239
Nutt JG (1987) On-off phenomenon: relation to levodopa pharmacokinetics and pharmacodynamics. Ann Neurol 22: 535–540
Nutt JG (1990) Levodopa-induced dyskinesia: Review, observations and speculations. Neurology 40: 340–345
Pincus JH, Barry K (1987) Protein redistribution diet restores motor function in patients with dopa-resistant "off" periods. Neurology 38: 481–483
Poewe WH, Lees AJ, Stern GM (1986) Treatment of motor fluctuations in Parkinson's disease with an oral sustained-release preparation of L-dopa: clinical and pharmacokinetic observations. Clin Neuropharmacol 9: 430–439
Rajput AM, Martin W, Saint-Hillaire MH, Dorflinger E, Pedder S (1997) Tolcapone improves motor function in parkinsonian patients with the "wearing off" phenom-

enon: a double-blind, placebo-controlled multicenter study. Neurology 49: 1066–1071
Riley DE, Lang AE (1993) The spectrum of levodopa-related fluctuations in Parkinson's disease. Neurology 43: 1459–1464
Roberts JW, Cora-Locatelli G, Bravi D, Amantea MA, Mouradian MM, Chase TN (1994) Catechol-0-methyltransferase inhibitor tolcapone prolongs levodopa/carbidopa action in parkinsonian patients. Neurology 44: 2685–2688
Steiger NJ, Quinn NP (1992) Levodopa-based therapy. In: Koller WC (ed) Handbook of Parkinson's disease. Marcel Dekker, New York, pp 391–410
Stern MB (1997) Contemporary approaches to the pharmacotherapeutic management of Parkinson's disease: An overview. Neurology 49 [Suppl 1]: S2–S9
Stibe CMH, Lees AJ, Kempster PA, Stern GM (1988) Subcutaneous apomorphine in parkinsonian on-off oscillations. Lancet 1: 403–406
Stocchi F, Nordera G, Marsden CD (1997) Strategies for treating patients with advanced Parkinson's disease with disastrous fluctuations and dyskinesias. Clin Neuropharmacol 20: 95–115
Waters CH (1997) Managing the late complications of Parkinson's disease. Neurology 49 [Suppl 1] S49–S57
Wooten GF (1988) Progress in understanding the pathophysiology of treatment-related fluctuations in Parkinson's disease. Ann Neurol 24: 363–365

Authors' address: Eldad Melamed, M.D., Professor and Chairman, Department of Neurology, Rabin Medical Center, Beilinson Campus, Petah Tikva, 49100 Israel

Differentiation of dopamine agonists and their role in the treatment of Parkinson's disease

D. B. Calne

Vancouver Hospital and Health Sciences Centre, Vancouver, Canada

Summary. Since the pioneering work of Hornykiewicz and his colleagues, it has been recognized that dopaminergic cells die selectively in Parkinson's disease, and considerable improvement in symptoms can be achieved by administering levodopa, so that it may be converted to dopamine. However, levodopa has side-effects, and its duration of action is relatively brief. For these reasons, alternative approaches have been undertaken to stimulate the dopamine receptors. In particular, artificial agonists for dopamine receptors have been developed. The pioneer compound was bromocriptine, which stimulates the D2 family of receptors. Bromocriptine is an ergot derivative, and other compounds that are structurally related to ergot have been developed. In particular, lisuride and pergolide have been used for several years. Recently, an ergot derivative with an exceptionally long plasma half-life has been studied, cabergoline. Now there are also non-ergot derivatives that are D2 agonists, and are likely to have a role in the treatment of Parkinson's disease. Both ropinirole and pramipexole fall into this category, and each has been released in various countries for the treatment of Parkinson's disease. All of these compounds stimulate the D2 family of receptors, but they have varying actions on the D1 family of receptors. At present, there is no definite information on the role of the D1 family of receptors in either the therapeutic response to levodopa, or the development of adverse reactions. However, preliminary studies with a D1 agonist, ABT-431, are now in progress.

Introduction

It is now thirty years since Cotzias et al. (1967) reported the spectacular results of high doses of levodopa in the treatment of Parkinson's disease. The critical observation, that dopamine was selectively depleted in the brains of patients with Parkinson's disease, was published by Ehringer and Hornykiewicz in 1960. From this background, there emerged the realization, for the first time, that a chronic neurological disease could be dramatically improved over the long term. In the early levodopa era, magazine articles appeared with titles such as "How the wonder drug set me free." The results were truly spectacular, and now they are not readily appreciated by younger neurologists who did not witness them. What happened, in the late sixties, was unique because patients with *advanced* Parkinson's disease were treated with levodopa for the first time. This never occurs today because levodopa and related drugs are introduced as treatment at an early stage, so naturally the benefits that can occur are less striking.

Following the initial exciting reports of the benefits of levodopa, it was realized that the treatment could lead to significant side-effects. In particular, it became apparent that after several years of advancing illness and treatment with levodopa, the response became shorter after each dose, and the changes in mobility that ensued were referred to as "oscillations" or "fluctuations." Subsequently the terms "on-off reaction" and "wearing off reaction" were coined. In addition, prominent dyskinesia became a problem, and most recently we have also been seeing increasing numbers of patients who develop hallucinations and delusions as side-effects.

Under the impact of these drawbacks, new approaches to treatment were explored. It became apparent that the deterioration in mobility at the end of the interval between doses of levodopa derived, at least in part, from the pharmacokinetics of levodopa. The half-life of levodopa in the plasma is around one hour, so an improved result could be expected from medications that had a more prolonged duration of action. In addition, it became recognized that the profile of side-effects caused by levodopa was probably determined by stimulation of more than one kind of dopamine receptor. So neurologists began to ask whether artificial agonists for dopamine receptors could be developed with an improved profile of side-effects achieved through a selective action on the various categories of dopamine receptors. Better kinetics were also sought through the development of a longer plasma and tissue half-life.

Receptors

Before considering the individual dopamine agonists, it is appropriate to comment on synaptic receptors in general, because the role of the agonists is determined by a combination of their profile of activity on dopamine receptors, and their pharmacokinetic properties. A "receptor" to a neurotransmitter has the ability to recognize and respond to that transmitter in minute quantities, selectively, with a precise biochemical or biophysical reaction. Receptors are generally classified pharmacologically, according to how they mimic drugs which are known to stimulate or block the normal neurotransmitter. In addition, receptors can be classified according to their molecular structure, their physiology, or their anatomy. When all methods of characterization are taken together, it is possible to establish a hierarchical classification of receptors. For example, under the general umbrella term of dopamine receptors we can talk about dopamine type-2 receptors and even dopamine type-2A receptors.

Early studies on dopamine receptors were undertaken in vitro. It was shown that striatal tissue responded to dopamine, in vitro, with the stimulation of a dopamine sensitive adenylyl cyclase (Kebabian and Greengard, 1991). This led to the accumulation of cyclic AMP. It was also found that dopamine had distinct actions unrelated to any stimulation of adenylate cyclase, such as suppression of prolactin release, which were not associated with the stimulation of a dopamine sensitive adenylyl cyclase. This effect was elicited in the pituitary and dopamine was discovered to be the "prolactin inhibitory factor" released into the portal circulation. These cardinal pharmacological distinctions — stimulation of adenylyl cyclase or suppression of prolactin release — became the key criteria for the early separation of dopamine receptors into "D1" (stimulating adenylate cyclase), and "D2" (suppressing prolactin release) (Kebabian and Calne, 1979). After this classification of dopamine receptors into two major families there followed further pharmacological developments which allowed a more extensive understanding of the disposition of the receptors. Table 1 (Sibley and Monsma, 1992) (from Calne) shows the main location of the D1 and D2 family of receptors in the brain. In addition, the pharmacological criteria for separation became more elaborate, see Table 2 (from Strange, 1997).

Finally, a new era in the classification of dopamine receptors emerged as a result of the steady progress of molecular biology. Five different peptides were identified as recognition sites for dopamine. All of these recognition molecules conform to the same general pattern of structure (see Fig. 1, Civelli et al., 1991). While it was thought, at first, that each of these molecules might play a separate physiological role, it subsequently became apparent that they all really still fell within the older classification of the major families of D1 and D2 receptors. The D1 and D5 receptors became D1-A and D1B, the D2-3-4 became D2-A-B-C.

Table 1. Distribution, function, and nomenclature of dopamine receptors[a]

Characteristic	D1		D2		
Currently used term	A	B	A	B	C
Previously used term	D1	D5	D2	D3	D4
Location (high concentrations)	Striatum Nucleus accumbens Amygdala Olfactory bulb	Hippocampus Hypothalamus	Striatum Nucleus accumbens Substantia nigra Olfactory bulb	Hypothalamus Nucleus accumbens Olfactory bulb	Frontal cortex Midbrain Medulla
Action (information is limited to biochemical indexes)	Increases cyclic AMP	Increases cyclic AMP	Reduces cyclic AMP Opens potassium channels Closes calcium channels	? Reduces cyclic AMP	Reduces cyclic AMP

[a] The analysis of the dopamine receptors is based on work in rodents

Table 2. Dopamine receptor subtypes defined from physiological, pharmacological, and biochemical studies

	D1	D2
Pharmacological characteristics		
Selective antagonists	SCH 23390	(−)-Sulpiride YM 09151-2
Selective agonists	SKF 38393	Quinpirole N-O437
Specific radioligands	[^{3}H]SCH 23390[b]	[^{3}H]YM 09151-2 [^{3}H]Spiperone
Physiological functions	Aspects of motor function (brain), cardiovascular function	Aspects of motor function and behavior (brain), control of prolactin and αMSH secretion from pituitary, cardiovascular function
Biochemical responses	Adenylate cyclase ↑ phospholipase C ↑	Adenylate cyclase ↓ K^+ channel ↑ Ca^{2+} channel ↓
Localization	Caudate nucleus, putamen, nucleus accumbens, olfactory tubercle, cerebral cortex (brain), cardiovascular system	Caudate nucleus, putamen, nucleus accumbens, olfactory tubercle, cerebral cortex (brain) anterior and neurointermediate lobes of pituitary gland, cardiovascular system
Size of protein (Daltons)		
Affinity labeling	72,000	85,000–150,000 (major species of 94,000, 118,000, 140,000)
Purification	—	92,000–95,000 (brain), 120,000 (pituitary)

Dopamine agonists

The story of artificial dopamine agonists goes back to 1954, when Sheleznyak suggested that ergot alkaloids suppressed the release of prolactin. From 1962, Flükiger and his colleagues began to study the inhibition of prolactin secretion by ergot derivatives, and in the course of their survey they tested a new derivative, bromocriptine. They were particularly impressed with how bromination appeared to increase the specificity of the inhibition of prolactin secretion. In 1967, bromocriptine was tested in human subjects. And then, as Flükiger reports (Flükiger et al., 1983):

There was no receptor concept to explain this new pharmacological action, but we were certain that serotonin receptors or adrenoceptors were not critically involved. We were also certain that prolactin secretion inhibition was achieved by a specific mechanism, as we found ergot compounds that stimulated prolactin release and antagonized the inhibitory action of bromocriptine (Flükiger, 1972; Flükiger et al., 1981). The existence of prolactin in humans, as a separate pituitary hormone, was discredited by most endocrinologists (Geschwind, 1972) — with the important exception of Pasteels (1972,

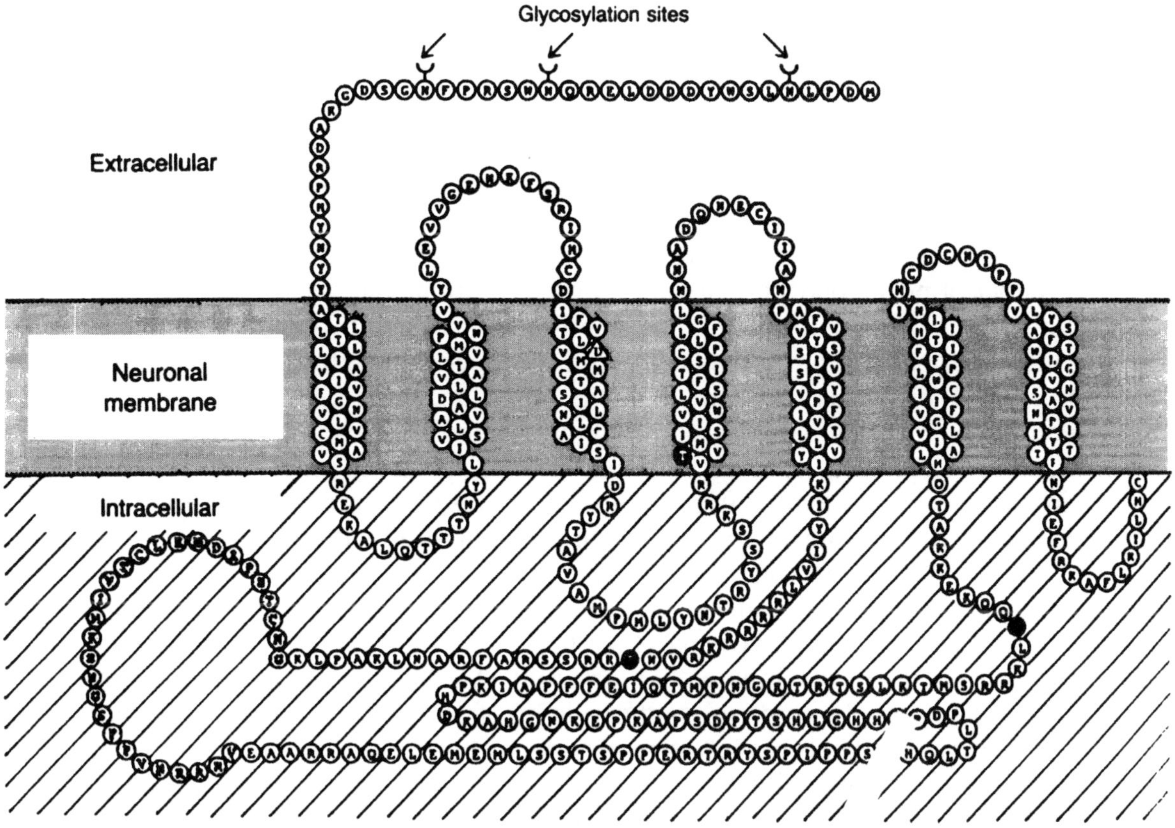

Fig. 1. Structure of the Dopamine D2A Receptor. Each link in the chain is an amino acid. The black components are thought to be phosphorylation sites. The square and triangular components are probably involved in ligand binding. The stippled components represent alternative splicing sites. Reprinted from Civelli et al., 1991

1973), whose positive evidence for the existence of human prolactin made us confident of the future of our project.

Ten years ago, the situation changed dramatically: Human prolactin (hPrl) was recognized as a separate hormone (Frantz and Kleinberg, 1970; Hwang et al., 1971a; Lewis et al., 1971; Loewenstein et al., 1971; Turkington, 1971) and a useful radio-immunoassay (RIA) for hPrl became available (Hwang et al., 1971b). Bromocriptine was shown to be therapeutically effective in galactorrhea (Lutterbeck et al., 1971), in inhibiting prolactin release in vitro from rat and human (fetal) pituitaries (Pasteels et al., 1971), and in reducing dopamine (DA) turnover in hypothalamic and neostriatal dopaminergic neurons (Fuxe and Hokfelt, 1970; Hokfelt and Fuxe, 1972), indicating a central dopaminomimetic property (Corrodi et al., 1973; Johnson et al., 1973). Hence, the new quality of DA receptor stimulation was recognized in ergot compounds, the clinical relevance of which was soon demonstrated in patients with Parkinson's disease (Calne et al., 1974a,b).

After the report of the efficacy of bromocriptine, studies were undertaken to determine what role it would have, if any, in the routine management of patients with Parkinson's disease who also had levodopa available to them. The results of these studies established a fairly clear guide on when and how to use bromocriptine. First, it was shown that bromocriptine caused less dyskinesia than levodopa. In particular, in patients who had never taken levodopa, bromocriptine hardly ever caused dyskinesia. In patients who had previously taken levodopa and had developed dyskinesia, it seemed that some kind of "kindling" phenomenon had taken place so that the bromocriptine could then cause dyskinesia. Nevertheless, there was certainly less dyskinesia in patients who were taking bromocriptine and levodopa combined in low dosage, compared with levodopa alone in high dosage. Another benefit, from bromocriptine, was a more prolonged duration of action. Thus bromocriptine helped patients who were experiencing "wearing off" reactions. The plasma half-life of pharmacotherapy increased from about an hour with levodopa to about three hours with bromocriptine. However, as is usual in clinical pharmacology, the benefits were accompanied by drawbacks. Compared with levodopa, bromocriptine tended to cause more psychiatric reactions. In particular, bromocriptine was more prone to induce visual hallucinations which were often frightening. There was also a greater likelihood for bromocriptine to cause paranoid delusions, compared with levodopa.

Some fifteen years ago the current practice in managing patients with Parkinson's disease became established. Once patients require treatment, they are started with levodopa (together with a peripheral decarboxylase inhibitor such as carbidopa or benserazide). As their illness progresses, the dose of levodopa is increased, and at a level of about 800 mg daily, or earlier if dyskinesia occurs, bromocriptine is added. Thus most patients, for most of the duration of their illness, take a combination of levodopa and bromocriptine. The principle behind this combined treatment is that of traditional therapeutics, as developed, particularly, in cancer chemotherapy. This principle ordains that when two drugs are available with similar efficacy and a different profile of side-effects, optimal treatment can be achieved by combining the two in low doses, rather than giving one drug in high doses. This approach to treatment has been the "gold standard" for the management of Parkinson's disease up to the present.

Studies on the pharmacology of bromocriptine, at synaptic level, showed that it was a potent D2 agonist, but had no D1 stimulating effect. In fact, bromocriptine appeared to be a mild inhibitor of D1 receptors. In the continuing search for alternative medications, two other artificial dopamine agonists were synthesized, lisuride and pergolide, which have been on the market in various countries for over a decade. Lisuride has a profile of pharmacological action similar to bromocriptine. Pergolide differs in that its duration of action seems to be longer, and it appears to stimulate the D1 receptors rather than inhibit them. All of these drugs are ergot derivatives, and there is no obvious difference in their clinical efficacy or their profile of adverse effects.

Recently, a new dopamine agonist with a very long half-life has been developed. This is cabergoline, whose half-life is in the region of seventy hours. Experience is too limited to draw any firm conclusion on what role this agent will achieve in the routine management of Parkinson's disease. Certainly, less frequent dosage should be an advantage. It can also be anticipated that "wearing off reactions" will be decreased. However, there is a potential problem in that side-effects will also persist longer with cabergoline. Thus it will take a few years to determine how, exactly, it will fit into the usual regimen for treating patients with Parkinson's disease.

Two new dopamine agonists have been developed which are not ergot derivatives. Ropinirole has a high affinity for the D2 family of receptors, binding avidly with the D2-A-B-C subcategories. It has a very low affinity for the D1 family of receptors. The elimination half-life averages six hours.

Pramipexole is a synthetic amino-benzathiazol derivative. It binds to the D2 family of receptors with highest affinity for the D2-B (old nomenclature D3 receptor). The elimination half-life varies from 8-12 hours, with older patients having clearance at the slower end of this range.

It is too early to say how these two new non-ergot dopamine agonists, ropinirole and pramipexole, will be used in the routine management of Parkinson's disease. However, it is apparent that they will certainly replace the ergot derivatives when idiosyncratic reactions associated with the ergot structure are encountered — that it to say, in patients who develop pleuropulmonary fibrosis, or pleural effusions, the non-ergot agonists should prove useful. Also, in patients who develop erythromelalgia there should be a benefit in switching from an ergot agonist to a non-ergot derivative.

ABT-431 is the diacetyl pro drug of A-86929, a potent selective D1 agonist. Preliminary studies indicate that ABT-431 produces less dyskinesia than levodopa in monkeys with MPTP lesions, who have been primed with levodopa (Grondin et al., 1997). Unfortunately, the drug is not readily absorbed, so it is not clear whether it will be suitable for chronic treatment in patients with Parkinson's disease.

References

Calne DB (1993) Treatment of Parkinson's disease. N Engl J Med 329: 1021–1027

Calne DB, Teychenne PF, Claveria LE, Eastman R, Greenacre JK, Petrie A (1974a) Bromocriptine in parkinsonism. Br Med J 4: 442–444

Calne DB, Teychenne PF, Leigh PN, Bamji AN, Greenacre JK (1974b) Treatment of parkinsonism with bromocriptine. Lancet 2: 1355–1366

Civelli O, Bunzow JR, Grandy DK, Zhou Q-Y, Van Tol HHM (1991) Molecular biology of the dopamine receptors. Eur J Pharmacol 207: 277–286

Corrodi H, Fuxe K, Hokfelt T, Lidbrink P, Ungerstedt U (1973) Effect of ergot drugs on central catecholamine neurons: Evidence for a stimulation of central dopamine neurons. J Pharm Pharmacol 25: 409–411

Cotzias GC, Van Woert MH, Schiffer LM (1967) Aromatic amino acids and modification of parkinsonism. New Engl J Med 276: 374–379

Ehringer H, Hornykiewicz O (1960) Verteilung von Noradrenalin und Dopamin (3-Hydroxytyramin) im Gehirn des Menschen Wien Klin Wochenschr 38: 1236–1239

Fluekiger E (1972) Drugs and the control of prolactin secretion. In: Boyns AR, Griffiths K (eds) Prolactin and carcinogenesis. Alpha Omega Alpha Publishing, Cardiff: 162–171

Fluekiger E, Briner U, Kovacs E, et al (1981) Prolactin secretion stimulation by an ergopeptine. Experientia 37: 669

Fluekiger E, Briner U, Enz A, Markstein R, Vigouret JM (1983) Dopaminergic compounds: An overview. In: Calne DB, Horowski R, McDonald RJ, Wuttke W (eds) Lisuride and other dopamine agonists. Raven Press, New York: 1–9

Frantz AG, Kleinberg DL (1970) Prolactin: Evidence that it is separate from growth hormone in human blood. Science 170: 745–747
Fuxe K, Hokfelt T (1970) Central monoaminergic systems and hypothalamic function. In: Martini L, Motta M, Fraschini F (eds) The hypothalamus. Academic Press, New York: 123–138
Geschwind II (1972) Introduction. In: Boyns AR, Griffiths K (eds) Prolactin and cardiogenesis. Alpha Omega Alpha Publishing, Cardiff: 1–3
Grondin R, Bedard PJ, Britton DR, Shiosaki K (1997) Potential therapeutic use of the selective dopamine D1 receptor agonist, A-86929: an acute study in parkinsonian levodopa-primed monkeys. Neurology 49: 421–426
Hokfelt T, Fuxe K (1972) On the morphology and the neuroendocrine role of the hypothalamic catecholamine neurons. In: Knigge KM, Scott DE, Weindl A (eds) Brain-endocrine interaction. Median eminence: structure and function. Karger, Basel: 181–223
Hwang P, Friesen H, Hardy J, Wilansky D (1971a) Biosynthesis of human growth hormone and prolactin by normal pituitary glands and pituitary adenomas. J Clin Endocrinol 33: 1–7
Hwang P, Guyda H, Friesen HG (1971b) A radioimmunoassay for human prolactin. Proc Natl Acad Sci USA 68: 1902–1906
Johnson AM, Vigouret JM, Loew DM (1973) Central dopaminergic actions of ergotoxine alkaloids and some derivatives. Experientia 29: 763
Kebabian JW, Calne DB (1979) Multiple receptor mechanisms for dopamine. Nature 227: 93–96
Kebabian JW, Greengard P (1991) Dopamine-sensitive adenyl cyclase: possible role in synaptic transmission. Science 174: 1346–1349
Lewis UJ, Singh RNP, Sinha YN, VahderLaan WP (1971) Electrophoretic evidence for human prolactin. J Clin Endocrinol 33: 153–156
Loewenstein JE, Mariz IK, Peake GT, Daughaday WH (1971) Prolactin bioassay by induction of N-acetyllactosamine synthetase in mouse mammary explants. J Clin Endocrinol 33: 217–224
Lutterbeck PM, Pryor S, Varga L, Wenner R (1971) Treatment of non-puerperal galactorrhea with an ergot alkaloid. Br Med J 3: 228–229
Pasteels JL (1972) Tissue culture of human hypophyses: evidence of a specific prolactin in man. In: Wolstenholme GEW, Knight J (eds) Lactogenic hormones. Churchill Livingstone, Edinburgh: 269–277
Pasteels JL (1973) Introduction. In: Pasteels JL, Robyn C (eds) Human prolactin. Excerpta Medica, Amsterdam: xi–xiii
Pasteels JL, Danguy A, Frerotte M, Ectors F (1971) Inhibition de la secretion de prolactine par l'ergocornine et la 2-Br-alpha-ergokryptine: action directe sur l'hypophyse en culture. Ann Endocrinol (Paris) 32: 188–192
Sibley DR, Monsma FJJ (1992) Molecular biology of dopamine receptors. TIPS 131(13): 61–69
Strange P (1997) Biochemical characterization of dopamine receptors. In: Neve KA, Neve RL (eds) The dopamine receptors. Humana Press, Totowa, NJ: 3–26
Turkington RW (1971) Measurement of prolactin activity in human serum by the induction of specific milk proteins in mammary gland in vitro. J Clin Endocrinol 33: 210–216

Author's address: Donald B. Calne, D.M., F.R.C.P.C. Vancouver Hospital and Health Sciences Centre, Purdy Pavilion, Room M36, 2221 Wesbrook Mall, Vancouver, BC, Canada V6T 2B5

Free radical scavengers: chemical concepts and clinical relevance

M. Gassen[1] and **M. B. H. Youdim**[2]

[1] Merck KGaA, Biomedical Research CNS, Darmstadt, Federal Republic of Germany
[2] Department of Pharmacology, Eve Topf and National Parkinson's Foundation Centers for Neurodegenerative Diseases, Bruce Rappaport Family Research Institute, Faculty of Medicine, Technion, Israel

Summary. Free radicals are involved in the pathology of many CNS disorders, like Parkinson's disease, Alzheimer's disease, or stroke. This discovery lead to the development of many radical scavengers for the clinical treatment of neurodegenerative diseases. In this review, the different chemical concepts for free radical scavenging will be discussed: nitrons, thiols, iron chelators, phenols, and catechols. Especially catechols, like the naturally occurring flavonols, the synthetic drug nitecapone, or the endogenous catacholamines and their metabolites, are of great interest, as they combine iron chelating with radical scavenging activity. We present data on the radical scvenging activity of dopamine and apomorphine, which prevent lipid peroxidation in rat brain mitochondria and protect PC12 cells against H_2O_2-toxicity.

Free radicals as mediators of neurological disorders

Free radicals and other reactive oxygen species (ROS) are formed as side products of the oxygen metabolism in every aerobic organism. The controlled enzymatic torch, that is the main source of metabolic energy in the cell, is based on the stepwise transfer of four electrons (taken from two molecules of NADH) to one molecule of dioxygen (O_2), leading to water as the end product. Between five and ten percent of the oxygen that enters cellular respiration receives less than the four electrons necessary for the complete reduction and is converted into superoxide ($O_2^{\bullet-}$) or hydrogen peroxide (H_2O_2) (Esterbauer, 1980). Efficient antioxidant systems have been developed to detoxify these reactive species before they can cause damage to cellular structures. Superoxide dismutase converts $O_2^{\bullet-}$ into H_2O_2, which is rapidly reduced by catalase and glutathione peroxidase (Halliwell and Gutteridge, 1985; Götz et al., 1994; Gerlach et al., 1996). Excessive cell damage can be caused by ROS if pathological conditions increase free radical formation or exhaust the antioxidative defense mechanisms. ROS-induced cell and tissue damage has been associated with a variety of disorders (see Table 1).

Table 1. Neurological disorders involving free radicals

Ischemia/reperfusion injury, e.g. after stroke	Oliver et al. (1990); Hensley et al. (1997)
Traumatic brain damage	Hall et al. (1993)
Parkinson's disease	Fahn and Cohen (1992); Youdim et al. (1994)
Alzheimer's disease	Smith et al. (1991); Shinobu and Beal (1997)
Neurotoxic insults,	
e.g. by MPTP	Gerlach et al. (1991)
or methamphetamine	Cadet et al. (1994)
Neuroinflammation	Yamasaki et al. (1996); Hallenbeck (1996)
Multiple sclerosis	Hunter et al. (1985)

Table 2. Neurotoxic action of free cellular iron

Fenton-reaction: promotion of the formation of reactive Fenton-products	Minotti and Aust (1992)
Depletion of cellular GSH	Riederer et al. (1989)
Alteration of Ca^{2+}-homeostasis	Anghileri et al. (1994)
DNA-damage and laddering	Spear and Aust (1995)
Accelerates autoxidation of dopamine to isoquinolines and melanin	Ben-Shachar et al. (1991)
Promotion of endogenous 6-OHDA formation	Ben-Shachar et al. (1991); Jellinger et al. (1996)
Release of glutamate	Gerlach et al. (1996)
Impairment of the mitochondral electron transport	Harley et al. (1993)

A rapid metabolism of $O_2^{\bullet-}$ and H_2O_2 is of importance for the cell, because competing mechanisms can lead to an activation of these molecules, giving rise to much more reactive and therefore more toxic radical species. H_2O_2 can react with ferrous iron (Fe^{2+}) to undergo Fenton-type activation. There is still a debate about the nature of the species formed here, most investigators, however refer to them as hydroxyl radicals ($HO^{\bullet}$). More likely than an entirely free $HO^{\bullet}$-radical are oxygen species coordinated to a transition metal. Minotti and Aust (1992) postulated a reactive mixed Fe^{2+}/Fe^{3+}-oxygen-complex as the main product of the Fenton-reaction under physiological conditions (see Table 2 for more neuropathological mechanisms involving iron). Nitric oxide (NO) combines with $O_2^{\bullet-}$ to form peroxynitrite $ONOO^-$. This molecule, already toxic by itself, can decompose in the presence of iron to form Fenton-like products (Cerruti et al., 1995). These highly reactive species account for most of the structural damage to unsaturated phospholipids, proteins or DNA. There are no enzymes, which can cope with the more aggressive ROS, and the organism relies on chemical scavenging strategies instead. Among the low molecular weight scavengers, which are sacrificed to protect important cellular structures, are glutathione and ascorbic acid in the aqueous cytosol and α-tocopherol in the lipid membranes.

Chemical strategies of free radical scavenging

Iron chelators

The first and still strongest indication of an involvement of iron as a primary or secondary phenomenon in the development and/or progression of neurodegenerative diseases is the finding of a marked increase in the concentration of iron in the affected brain areas. This has been confirmed for Parkinson's disease, Huntingtons's disease, supranuclear palsy, multisystem atrophy, as well as Alzheimer's disease. All neurotoxins, inculding NO, 6-hydroxydopamine (6-OHDA), or methylphenyltetrahydropyridine (MPTP) produce a similar increase of iron in the substantia nigra pars compacta of rats and mice, as can be observed in Parkinson's disease patients (Gerlach et al., 1994). Factors different from the classical neurodegenerative diseases can also lead to increased iron concentrations in the brain: hereditary hemochromatosis, a disorder, which causes exceeding intestinal iron uptake, or mechanical or drug induced defects of the blood-brain barrier weakening the normally very tight regulation of the iron entry into the central nervous system. Chronic treatment of rats with the neuroleptics chlorpromazine and haloperidol opens the blood-brain barrier and facilitates the uptake of Fe^{3+} into the brain (Ben-Shachar et al., 1993; Leenders et al., 1994; Ben-Shachar et al., 1994). Recent studies in dietary iron-deficient rats have shown that kainate and 6-OHDA do not induce neurodegeneration in the hippocampus and substantia nigra (Youdim and Shoham, personal communication). This strongly suggests that free iron is involved in the process of neurodegeneration induced by these toxins.

Chelation of redox active transition metals to prevent Fenton-type activation of ROS is of interest not only in neurodegenerative diseases, but in any free radical mediated disorder, as well as in some infectious diseases and in cancer (for a review see Hershko, 1994). Studies with desferal in various in vitro and in vivo models also indicated a possible clinical application in the therapy of ischemia/reperfusion (Marx and van Asbeck, 1996). Given the high potential of desferal, other iron chelators (Fig. 1) have been developed and tested but so far only for non-CNS indications. Deferiprone (CP20) and CP22 are representatives of the novel hydroxypyridone-type iron-chelators, which efficiently prevent radical reactions in in vitro (Morel et al., 1992). In clinical trials, deferiprone has been shown to reduce serum iron concentrations (Olivieri, 1996) in thalassemia major patients. With hydroxypyridone derived iron chelators as well as with desferal, promising preclinical results have been obtained in models of rheumatoid arthritis (Trenam et al., 1992).

Important considerations for the design of iron chelating drugs have been summarized in a recent review by Hider et al. (1996). As it is almost impossible to construct a chelator for Fe^{2+}, which does not interact with other divalent transition metal cations such as Cu^{2+}, Co^{2+}, or Zn^{2+}, chelators for Fe^{3+} are likely to be much more selective. Hexadentate ligands normally form the most stable complexes with metal ions, but, due to their high molecular weight, the bioavailabilty can be low after oral application. Among the

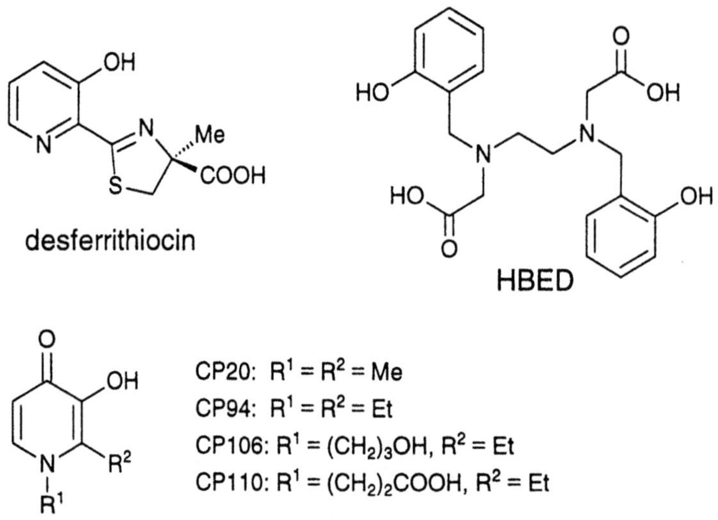

Fig. 1. Structures of experimental iron chelating drugs. *CP20* 1,2-dimethyl-3-hydroxypyrid-4-one, *CP94* 1,2-diethyl-3-hydroxypyrid-4-one, *CP106* 2-ethyl-3-hydroxy-1-(3-hydroxypropyl)pyrid-4-one, *CP110* 3-(2-ethyl-3-hydroxypyrid-4-on-1-yl)propionic acid, *HBED* *N,N'*-di(2-hydroxybenzyl) ethylene-1,2-diamine *N,N'*-diacetic acid

4-hydroxypyridones, however, bidentate ligands are the better chelators (Xiao et al., 1996) and are just as effective in preventing iron-catalyzed radical reactions (Morel et al., 1992). Another problem that arises especially with iron chelators targeted at the central nervous system is the blood-brain barrier permeability. Iron chelators are generally polar molecules with many functional groups, which makes an effective passage through the blood-brain barrier quite difficult. These problems can be solved, if, in a prodrug, these functional groups are chemically masked (Fig. 2a,b), or if they are introduced by enzymatic or nonenzymatic reactions at the target organ (Fig. 2c). This considerations motivated the development of bioactivatable iron chelators, non-chelating prodrugs, which are metabolically activated. The bispiperazonedione dexrazoxane undergoes enzymatic hydrolysis to form an iron chelator which resembles ethylenediaminetraacetic acid (EDTA, Fig. 2a) (Hasinoff et al., 1994) and decreases anthracycline-induced radical formation (Hershko et al., 1993). (Galey et al., 1995) introduced an iron chelator which is triggered by oxidative stress. In vitro, the prodrug *N,N'*-bis-dibenzylethylenediaminediacetic acid (DBED), is converted into the active compound *N*-(2-hydroxybenzyl)-*N'*-benzylethylenediaminediacetic acid (HBBED) by treatment with H_2O_2 (Fig. 2c). For compounds specifically targeted at the CNS, it may be worth while to envisage a bioactivation by CNS-specific enzymes, e.g. monoamine oxidase-A, tyrosine hydroxylase, dihydroxyphenylalanine decarboxylase.

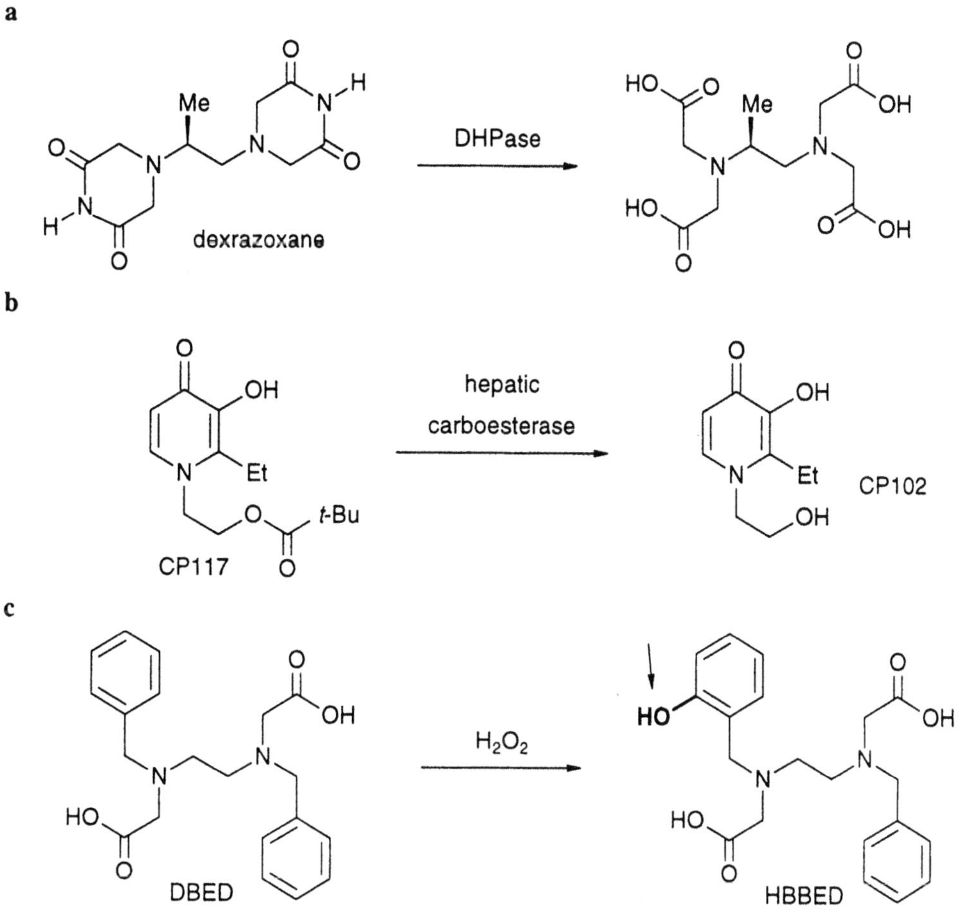

Fig. 2. In situ formation of active chelators from prodrugs: by hydrolytic activation of chemically protected functional groups (**a,b**) or by de novo formation of a functional group (**c**) *CP102* 2-ethyl-3-hydroxy-1-(2-hydroxyethyl)pyrid-4-one, *CP117* 2-ethyl-3-hydroxy-1-(2-pivolyloxyethyl)pyrid-4-one, *DHPase* dihydropurinase

Thiol-antioxidants

Glutathione is the parent compound for the whole class of thiol antioxidants. The most important representatives are the glutathione ethyl and isopropyl esters (prodrugs of glutathione), acetylcysteine (a prodrug of cysteine), D-penicillamine, and α-lipoic acid, which is reduced to dihydrolipoate, the dithiol form. Thiols support the cellular defense by directly increasing the availability of reduced glutathione (Packer et al., 1997).

Glutathione is a key molecule for the detoxification of ROS and the repair of ROS-induced damage in the CNS. As catalase-activity is very low in neurons, glutathione peroxidase and its key substrate, reduced glutathione (GSH) are of great importance to metabolize H_2O_2. GSH can also directly interact with Fe^{2+} and prevent Fenton-type activation of ROS (Ben-Shachar et al., 1994). It prevents the crosslinking and desactivation of enzymes and proteines with active thiol groups (Panigrahi et al., 1996). Furthermore, GSH is required for the conjugation and clearance of potentially toxic organic molecules formed through radical reactions in the cell, like the α,β-unsaturated aldehydes formed from peroxidized membrane lipids (Esterbauer, 1980). NADH is required for the reduction of oxidized glutathione (GSSG) by glutathione reductase. During reperfusion after cerebral ischemia, NADH and consequently GSH-stores are depleted, leading to increased protein crosslinking (Panigrahi et al., 1996). Pretreatment with α-lipoic was protective against transient global iscemia in rats (Panigrahi et al., 1996) and gerbils (Cao and Phillis, 1995), as well as against focal ischemia in mice (Wolz and Krieglstein, 1996). α-Lipoic acid potects against lipid peroxidation and regenerates the cellular stores of GSH, α-tocopherol and ascorbic acid (Packer et al., 1997).

N-Acetylcysteine has no antioxidant properties by itself but increases the availability of GSH. It reduces lead poisoning in mice (Ercal et al., 1996) and protects hippocampal neurons after forebrain ischemia in rats (Knuckey et al., 1995). In vitro, *N*-Acetylcysteine shows pronounced anti-apoptotic potential: In PC12 cells it prevents apoptosis induced by glutamate (Froissard et al., 1997) or trophic withdrawal (Ferrari et al., 1995). It also protects cultured forebrain neurons from dopamine toxicity (Hoyt et al., 1997) and enhances the survival of mesencephalic dopaminergic neurons (Colton et al., 1995).

D-Penicillamine is a antioxidant with metal chelating properties, mainly against copper. It also ameliorates mercury-poisoning in rats (Benov et al., 1990). After seven days pretreatment, it reduced the iron content of rat brains with a significant effect on TBARS-formation as indicators of free radical damage (Ciuffi et al., 1992).

Nitrone spin traps

Nitrone spin traps originally have been developed to make free radicals accessible for spectroscopic measurement (Janzen and Blackburn, 1968). The relatively long halflife of the nitrone-radical-adducts allows the quantification of these species by electron paramagnetic resonance specroscopy (ESR) (Janzen et al., 1992). However, the ability to convert ROS into much less reactive products prompted many investigators to look also at the pharmacological properties of nitrones in various models of potentially free radical mediated diseases. Since 1990, phenyl-*t*-butyl nitrone (PBN) and other nitrones were found to be protective against ischemia and reperfusion in various organs, amyloid-β-protein and MPTP neurotoxicity, tetrachloromethane hepatotoxicity, age related striatal desensitation (for a recent review see Hensley et al., 1997).

The neuroprotective profile of PBN in models of cerebral ischemia is especially impressive. Zhao et al. (1994) demonstrated a marked attenuation of the brain infarct volume after transient occlusion of the middle cerebral artery in rats, if PBN was applied as late as three hours post ischemia. Other work (Folbergrova et al., 1995) confirms that PBN is able to promote the recovery of the ATP concentration as well as the removal of excess lactate. Excessive formation of ROS is obvious primarily during and after reperfusion, a pharmacological intervention with a radical scavenger therefore appears feasible. The antioxidant action of PBN during ischemia/reperfusion in gerbils is well documented: it attenuates protein carbonyl formation and protects glutamine synthetase, an enzyme sensitive to ROS (Oliver et al., 1990).

PBN can reach concentrations of several hundred micromolar in the brain after systemic application (Cheng et al., 1993). Thus it appears that straightforward radical scavenging is the basis for the neuroprotection by PBN. However, PBN does not only show overall tissue protection in cerebral ischemia/reperfusion. The expression of the immediate early genes *c-fos*, *c-jun*, as well as other proapoptotic factors like heat-shock protein (HSP)-70, inducible NO-synthase, and ornitine decarboxylase, is characteristic of neuronal death after ischemia (Carney et al., 1994). PBN attenuates the induction of *c-fos* and HSP-70 in gerbils after ischemia. It cannot be ruled out that PBN indirectly stimulates these genes on the transcriptional level, possibly via oxidatively sensitive pathways involving the tarnscription factor NFκB and a redox sensitive G-protein encoded by $p21^{ras}$ (Hensley et al., 1997).

Aminosteroids (lazaroids)

21-aminosteroids (lazaroids) constitute another new class of drugs with high therapeutic potential in models of focal cerebral ischemia, brain injury and excitotoxic neuronal damage (Hall and McCall, 1993). Some representatives have a dual mode of action and act as iron chelators and show radical scavenging activity. It is, however still very controversial whether iron chelation really makes a significant contribution to their pharmacologic profile. Due to their ambiphilic character, they can integrate into cell membranes and protect unsaturated lipids from peroxidation. Simultaneously they are able to maintain membrane stability and phospholipid fluidity (Hall, 1997).

Phenols and catechols

Hydroxylated aromatic compounds are easily oxidizable and can therefore react with ROS. The primary products of this reaction are phenolic radicals (or semiquinone radicals if formed from catechols), which can undergo a wide variety of secondary oxidation and polymerization reactions (Terland et al., 1997). These secondary products may have pro- or antioxidant properties, depending on the concentration, pH, availability of transition metals, and the type of ROS involved. Therefore, it difficult to define the overall redox characteristics of phenolic compounds (see Table 3).

The most important endogenous phenolic free radical scavenger is α-tocopherol (Vatessery, 1997). It is almost equally distributed throughout the

Table 3. Redox-active phenol- and catechol-derivatives

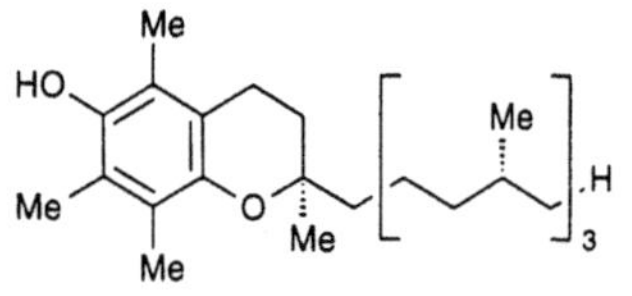 α-Tocopherol	Most important lipid soluble radical scavenger (Vatessery, 1997). Occurs throughout the CNS (McCay, 1985).
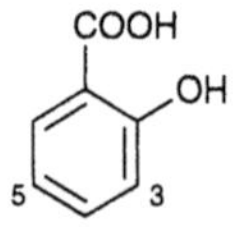 Salicylic Acid	Scavenger of "HO·", in vivo probe for ROS-formation (Chiueh et al., 1992; Giovanni et al., 1995).
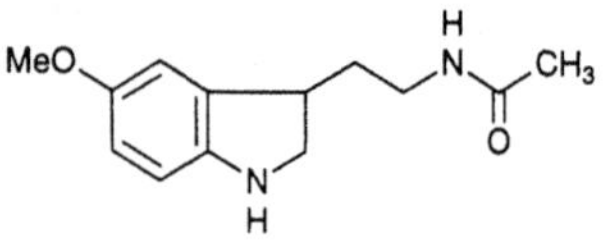 Melatonin	Prevents autoxidation of catecholamines (Miller et al., 1996).
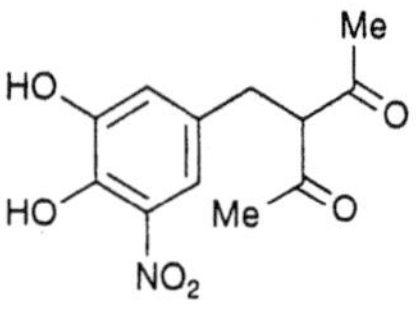 Nitecapone	Synthetic catechol active against cardiac ischemia (Haramaki et al., 1995). Prevents oxidative LDL-modification (Pentikainen et al. 1995).
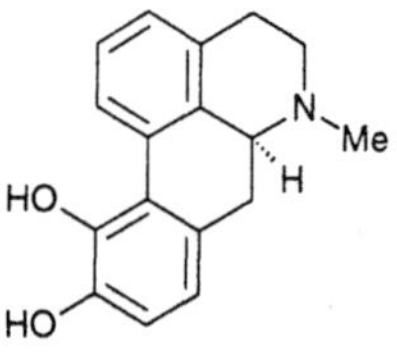 Apomorphine	Antioxidant in ascorbate/Fe^{2+} stimulated ROS in rat brain mitochondria (Gassen et al., 1996). Prevents H_2O_2-induced degeneration of PC12 cells (Gassen et al., 1997).
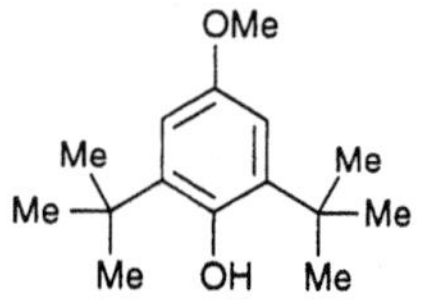 BHA	BHA and other "hindered phenols" are free radical scavengers and Ca^{2+}-antagonists (Sgaragli et al., 1993).
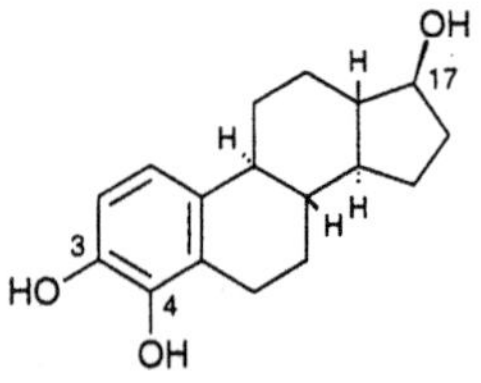 4-Hydroxyestradiol	Antioxidant in vitro (Hess et al., 1996). May cause breast cancer (Liehr, 1997a; Liehr, 1997b).
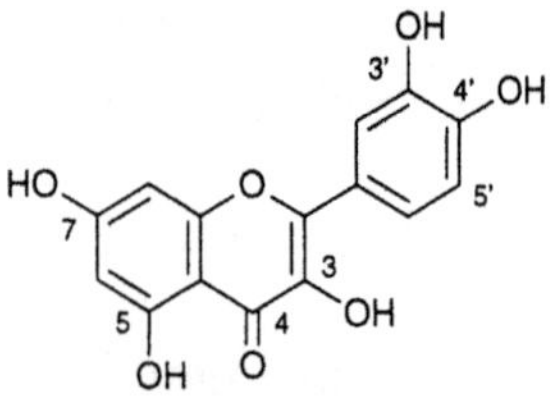 Quercitin	Naturally occurring flavonoid. Iron chelator and antioxidant (Morel et al., 1993). Peroxynitrite scavenger (Haenen et al., 1997).

CNS, with slightly lower concentrations in the cerebellum and spinal cord. Its main purpose is the protection of the sensitive polyunsaturated fatty acids (arachidonic acid, linolic acid, or linolenic acid) in the phospholipids of biological membranes. ROS-induced lipid peroxidation leads to altered membrane fluidity, pore formation and, through the breakdown of oxidized fatty acids, to the release of toxic fragments like malonic dialdehyde or 4-hydroxynonenal (Esterbauer, 1980). In the presence of ROS, α-tocopherol forms a stable radical, which can be recycled by ascorbic acid and GSH (McCay, 1985), as well as by dihydro-α-lipoic acid (Packer et al., 1997).

Phenols react primarily with Fenton-like ROS. The most prominent example is salicylic acid, which can be used as a probe for "$HO^•$" in vivo (Chiueh et al., 1992). Catechols (1,2-dihydroxyphenols), like the catecholamines, are also antioxidants (Liu and Mori, 1993; Sam and Verbeke, 1995; Gassen et al., 1996) and additionally excellent chelators for Fe^{3+} (Kawabata et al., 1996). Liu and Mori (1993) established the antioxidant properties of the dopamine metabolites with dopamine = norepinephrine > dihydroxyphenylacetic acid > homovanillic acid. The autoxidation of catechols leads to semiquinone radicals, which are frequently toxic. There is evidence that the semiquinone radicals of dopamine and L-DOPA are responsible for the in vitro neurotoxicity of these compounds (Haenen et al., 1997), while parent catechols are excellent free radical scavengers. The 3-*O*-methyl-derivative of dopamine is protected from autoxidation and has antioxidant properties.

Another example for such bipolar effects are the catecholamines formed by hydroxylation of estrogens. While catechol estrogens seem to have antioxidant properties in vitro, the most important representative 4-hydroxyestradiol has been implied in hormone-associated breast cancer (Liehr, 1997a; Liehr, 1997b).

As the toxicity of phenolic compounds is correlated with the reactivity of the corresponding semiquinone radicals (Hastings et al., 1996; Haenen et al., 1997), "hindered phenols" have been designed, where the phenolic hydroxyl group is sterically shielded by a bulky substituent (like *t*-butyl) in 2-position. The parent compound are the technical antioxidants di-*t*-butylhydroxytoluene (BHT) or di-*t*-butyl-4-hydroxyanisol (BHA). Gossypol, a naturally occurring sesquiterpene with antioxidant properties, is a structurally related compound (Sgaragli et al., 1993). Flavonoids, especially the flavonols with a hydroxy group at C-3 are potent antioxidants, scavengers of Fenton-like products and $ONOO^-$, and iron chelators of plant origin (Morel et al., 1993). Rutin, the quercitin derived rhamnoglucoside from Ruta graveolens, also known as vitamin P, has a beneficial effect on the incidence of coronary heart disease (Haenen et al., 1997).

The great therapeutic potential of these naturally occurring compounds has prompted the development of the synthetic catechol derivative nitecapone. It has been shown to ameliorate cardiac damage after ischemia/reperfusion in rats (Haramaki et al., 1995) and to prevent the oxidative modification of isolated low density lipoprotein from mice (Pentikainen et al., 1995).

Apomorphine: pro- and antioxidant properties

During the last years, the mixed type dopamine D_1-D_2-receptor agonist apomorphine has had an impressive career as a replacement for L-DOPA in the therapy for Parkinson's disease in the late stage (Gancher et al., 1995). Earlier attempts to introduce apomorphine as dopamine replacement had failed due to its extremely short duration of action and due to serious gastrointestinal side effects. More recently, the pharmacokinetic problems have been overcome with more sophisticated drug delivery devices like portable infusion pumps or insulin pens. Coadministration of the peripherally acting dopamine receptor antagonist domperidone prevents the undesired non-central effects of apomorphine, which has become the medication of choice for patients suffering from on-off fluctuations after years of L-DOPA administration (for a review see Lees, 1993).

Apomorphine protects rat brain mitochondrial membranes from ascorbate/Fe^{2+}-induced radical damage

We recently examined the effect of dopamine and apomorphine on the formation from radical induced lipid and DNA oxidation. Incubation of rat brain mitochondria with ascorbic acid (50 μM) and $FeSO_4$ (1–10 μM) leads to a rapid increase in the formation of thiobarbituric acid reactive substances (TBARS), with maximum concentrations reached after 90 min. This effect can be almost completely abolished by addition of 0.6 μM apomorphine (Gassen et al., 1996). The addition of 0.3 μM apomorphine, approximately the EC_{50} for 2.5 μM Fe^{2+}, slows down the formation of free radical products (Fig. 3a). The concentration of apomorphine required for an effective protection depends on the Fe^{2+} concentration (see Table 1). Oxidation of apomorphine, which eventually leads to a dark melanin-like polymer, was monitored by photometric determination of the strong chromophor of the oxidation products at λ = 619 nm. Autoxidation of apomorphine is slow in the presence of 5 μM Fe^{2+} and 50 μM ascorbic acid (the late increase is due to ascorbate autoxidation), but markedly accelerated in the presence of mitochondria. This accelerated oxidation of apomorphine reflects the protection of mitochondrial lipids (Fig. 3b). These kinetic experiments revealed a negative correlation between apomorphine oxidation and TBARS formation. The former reaction occurs at a high rate during early incubation, completely suppressing TBARS. When apomorphine oxidation slows down later on, increasing amounts of TBARS are generated in the system. Apomorphine showed maximum activity even if added at much lower concentrations than the metal ion. Consequently it appears that chelation of iron does not contribute to this effect to a significant extent, but further work is required to enforce that point. This interpretation is confirmed by the latest results by Sam and Verbeke (1995), who demonstrated that the *R*(−)- and the *S*(+)-enantiomer of apomorphine, as well as dopamine, inhibit the oxidation of the two polyunsaturated fatty acids cholesteryl linoleate and trilinolein by free radicals, which were generated in

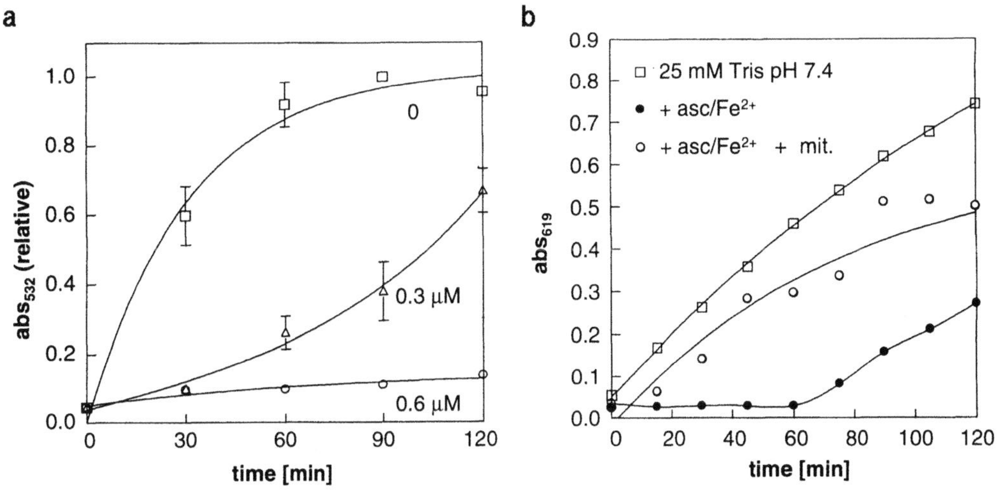

Fig. 3. Modified from Gassen et al. 1996. **a** Apomorphine prevents ascorbate/Fe^{2+}-induced radical damage. Mitochondria were prepared from rat brain and incubated with 2.5 μM $FeSO_4$ and 50 μM ascorbic acid at 37°C. Apomorphine was added at the beginning on the incubation. Free radical damage was assayed as TBARS-formation. **b** The autoxidation of apomorphine (10 μM in TRIS-buffer, +1% digitonin) was monitored photometrically. Additions of $FeSO_4$, ascorbic acid, or mitochondria (1 $mg_{Protein}$/ml) at the beginning of the incubation

an iron free system through thermal decomposition of 2,2′-azobis(2-methylpropionitrile).

Dopamine also shows antioxidant properties in rat brain mitochondrial system, although it was less effective by a factor 20 (Fig. 4). On the basis of this data, iron chelation by dopamine may be a major contribution to the observed inhibition of TBARS formation. Only apomorphine, and neither dopamine, desferal, or pergolide were able to afford full protection from lipid peroxidation induced by ascorbate/Fe^{2+} (Fig. 4).

Apomorphine also protects against oxidation of proteins. ROS induce cysteine-cysteine and tyrosine-tyrosine cross links and react with proline, arginine, lysine, and threonine leading to the formation of new keto- and aldehyde-functional groups (Stadtman, 1993). In order to detect these, we labeled the carbonyl groups with the specific reagent 2,4-dinitrophenylhydrazine. We found that forcing conditions (250 μM Fe^{2+}, 15 μM ascorbic acid) were needed to induce a threefold increase of protein carbonyls in the mitochondrial proteins. This effect could be reduced by 50% in the presence of 100 μM apomorphine (Gassen et al., 1996).

Apomorphine protects Pheochromocytoma (PC12) cells against H_2O_2 and 6-hydroxydopamine

A key question that remained after these results concerns the balance between catecholamine toxicity and possible beneficial effects due to

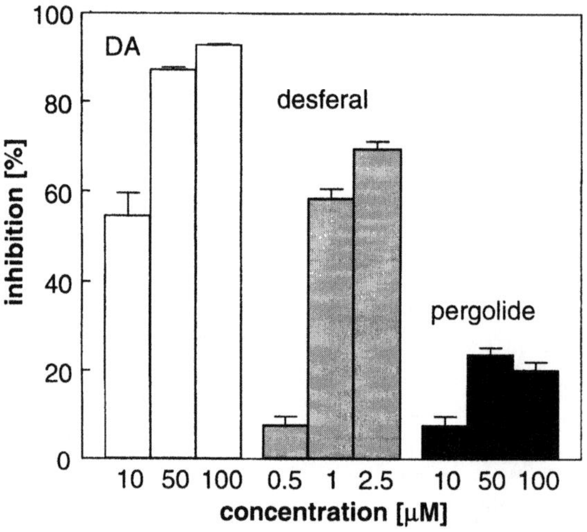

Fig. 4. Modified from Gassen et al. 1996. Antioxidant effects of dopamine (*DA*), desferal, and pergolide against ascorbate/Fe^{2+}-induced radical damage (for details see the caption of Fig. 3a)

antioxidation. We looked at this problem in PC12 cell culture, a well established system to study apoptotic and necrotic cell death (Vimard et al., 1996). Oxidative stress can be induced by various agents like H_2O_2, organic hydroperoxides, or 6-OHDA.

We treated PC12 cells with H_2O_2 and 6-OHDA and observed cell death in a concentration dependent manner within 24h. There was no significant difference of the sensitivity between cells that were grown in medium containing 15% serum (1/3 fetal calf serum, 2/3 horse serum) and those that had been differentiated for six days with additional 100μg/ml 7S-NGF (Greene et al., 1987), if all the NGF had been washed out prior to the experiment. Although it takes 24h to observe the maximum cell death, only two hours exposure to the toxic agent is sufficient to induce the full damage (Gassen et al., 1997). The viability of the cells has been alternatively measured by counting cells after trypane blue exclusion or by measurement of metabolic conversion of 3-(4,5-dimethylthiozol-2-yl)-2,5-diphenyltetrazolium bromide (MTT) into a colored formazane derivative. Both methods produced equivalent results and we used the faster MTT-procedure for routine experiments. Exact EC_{50} values were obtained: 400μM for H_2O_2 and 150μM for 6-OHDA were necessary to kill 50% of the cultured cells.

In this system, apomorphine was tested for its ability to protect PC12 cells from the oxidative insults with H_2O_2 and 6-OHDA. We found also in cell culture that apomorphine is a potent antioxidant. When PC12 cells were treated with H_2O_2 or 6-OHDA at LD_{50}-concentrations, apomorphine provided concentration dependent protection (Fig. 5a and b). The maximum effects were observed with 5μM apomorphine, which improved the cellular survival to 85% after the H_2O_2-insult (Fig. 5a), while the survival rate after 150μM 6-OHDA was improved to 70% with only 1μM apomorphine

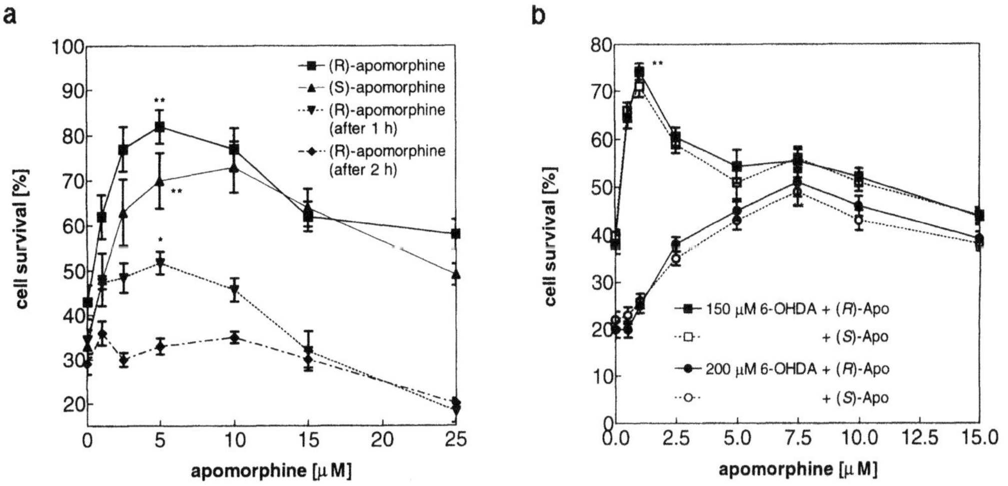

Fig. 5. Modified from Gassen et al. 1997. Cytoprotective effects of apomorphine against **a** H_2O_2-induced cell death; **b** 6-OHDA-induced cell death. PC12 cells (passages 5–10) were grown in Dulbecco's Modified Eagles Medium with 5% fetal calf serum and 10% horse serum. On confluence, the cells were harvested and placed at 10000/well into 96 well microtiter plates precoated with collagen. Cells were used for experiments after 24 h. Cell viability was assayed 24 h after addition of toxin and/or drug with MTT, after careful washing to completely remove catecholamines

(Fig. 5b). This is the first example a catecholamine to attenuate the toxicity of 6-OHDA in cell culture. Any protection against H_2O_2 by apomorphine depended on the presence of the drug during the insult. Preincubation with apomorphine and washout prior to H_2O_2 addition or addition of H_2O_2 one hour after the toxin did not improve the survival as compared with controls only treated with the oxidant.

There was no significant difference between the antioxidant effects of *R*(−)- and the *S*(+)-enatiomers with respect to their antioxidant effects, although *R*(−)-apomorphine, which is the isomer binding the dopamine receptor, appeared to be slightly more active.

Apomorphine was also cytotoxic by itself with LD_{50} = 50 μM. This leaves only a narrow window for the cytoprotective effect. Further work is required to elucidate, whether the antioxidant effect of apomorphine can be demonstrated in vivo, before any conclusions about any therapeutic implications can be drawn.

At subtoxic concentrations, even dopamine, which is generally regarded as a pro-oxidant, potentially neurotoxic catecholamine, shows antioxidant effects against H_2O_2-, but not against 6-OHDA-toxicity. Against H_2O_2, dopamine was significantly less active than apomorphine. With 50 μM dopamine, the survival of PC12 cells could be improved from 45% to 58%. The chelating agents desferal and EDTA and the NO-synthase inhibitor L-NAME showed no effect (Gassen et al., 1997). Different from apomorphine, dopamine was not toxic in PC12 cell cultures below 200 μM concentration.

Conclusion

Since free radicals have been identified as a key factor in the pathology of neurodegenerative diseases, as well as many other disorders, free radical scavengers have been in the focus of interest. The chemical concepts for free radical inactivation lead to several different classes of compounds: nitrones, thiols, iron chelators, as well as phenolic and chatechol radical scavengers. Through the interaction with biogenic free radicals, the scavengers are themselves converted into radical species, which should be, however, less reactive. Many catecholamines, like dopamine, form toxic semiquinone radicals after autoxidation. We demonstrated for dopamine and for the dopamine receptor agonist apomorphine that they can be potent antioxidants in vitro. This should be an encouragment to have a close look at the role of catecholamines and their metabolites in the pathogenesis of neurodegenerative diseases, as well as at the potential of catechol-derived compounds as neuroprotective drugs.

References

Anghileri LJ, Maincent P, Cordova MA, Thouvenot P (1994) The role of ATP as a mediator in the action of iron complexes on cellular calcium homeostasis. Biol Trace Elem Res 46: 103–111

Ben-Shachar D, Riederer P, Youdim MBH (1991) Iron-melanin interaction and lipid peroxidation: implications for Parkinson's disease. J Neurochem 57: 1609–1614

Ben-Shachar D, Livne E, Spanier I, Zuk R, Youdim MBH (1993) Iron modulates neuroleptic-induced effects related to the dopaminergic system. Isr J Med Sci 29: 587–592

Ben-Shachar D, Livne E, Spanier I, Leenders KL, Youdim MBH (1994) Typical and atypical neuroleptics induce alteration in blood-brain barrier and brain $59FeCl_3$ uptake. J Neurochem 62: 1112–1118

Benov LC, Benchev IC, Monovich OH (1990) Thiol antidotes effect on lipid peroxidation in mercury-poisoned rats. Chem Biol Interact 76: 321–332

Cadet JL, Ali S, Epstein C (1994) Involvement of oxygen-based radicals in methamphetamine-induced neurotoxicity: evidence from the use of CuZnSOD transgenic mice. Ann N Y Acad Sci 738: 388–391

Cao X, Phillis JW (1995) The free radical scavenger, α-lipoic acid, protects against cerebral ischemia-reperfusion injury in gerbils. Free Radic Res 23: 365–370

Carney JM, Kindy MS, Smith CD, Wood K, Tatsuno T, Wu JF, Landrum WR, Floyd RA (1994) Gene expression and functional changes after acute ischemia: age-related differences in outcome and mechanisms. In: Hartmann A, Yatsu F, Kuschinsky W (eds) Cerebral ischemia and basic mechanisms. Springer, Berlin Heidelberg New York Tokyo, pp 301–311

Cerruti C, Sheng P, Ladenheim B, Epstein CJ, Cadet JL (1995) Involvement of oxidative and L-arginine-NO pathways in the neurotoxicity of drugs of abuse in vitro. Clin Exp Pharmacol Physiol 22: 381–382

Cheng HY, Liu T, Feuerstein G, Barone FC (1993) Distribution of spin-trapping compounds in rat blood and brain: in vivo microdialysis determination. Free Radic Biol Med 14: 243–250

Chiueh CC, Krishna G, Tulsi P, Obata T, Lang K, Huang SJ, Murphy DL (1992) Intracranial microdialysis of salicylic acid to detect hydroxyl radical generation through

dopamine autooxidation in the caudate nucleus: Effects of MPP^+. Free Radic Biol Med 13: 581–583
Ciuffi M, Gentilini G, Franchi Micheli S, Zilletti L (1992) D-penicillamine affects lipid peroxidation and iron content in the rat brain cortex. Neurochem Res 17: 1241–1246
Colton CA, Pagan F, Snell J, Colton JS, Cummins A, Gilbert DL (1995) Protection from oxidation enhances the survival of cultured mesencephalic neurons. Exp Neurol 132: 54–61
Ercal N, Treeratphan P, Hammond TC, Matthews RH, Grannemann NH, Spitz DR (1996) In vivo indices of oxidative stress in lead-exposed C57BL/6 mice are reduced by treatment with meso-2,3-dimercaptosuccinic acid or N-acetylcysteine. Free Radic Biol Med 21: 157–161
Esterbauer H (1980) Aldehydes of lipid peroxidation. In: McBrien DCH, Slater TF (eds) Free radicals, peroxidation, and cancer. Academic Press, London, pp 101–122
Fahn S, Cohen G (1992) The oxidant stress hypothesis in Parkinson's disease: evidence supporting it. Ann Neurol 32: 804–812
Ferrari G, Yan CY, Greene LA (1995) N-acetylcysteine (D- and L-stereoisomers) prevents apoptotic death of neuronal cells. J Neurosci 15: 2857–2866
Folbergrova J, Zhao Q, Katsura K, Siesjö BK (1995) N-tert-butyl-α-phenylnitrone improves recovery of brain energy state in rats following transient focal ischemia. Proc Natl Acad Sci U S A 92: 5057–5061
Froissard P, Monrocq H, Duval D (1997) Role of glutathione metabolism in the glutamate-induced programmed cell death of neuronal-like PC12 cells. Eur J Pharmacol 326: 93–99
Galey JB, Dumats J, Beck I, Fernandez B, Hocquaux M (1995) N,N′-bis-dibenzyl ethylenediaminediacetic acid (DBED): a site-specific hydroxyl radical scavenger acting as an "oxidative stress activatable" iron chelator in vitro. Free Radic Res 22: 67–86
Gancher ST, Nutt JG, Woodward WR (1995) Apomorphine infusional therapy in Parkinson's disease: clinical utility and lack of tolerance. Mov Disord 10: 37–43
Gassen M, Glinka Y, Pinchasi B, Youdim MBH (1996) Apomorphine is a highly potent free radical scavenger in rat brain mitochondrial fraction. Eur J Pharmacol 308: 219–225
Gassen M, Gross A, Pinchasi B, Youdim MBH (1997) Apomorphine enantiomers protect cultured pheochromocytoma (PC12) cells from oxidative stress induced by H_2O_2 and 6-hydroxydopamine. Mov Disord in press:
Gerlach M, Riederer P, Przuntek H, Youdim MBH (1991) MPTP mechanisms of neurotoxicity and their implications for Parkinson's disease. Eur J Pharmacol 208: 273–286
Gerlach M, Ben-Shachar D, Riederer P, Youdim MBH (1994) Altered brain metabolism of iron as a cause of neurodegenerative diseases? J Neurochem 63: 793–807
Gerlach M, Riederer P, Youdim MBH (1996) Molecular mechanisms for neurodegeneration. Synergism between reactive oxygen species, calcium, and excitotoxic amino acids. Adv Neurol 69: 177–194
Giovanni A, Liang LP, Hastings TG, Zigmond MJ (1995) Estimating hydroxyl radical content in rat brain using systematic and intraventricular salicylate: impact of methamphetamine. J Neurochem 64: 1819–1825
Götz ME, Kunig G, Riederer P, Youdim MBH (1994) Oxidative stress: free radical production in neural degeneration. Pharmacol Ther 63: 37–122
Greene LA, Aletta JM, Rukenstein A, Green SH (1987) PC12 pheochromocytoma cells: culture, nerve growth factor treatment, and experimental exploitation. Methods Enzymol 147: 207–216
Haenen GR, Paquay JB, Korthouwer RE, Bast A (1997) Peroxynitrite scavenging by flavonoids. Biochem Biophys Res Commun 236: 591–593
Hall ED (1997) Antioxidant Therapeutic Strategies in CNS Disorders. In: Connor JR (ed) Metals and oxidative damage in neurological disorders. Plenum Press, New York, pp 325–339

Hall ED, McCall JM (1993) Lazaroids: potent inhibitors of iron-dependent lipid peroxidation for neurodegenerative disorders. In: Riederer P, Youdim MBH (eds) Iron in central nervous system disorders. Springer, Wien New York, pp 173–188

Hall ED, Andrus PK, Yonkers PA (1993) Brain hydroxyl radical generation in acute experimental head injury. J Neurochem 60: 588–594

Hallenbeck JM (1996) Significance of the inflammatory response in brain ischemia. Acta Neurochir [Suppl] Wien 66: 27–31

Halliwell B, Gutteridge JMC (1985) Oxygen radicals and the nervous system. Trends Neurosci 8: 22–29

Haramaki N, Stewart DB, Aggarwal S, Kawabata T, Packer L (1995) Role of ascorbate in protection by nitecapone against cardiac ischemia-reperfusion injury. Biochem Pharmacol 50: 839–843

Harley A, Cooper JM, Schapira AH (1993) Iron induced oxidative stress and mitochondrial dysfunction: relevance to Parkinson's disease. Brain Res 627: 349–353

Hasinoff BB, Venkataram S, Singh M, Kuschak TI (1994) Metabolism of the cardioprotective agents dexrazoxane (ICRF-187) and levrazoxane (ICRF-186) by the isolated hepatocyte. Xenobiotica 24: 977–987

Hastings TG, Lewis DA, Zigmond MJ (1996) Reactive dopamine metabolites and neurotoxicity: implications for Parkinson's disease. Adv Exp Med Biol 387: 97–106

Hensley K, Carney JM, Steward CA, Tabatabaie T, Pye Q, Floyd RA (1997) Nitron-based free radical traps as neuroprotective agents in cerebral ischemia and other pathologies. In: Green AR, Cross AJ (eds) Neuroprotective agents in cerebral ischemia. Academic Press, San Diego, pp 299–317

Hershko C (1994) Control of disease by selective iron depletion: a novel therapeutic strategy utilizing iron chelators. Baillieres Clin Haematol 7: 965–1000

Hershko C, Link G, Tzahor M, Pinson A (1993) The role of iron and iron chelators in anthracycline cardiotoxicity. Leuk Lymphoma 11: 207–214

Hess DC, Thompson Y, Sprinkle A, Carroll J, Smith J (1996) E-selectin expression on human brain microvascular endothelial cells. Neurosci Lett 213: 37–40

Hider RC, Choudhury R, Rai BL, Dehkordi LS, Singh S (1996) Design of orally active iron chelators. Acta Haematol 95: 6–12

Hoyt KR, Reynolds IJ, Hastings TG (1997) Mechanisms of dopamine-induced cell death in cultured rat forebrain neurons: interactions with and differences from glutamate-induced cell death. Exp Neurol 143: 269–281

Hunter MIS, Nlemadim BC, Davidson DLW (1985) Lipid peroxidation production and antioxidant properties in plasma and cerebrospinal fluid of multiple sclerosis patients. Neurochem Res 10: 1645–1652

Janzen EG, Blackburn BJ (1968) Detection and identification of short-lived free radicals by an electron spin resonance trappind technique. J Am Chem Soc 90: 5909–5915

Janzen EG, Kotake Y, Hinton RD (1992) Stabilities of hydroxyl radical spin adducts of PBN-type spin traps. Free Radic Biol Med 12: 169–173

Jellinger K, Linert L, Kienzl E, Herlinger E, Youdim MBH, Ben-Shachar D, Riederer P (1996) Chemical evidence for 6-hydroxydopamine to be an endogenous toxic factor in the pathogenesis of Parkinson's disease Iron-melanin interaction and lipid peroxidation: implications for Parkinson's disease. J Neural Transm [Suppl] 57: 1609–1614

Kawabata T, Schepkin V, Haramaki N, Phadke RS, Packer L (1996) Iron coordination by catechol derivative antioxidants. Biochem Pharmacol 51: 1569–1577

Knuckey NW, Palm D, Primiano M, Epstein MH, Johanson CE (1995) N-acetylcysteine enhances hippocampal neuronal survival after transient forebrain ischemia in rats. Stroke 26: 305–310

Leenders KL, Antonini A, Schwarzbach R, Smith-Jones P, Reist H, Ben-Shachar D, Youdim MBH, Henn V (1994) Blood to brain iron uptake in one rhesus monkey using [Fe-52]-citrate and positron emission tomography (PET): influence of haloperidol. J Neural Transm [Suppl] 43: 123–132

Lees AJ (1993) Dopamine agonists in Parkinson's disease: a look at apomorphine. Fundam Clin Pharmacol 7: 121–128
Liehr JG (1997a) Dual role of oestrogens as hormones and pro-carcinogens: tumour initiation by metabolic activation of oestrogens. Eur J Cancer Prev 6: 3–10
Liehr JG (1997b) Hormone-associated cancer: mechanistic similarities between human breast cancer and estrogen-induced kidney carcinogenesis in hamsters. Environ Health Perspect 105 [Suppl] 3: 565–569
Liu J, Mori A (1993) Monoamine metabolism provides an antioxidant defense in the brain against oxidant- and free radical-induced damage. Arch Biochem Biophys 302: 118–127
Marx JJ, van Asbeck BS (1996) Use of iron chelators in preventing hydroxyl radical damage: adult respiratory distress syndrome as an experimental model for the pathophysiology and treatment of oxygen-radical-mediated tissue damage. Acta Haematol 95: 49–62
McCay PB (1985) Vitamin E: Interactions with free radicals and ascorbate. Ann Rev Nutr 5: 323–340
Miller JW, Selhub J, Joseph JA (1996) Oxidative damage caused by free radicals produced during catecholamine autoxidation: protective effects of O-methylation and melatonin. Free Radic Biol Med 21: 241–249
Minotti G, Aust SD (1992) Redox cycling of iron and lipid peroxidation. Lipids 27: 219–226
Morel I, Cillard J, Lescoat G, Sergent O, Pasdeloup N, Ocaktan AZ, Abdallah MA, Brissot P, Cillard P (1992) Antioxidant and free radical scavenging activities of the iron chelators pyoverdin and hydroxypyrid-4-ones in iron-loaded hepatocyte cultures: comparison of their mechanism of protection with that of desferrioxamine. Free Radic Biol Med 13: 499–508
Morel I, Lescoat G, Cogrel P, Sergent O, Pasdeloup N, Brissot P, Cillard P, Cillard J (1993) Antioxidant and iron-chelating activities of the flavonoids catechin, quercetin and diosmetin on iron-loaded rat hepatocyte cultures. Biochem Pharmacol 45: 13–19
Oliver CN, Starke-Reed PE, Stadtman ER, Liu GJ, Carney JM, Floyd RA (1990) Oxidative damage to brain proteins, loss of glutamine synthetase activity, and production of free radicals during ischemia/reperfusion-induced injury to gerbil brain. Proc Natl Acad Sci U S A 87: 5144–5147
Olivieri NF (1996) Long-term therapy with deferiprone. Acta Haematol 95: 37–48
Packer L, Tritschler HJ, Wessel K (1997) Neuroprotection by the metabolic antioxidant α-lipoic acid. Free Radic Biol Med 22: 359–378
Panigrahi M, Sadguna Y, Shivakumar BR, Kolluri SV, Roy S, Packer L, Ravindranath V (1996) α-Lipoic acid protects against reperfusion injury following cerebral ischemia in rats. Brain Res 717: 184–188
Pentikainen MO, Lindstedt KA, Kovanen PT (1995) Inhibition of the oxidative modification of LDL by nitecapone. Arterioscler Thromb Vasc Biol 15: 740–747
Riederer P, Sofic E, Rausch WD, Schmidt D, Reynolds GP, Jellinger K, Youdim MBH (1989) Transition metals, ferritin, glutathione, and ascorbic acid in parkinsonian brains. J Neurochem 52: 512–520
Sam EE, Verbeke N (1995) Free radical scavenging properties of apomorphine enantiomers and dopamine: Possible implication in their mechanism of action in parkinsonism. J Neural Transm PD Sect 10: 115–127
Sgaragli GP, Valoti M, Gorelli B, Fusi F, Palmi M, Mantovani P (1993) Calcium antagonist and antiperoxidant properties of some hindered phenols. Br J Pharmacol 110: 369–377
Shinobu LA, Beal MF (1997) The role of oxidative processes and metal ions in aging and Alzheimer's disease. In: Connor JR (ed) Metals and oxidative damage in neurological disorders. Plenum Press, New York, pp 237–275
Smith CD, Carney JM, Starke-Reed PE, Oliver CN, Stadtman ER, Floyd RA, Markesbery WR (1991) Excess brain protein oxidation and enzyme dysfunction

in normal aging and in Alzheimer disease. Proc Natl Acad Sci USA 88: 10540–10543

Spear N, Aust SD (1995) Effects of glutathione on Fenton reagent-dependent radical production and DNA oxidation. Arch Biochem Biophys 324: 111–116

Stadtman ER (1993) Oxidation of free amino acids and amino acid residues in proteins by radiolysis and by metal-catalyzed reactions. Annu Rev Biochem 62: 797–821

Terland O, Flatmark T, Tangeras A, Gronberg M (1997) Dopamine oxidation generates an oxidative stress mediated by dopamine semiquinone and unrelated to reactive oxygen species. J Mol Cell Cardiol 29: 1731–1738

Trenam CW, Winyard PG, Morris CJ, Blake DR (1992) Iron promotes oxidative damage in rheumatic diseases. In: Lauffer RB (ed) Iron and human disease. CRC Press, Boca Raton, pp 395–418

Vatessery GT (1997) Vitamin E: Neurochemical aspects and relevance to nervous system disorders. In: Connor JR (ed) Metals and oxidative damage in neurological disorders. Plenum Press, New York, pp 175–188

Vimard F, Nouvelot A, Duval D (1996) Cytotoxic effects of an oxidative stress on neuronal-like pheochromocytoma cells (PC12). Biochem Pharmacol 51: 1389–1395

Wolz P, Krieglstein J (1996) Neuroprotective effects of α-lipoic acid and its enantiomers demonstrated in rodent models of focal cerebral ischemia. Neuropharmacology 35: 369–375

Xiao G, van der Helm D, Hider RC, Rai BL (1996) Molecular modeling studies of a ferric hexadentate 3-hydroxy-2(1H)-pyridone complex and an analog by molecular mechanics, molecular dynamics, and free energy pertubation simulations. J Phys Chem 100: 2345–2352

Yamasaki Y, Itoyama Y, Kogure K (1996) Involvement of cytokine production in pathogenesis of transient cerebral ischemic damage. Keio J Med 45: 225–229

Youdim MBH, Lavie L, Riederer P (1994) Oxygen free radicals and neurodegeneration in Parkinson's disease: a role for nitric oxide. Ann N Y Acad Sci 738: 64–68

Zhao Q, Pahlmark K, Smith ML, Siesjö BK (1994) Delayed treatment with the spin trap α-phenyl-N-tert-butyl nitrone (PBN) reduces infarct size following transient middle cerebral artery occlusion in rats. Acta Physiol Scand 152: 349–350

Authors' address: Dr. Michael Gassen, Merck KGaA, Frankfurter Strasse 250, D-64271 Darmstadt, Federal Republic of Germany

Differential diagnosis of tremor*

G. Deuschl

Department of Neurology, Christian Albrechts University Kiel, Federal Republic of Germany

Summary. The differential diagnosis of tremor is based on the clinical distinction of rest, postural and intention tremor and the presence of additional clinical signs and data from the medical history. The most common pathological tremors are essential tremor and the tremors of Parkinson's disease. Among the patients with essential tremor those with intention tremor are often misdiagnosed as cerebellar tremors. Patients with monosymptomatic resting tremors represent a special subgroup of Parkinson's disease. Primary orthostatic tremor and dystonic tremor are rare clinical syndromes which have recently been well defined. Holmes' tremors are defined by their low frequency and the occurence of resting and intention tremor. Palatal tremor can be separated into two subgroups. Psychogenic tremor can be diagnosed on the basis of clinical criteria. The gold standard of tremor differential diagnosis is still based on clinical criteria.

Phenomenological definition

Tremor is defined as an involuntary rhythmically oscillating movement of at least one functional region. It is either present alone or part of a neurologic syndrome. Various types of tremor are distinguished clinically. Their distinction is based on the activation condition, frequency and topographical distribution. Associated neurologic symptoms (e.g. spasticity, akinesia, dystonia) sometimes help to pinpoint the underlying syndrome or the etiology.

Tremor may be elicited at rest or during action. *Resting tremor* is suspected, if it occurs with the patient sitting with his arms firmly supported (no voluntary activity), if it increases with mental stress (counting backwards) and if it is suppressed by voluntary movements. It may defy differentiation from

---

*This contribution is based on two recently published overviews (Deuschl et al., 1998; Krack et al., 1998).

other types of tremor, if the patient cannot completely relax. *Action tremor* may take one of 3 forms, i.e. static tremor, kinetic tremor during undirected movements (simple kinetic tremor) and intention tremor. *Static tremor* occurs when a tonic movement is attempted against gravity. *Simple kinetic* tremor is tested during undirected flexion and extension. *Intention tremor* is characterized by visibly increased trembling during visually guided movements towards an object and shows typical arrhythmic (myocloniform) amplitude variations the closer the object comes. *Task-specific tremors* occur during certain activities (writing, playing a musical instrument) and occupy a special place. *Isometric tremor* is sometimes seen when a heavy object is held and the muscles involved are tonically innervated.

Tremors can be visually classified by their frequency in low-frequency (<4 Hz), intermediate-frequency (4–7 Hz) and high-frequency (>7 Hz) trembling.

Clinically, tremors are described by noting precisely which topographical regions of the body (limbs, head, voice, etc.) are involved.

Etiology and classification by syndromes

In clinical routine several tremor syndromes are distinguished by the signs and symptoms present. Classifying tremors by syndromes (Fig. 1) either brings to light the true etiology or suggests a number of etiologies to be considered in the further differential diagnostic work-up.

In 1997 the Movement Disorder Society proposed a classification which differentiates various tremor syndromes clinically (Deuschl et al., 1998). These are:

Physiologic tremor

Physiologic tremor is seen in all normal individuals when muscles are activated. Its rate is between 6 and 12 Hz in the hands, varies in different joints and decreases slightly above age 60 years, while its amplitude may increase somewhat. It is caused by mechanical oscillations of the activated limb in most cases.

Enhanced physiologic tremor

This is seen in patients with predominantly static tremor. It is characterized by a high tremor frequency (like physiologic tremor), a short history (<2 years) and the absence of other clinical signs of a neurologic disease associated with tremor. Its etiologies vary widely. Endogenous and exogenous intoxifications (e.g. hyperthyroidism or drugs) are typical examples. The pathophysiology varies depending on the cause and can be due to mechanical and reflex factors as well as central oscillators.

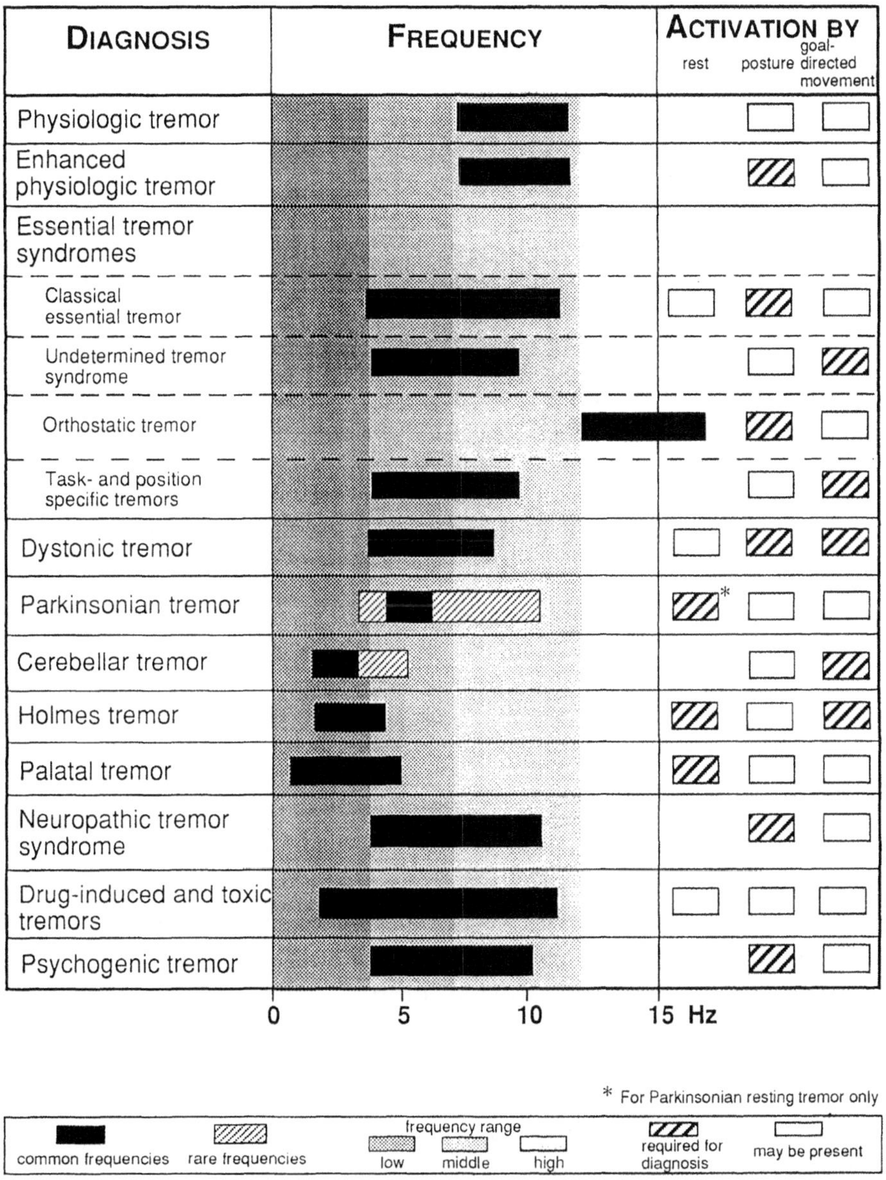

Fig. 1. Overview of various tremor variants (modified from Deuschl et al., 1998). Frequency and eliciting conditions are helpful criteria in the differential diagnosis. But if used alone, their discriminating power is limited to just a few tremor variants

Essential tremor syndromes

This term mainly refers to a syndrome which is often hereditary and manifests itself by no symptoms other than tremor. For the purposes of this contribution it is called classical essential tremor. However, a number of tremor variants of unknown origin have also been referred to as "essential", although they are clinically distinguishable from classical essential tremor. These subgroups are described below:

Classical essential tremor

Essential tremor (Brain et al., 1994; Findley and Koller, 1995; Koller et al., 1994) is a slowly or sometimes rapidly progressive monosymptomatic disease. Its major manifestations are static and action tremor. While mostly static, the tremor may be intentional in more than 20% of patients and may cause severe disability. Patients with intention tremor are often misdiagnosed. Very rarely resting tremor is also seen. Above age 40 years the prevalence is between 0.4 and 5.6%. The onset may either be during adolescence or at a more advanced age (mean age at onset, about 40 years). In 60% of cases there is evidence suggesting that the condition is inherited as an autosomal dominant trait. The gene involved is still unknown. Ingestion of alcohol reduces the tremor intensity in 50 to 70% of patients. Body regions are variably affected: hands — 94%; head — 33%; voice — 16%; face — 3%; legs — 12%; trunk — 3%. At times, the tremor is limited to a single region (prototype: isolated head or vocal tremor). Almost all patients are handicapped socially. Up to 25% of them are forced to change their occupations or seek early retirement. A central oscillator is the most likely underlying pathophysiologic mechanism.

Primary writing tremor and other task-specific tremors

Tremor which occurs only when writing is called primary writing tremor, if other neurological manifestations are absent. Whether it is dystonic in nature or represents a focal variant of classical essential tremor is still controversial. Task or posture-specific tremors have also been described during other activities. Professional musicians or athletes are most often affected. The common denominator is a highly specialized motor strain close to the limits of coordination.

Isolated vocal tremor

This diagnosis is made when the tremor is limited to vocal production. As for writing tremors, it is still not known whether the patients affected by it have focal dystonia or a focal variant of essential tremor. Both is conceivable. Dystonic vocal tremor should be suspected if emotional vocal productions or singing, which are invariably involved in focal essential tremor, are spared. Vocal tremor more commonly occurs in the context of a generalized tremor syndrome (e.g. cerebellar tremor or essential tremor) than in isolation.

Primary orthostatic tremor

This tremor syndrome is characterized by subjectively experienced unsteadiness while standing. Walking, sitting and lying down are unaffected. Just barely visible or at best palpable vibration of the leg muscles while standing is the only typical finding. Confirmation of the diagnosis can only be obtained by electromyography which typically shows a 13–18 Hz pattern in the leg muscles with the patient standing. Although other tremors may equally cause discomfort while standing, they all show lower frequencies.

Unclassifiable postural tremor

Some patients present with a predominantly static tremor associated with other neurologic abnormalities (e.g. slight hypomimia, cogwheel phenomenon, mild akinesia) which are, however, not severe enough to confirm the presence of another neurologic entity, i.e. Parkinson's disease or dystonia. Some of these patients defy classification. In these, it is wiser to describe the condition by a term implying the need for waiting rather than to attach a wrong label to it.

Dystonic tremor

This variant is present if the tremor involves a body region that is dystonic (Jankovic et al., 1988; Rosenbaum and Jankovic, 1988). It is often focally limited with highly variable amplitudes and a frequency of less than 7 Hz. In most cases it is elicited by static innervation or action and not detectable at rest, although exceptions to this general rule are known.

Focal tremors without clear-cut signs of dystonia have also been classified among the dystonic tremors, because they involve circumscribed areas (Rivest and Marsden, 1990). Many patients affected by dystonic tremors use tricks to reduce the tremor amplitude. These together with the absence of attempts at suppressing the tremor by voluntary muscle contractions is a fairly reliable diagnostic sign. It is particularly common in patients with dystonic head tremor and has been described as "geste antagoniste". Dystonic tremors should be distinguished from static tremors associated with dystonia which occur unspecifically in regions unaffected by dystonia (e.g. tremor of the hands in patients with isolated spasmodic torticollis).

Parkinsonian tremor

Tremor is almost always present in patients with parkinsonism. Resting tremor is the typical variant seen, but static and action tremors also occur either alone or in combination. The tremor is thought to be parkinsonian, if the patient shows the clinical manifestations of Parkinson's disease (akinesia and one of the cardinal symptoms, i.e. rigidity, resting tremor and abnormal postural reflexes) and presents with any one variant of tremor. Various mechanisms — probably co-existent — have been implicated in the development of action tremor: (1) a continuation of resting tremor under static conditions; (2) the additional presence of an increased physiologic tremor and (3) an associated tremor of the essential static tremor type.

Clinical manifestations are further characterized phenomenologically. This ultimately dictates the therapeutic approach.

Resting tremor or resting and static tremor of the same frequency (classical parkinsonian tremor)

The usual frequency of the resting tremor is between 4 and 6 Hz, but higher frequencies may be recordable in the early stages. If postural tremor is also present, the resting and postural/action tremors have the same frequencies. In such cases the tremor pattern is thought to be a continuation of the resting tremor under conditions of tonic muscle innervation.

Resting and postural tremor of different frequency

In these cases the postural or kinetic tremor is of higher frequency (>1.5 Hz) than the resting tremor. Some of these patients have a long history of postural tremor (essential tremor?) before the onset of unequivocal parkinsonian symptoms.

Action tremor

Action tremors are of high frequency, become most obvious during slow flexion and extension of the hands and may be closely related with parkinsonian rigidity.

Monosymptomatic resting tremor

Isolated resting tremor may be present for years without any definite signs of bradykinesia or rigidity. Hence the proposal to use the term monosymptomatic resting tremor for patients without convincing evidence of Parkinson's syndrome (akinesia rigidity), but a history of at least 2 years (Deuschl et al., 1998). Case reports as well as recent PET studies (Brooks, 1993) suggest that these patients suffer from a relatively slowly progressive variant of Parkinson's disease as far as akinesia and rigidity are concerned.

In contrast to other tremors, resting tremor is the only specific sign of Parkinson's disease. In patients with classical resting tremor there is a likelihood of >90% for Parkinson's syndrome to be idiopathic rather than secondary or symptomatic. Early in the course of Parkinson's disease the differentiation from essential tremor may be extremely difficult. Tremor of the legs, asymmetry and established sensitivity to dopa argue in favor of

Parkinson's disease. A positive family history, head or vocal tremor or an improvement of symptoms on alcohol ingestion are suggestive of essential tremor.

Cerebellar tremor syndromes

Classical cerebellar tremor is often described as intention tremor. Its frequency is typically below 5 Hz. Static tremor of the same frequency may also be present, while resting tremor is absent. This type of cerebellar tremor is due to structural or functional lesions in the ipsilateral cerebellum or its afferent and efferent projections so that the tremor is symptomatic in nature.

Titubation is another tremor variant. It is characterized by a low frequency (usually less than 4 Hz), is postural in nature, has an added volitional component and particularly affects the head and the upper half of the body. Its marked rhythmicity is, at times, the only sign distinguishing it from ataxia of the trunk.

In patients with cerebellar lesions other tremor variants (mostly postural) have also been described. These only qualify as cerebellar tremors if other clinical signs of cerebellar dysfunction are present.

Holmes tremor

There are some clinically and phenomenologically similar tremors which are referred to by quite different names, i.e. rubral tremor, mesencephatic tremor, Benedict's syndrome, myorhythmia. Most of these names imply a specific location. But as the underlying lesions may be located at widely different sites, this type of tremor has recently been named after G. Holmes who described some patients presenting with it (Holmes, 1904). Like cerebellar tremor, this type of tremor is almost always attributable to a circumscribed lesion. It is characterized by a resting and an intentional component. Often a postural component is also present. Its frequency is less than 4.5 Hz. In patients in whom the time of the brain lesion is known the tremor is seen to occur after a latency of 2 weeks to 2 years. Recently the clinical signs and symptoms described were found to be present in several patients with dorsolateral thalamic lesions ("thalamic tremor") (Mivoa et al., 1996). As they all but defy differentiation from Holmes tremor and sometimes have an additional dystonic element, this tremor variant is dealt with under this heading and the etiology of an underlying thalamic lesion is mentioned.

Palatal tremor

Previously called palatal myoclonus, palatal tremor may either be symptomatic or essential (Deuschl et al., 1990; Deuschl et al., 1994).

Symptomatic palatal tremor is characterized by a lesion in the brainstem or cerebellum followed by pseudohypertrophy of the inferior olive which can usually be detected by MRI. Phenomenologically its main features are rhythmic movements of the velum (levator veli palatini muscle) and other cranial nerves or limb muscles. In essential palatal tremor brainstem lesions and olivary hypertrophy are absent. The rhythmic movements of the velum are caused by an activation of the tensor veli palatini muscle and are almost always associated with audible rhythmic clicks in the ear. In symptomatic palatal tremor the rhythmic activity is believed to originate in the inferior olive. The origin of essential palatal tremor is unknown.

Drug or toxin induced tremors

Drugs and toxins may cause all types of tremor known clinically (Krack and Deuschl, 1998). Increased physiologic tremor is most commonly seen and is elicited by such drugs as sympathomimetics or antidepressants. Resting tremor may be elicited by dopamine receptor blockers, while intention tremor may occur in patients with lithium and other drug intoxications.

Tardive tremor following prolonged neuroleptic medication is a special variant which usually is static in nature, but may also occur at rest and on intentional movements.

Tremor associated with peripheral neuropathies

Tremor may also occur in some peripheral neuropathies. It is most commonly seen in demyelinating disease and may be static or kinetic (Bain et al., 1996; Dalakas et al., 1984). Its causes are still unknown. Often distal and proximal muscles of one and the same extremity show tremors of different frequencies (Brain et al., 1996).

Psychogenic tremor

The criteria suggestive of psychogenic tremor are sudden onset and/or complete remission. Resting and action components are usually present side by side in unusual combinations. When the patients are distracted, the amplitude often decreases or the frequency varies.

Psychogenic tremor shows palpable co-activation during passive limb movements which appears to be volitional. When the arms are outstretched, the trembling limb may assume an abnormal position. In terms of the underlying pathophysiology, psychogenic tremor is probably a variant of voluntarily induced clonus. The co-activation sign of psychogenic tremor is therefore a particularly important clinical sign. This common type of tremor should be distinguished from low-frequency tremor of the limbs or trunk which is apparently intentionally induced.

The above classification which is shown in Fig. 1 helps to diagnose the overwhelming majority of tremors.

Differential diagnosis

In the clinical differential diagnosis of tremor rhythmic myoclonus should invariably be considered. It is typically characterized by brief muscle twitches, has a frequency of lees than 5 Hz and is usually confined to one limb or to neighboring body regions. The symptoms are not always fully differentiable clinically (Silfverskiöld, 1986). But spike — wave complexes in the EEG or spinal lesions make the diagnosis probable (Deuschl et al., 1996).

Clonus is characterized by a rhythmic movement which is enhanced by muscle stretching.

Static tremor is often mistaken for asterixis. This is characterized by a sudden interruption of EMG activity in several muscles of an extremity. The diagnosis is confirmed by electromyography. Prolonged absence of EMG activity may be associated with "flapping" on abduction of the upper extremity. Briefer spells of EMG inactivity are associated with irregular static tremor. Unilateral asterixis is usually due to a focal lesion. Bilateral asterixis is, as a rule, caused by intoxication (Ikeda et al., 1990; Shibasaki, 1995; Young and Shahani; 1986).

Epilepsia partialis continua may be confused with low-frequency tremor (Thomas et al., 1977). The differential diagnosis is clinched by a negative history of tremor, the absence of other focal or multifocal lesions and an abnormal EEG or EMG.

Studies to confirm the diagnosis

In patients with increased physiologic tremor certain lab tests should be ordered. Depending on the clinical condition, these may include: T3, T4, TRH, Na^{+}, Ca^{++}, Cl, γ-GT, COT, GPT, cholinesterase, creatinine, urea, glucose. If suggested by clinical signs and symptoms, other tests may be required, i.e. cortisol, parathyroid hormone, copper excretion in 24-h urinary samples, ceruloplasmin and special toxicologic studies.

Multi-channel electromyography is mandatory for confirming the diagnosis of primary orthostatic tremor and asterixis and for instituting targeted botulinum toxin treatment, if indicated. It is also helpful in a number of other tremor syndromes (Deuschl et al., 1996). Time frequency analysis is widely used for quantitation and for distinguishing tremors of central or peripheral origin. Long-term tracings mainly have a place in therapeutic trials (Spieker et al., 1995). Functional imaging by fluorodopa-PET or SPECT with IBZM or β-CIT may be useful for differentiating parkinsonian syndromes. But despite the battery of helpful studies, the differential diagnosis of tremor is still primarily made on the basis of the patient's history and the clinical examination.

References

Bain PG, Findley LJ, Thompson PD, Gresty MA, Rothwell JC, Harding AE, Marsden CD (1994) A study of hereditary essential tremor. Brain 117: 805–824

Bain PG, Britton TC, Jenkins IH, Thompson PD, Rothwell JC, Thomas PK, Brooks DJ, Marsden CD (1996) Tremor associated with benign IgM paraproteinaemic neuropathy. Brain 119: 789–799

Brooks DJ (1993) PET studies on the early and differential diagnosis of Parkinson's disease. Neurology 43

Dalakas MC, Teravainen H, Engel WK (1984) Tremor as a feature of chronic relapsing and dysgammaglobulinemic polyneuropathies. Incidence and management. Arch Neurol 41: 711–714

Deuschl G, Mischke G, Schenck E, Schulte-Mönting J, Lücking CH (1990) Symptomatic and essential rhythmic palatal myoclonus. Brain 113: 1645–1672

Deuschl G, Toro C, Valls SJ, Zeffiro T, Zee DS, Hallett M (1994) Symptomatic and essential palatal tremor. 1. Clinical, physiological and MRI analysis. Brain 117: 775–788

Deuschl G, Krack P, Lauk M, Timmer J (1996) Clinical electrophysiology of tremor. J Clin Neurophysiol 13: 110–122

Deuschl G, Bain P, Brin M (1998) Tremor: Consensus-statement of the Movement Disorder Society. Mov Disord 13 (Suppl) 3: 2–23

Findley LJ, Koller WC (1995) Handbook of tremor disorders. Marcel Dekker, New York

Holmes G (1904) On certain tremors in organic cerebral lesions. Brain 27: 327–375

Ikeda A, Kakigi R, Funai N, Neshige R, Kuroda Y, Shibasaki H (1990) Cortical tremor: a variant of cortical reflex myoclonus. Neurology 40: 1561–1565

Jankovic J, Van der Linden C (1988) Dystonia and tremor induced by peripheral trauma: predisposing factors. J Neurol Neurosurg Psychiatry 51: 1512–1519

Koller WC, Busenbark K, Miner K (1994) The relationship of essential tremor to other movement disorders: report on 678 patients. Essential Tremor Study Group. Ann Neurol 35: 717–723

Krack P, Deuschl G (1998) Tremor. In: Hopf H, Diener GD H C, Reichmann H (eds) Handbuch der Neurologie. Thieme, Stuttgart

Miwa H, Hatori K, Kondo T, Imai H, Mizuno Y (1996) Thalamic tremor: case reports and implications of the tremor-generating mechanism. Neurology 46: 75–79

Rivest J, Marsden CD (1990) Trunk and head tremor as isolated manifestations of dystonia. Mov Disord 5: 60–65

Rosenbaum F, Jankovic J (1988) Focal task-specific tremor and dystonia: categorization of occupational movement disorders. Neurology 38: 522–527

Shibasaki H (1995) Myoclonus. Curr Opin Neurol 8: 331–334

Silfverskiöld BP (1986) Rhythmic myoclonias including spinal myoclonus. In: Fahn S, Marsden CD, Van Woert M (eds) Advances in neurology. Raven Press, New York, pp 275–285

Spieker S, Jentgens C, Boose A, Dichgans J (1995) Reliability, specificity and sensitivity of long-term tremor recordings. Electroencephalogr Clin Neurophysiol 97: 326–331

Thomas JE, Reagan TJ, Klass DW (1977) Epilepsia partialis continua. Arch Neurol 34: 266–275

Young RR, Shahani BT (1986) Asterixis: one type of negative myoclonus. In: Fahn S, Marsden CD, Van Woert M (eds), Myoclonus. Raven Press, New York, pp 137–156

Author's address: Prof. Dr. Günther Deuschl, Department of Neurology, Christian-Albrechts-University Kiel, Niemannsweg 147, D-2405 Kiel, Federal Republic of Germany

Springer-Verlag and the Environment

WE AT SPRINGER-VERLAG FIRMLY BELIEVE THAT AN international science publisher has a special obligation to the environment, and our corporate policies consistently reflect this conviction.

WE ALSO EXPECT OUR BUSINESS PARTNERS – PRINTERS, paper mills, packaging manufacturers, etc. – to commit themselves to using environmentally friendly materials and production processes.

THE PAPER IN THIS BOOK IS MADE FROM NO-CHLORINE pulp and is acid free, in conformance with international standards for paper permanency.

GPSR Compliance

The European Union's (EU) General Product Safety Regulation (GPSR) is a set of rules that requires consumer products to be safe and our obligations to ensure this.

If you have any concerns about our products, you can contact us on ProductSafety@springernature.com

In case Publisher is established outside the EU, the EU authorized representative is:

Springer Nature Customer Service Center GmbH
Europaplatz 3
69115 Heidelberg, Germany

Batch number: 10732722

Printed by Printforce, the Netherlands